MICROSTRIP CIRCUIT ANALYSIS

MICROSTRIP CIRCUIT ANALYSIS

DAVID H. SCHRADER
Washington State University

For book and bookstore information

http://www.prenhall.com
gopher to gopher.prenhall.com

Prentice Hall PTR
Upper Saddle River, New Jersey 07458

Library of Congress Cataloging-in-Publication Data

Schrader, David H.
Microstrip circuit analysis / David H. Schrader.
p. cm.
Includes index.
ISBN 0-13-588534-5
1. Microwave circuits 2. Strip transmission lines. 3. Electric circuit analysis. I. Title.
TK7876.S353 1995
621.381'32—dc20 95-9926
CIP

Acquisitions editor: Karen Gettman
Cover designer: Miguel Ortiz
Manufacturing buyer: Alexis R. Heydt
Compositor/Production services: Pine Tree Composition, Inc.

Prentice-Hall, Inc.
A Simon & Schuster Company
Upper Saddle River, New Jersey 07458

The publisher offers discounts on this book when ordered in bulk quantities.

For more information contact:
Corporate Sales Department
Prentice Hall PTR
One Lake Street
Upper Saddle River, New Jersey 07458

Phone: 800-382-3419
Fax: 201-236-7141
email: corpsales@prenhall.com

Printed in the United States of America
10 9 8 7 6 5 4 3 2 1

ISBN: 0-13-588534-5

Prentice Hall International (UK) Limited, *London*
Prentice Hall of Australia Pty. Limited, *Sydney*
Prentice Hall Canada, Inc., *Toronto*
Prentice Hall Hispanoamericana, S.A., *Mexico*
Prentice Hall of India Private Limited, *New Delhi*
Prentice Hall of Japan, Inc., *Tokyo*
Simon & Schuster Asia Pte. Ltd., *Singapore*
Editora Prentice Hall do Brasil, Ltda., *Rio de Janeiro*

CONTENTS

Preface

Although this book is intended to be a textbook for an upper division course on microstrip circuit analysis, I hope that it will be of interest to practicing engineers also. I have tried to include sufficient examples of microstrip circuits to illustrate the theory. However, I have not tried to describe all possible microstrip circuits because the aim was to present methods for linear analyses of microstrip circuits.

I have tried to provide coverage of the theory underlying circuit and amplifier design. I have included some optional topics which can be covered to make this text suitable for a first year graduate course as well. My experience is that the subject matter cannot be covered in a semester. Depending upon the background of the students, one can cover in one semester the essentials of the subject which are contained in Sections 1.0 through 1.7, Sections 2.0 through 2.11, Chapter 3, parts of Chapter 4, Sections 5.0 through 5.6, Chapter 6, and parts of Chapter 7.

Undergraduates using this text will require a background equivalent to junior level courses in electromagnetics (covering through Maxwell's equations and plane wave propagation) and circuit analysis (including Laplace transforms and transfer functions). Although complex contour integration is used briefly in Chapter 7, subsequent material does not require an understanding of it.

Frequent references have been given in Chapters 2, 4, 5, and 7 for those readers wishing to pursue some aspect of the subject further. This book is not a treatise on the subject so I have tried to furnish a representative sample of references but not an exhaustive list.

Chapter

1

Waveguides

1.0 INTRODUCTION

The step-by-step advances in our ability to design and fabricate circuits at higher frequencies have produced a technology for the use of microwaves. Although there is no set definition for the microwave frequency range, we shall take it to be that starting at approximately 1 GHz and extending to around several hundred GHz. A planar circuit technology has evolved and become dominant for low power circuits in this frequency range. The design techniques for these circuits differ significantly from those for circuits below 100 MHz. It is these techniques with which this book is concerned.

Early microwave equipment used rectangular waveguides and coaxial cables to interconnect bulky microwave devices, resulting in designs which were large and costly to construct. This began to change in the late 1940s and early 1950s, when the idea of using a flat strip of metal between two conducting planes as a waveguide for microwaves was conceived and promoted by R. M. Barrett.[1] Independently in 1942, a similar form of line was used by Wheeler.[2] These evolved into *strip lines* (see Figure 1–1). Shortly thereafter, a modification of that idea resulted in a *microstrip line* as shown in Figure 1–1. This configuration consists of a metallic strip on a dielectric slab which, in turn, is backed by a conducting ground plane. Because of its planar geometry, microstrip readily lends itself to fabrication. Further development of this waveguide has resulted in designs that are capable of integrated circuit construction with reduced size and cost. These are called microwave integrated circuits: *MIC.*

There are two types of MIC: The first type uses some discrete components, such as chip capacitors or resistors, mounted on an integrated circuit structure. These circuits are called *hybrid MICs.* Such circuits are used mainly in the lower microwave frequency

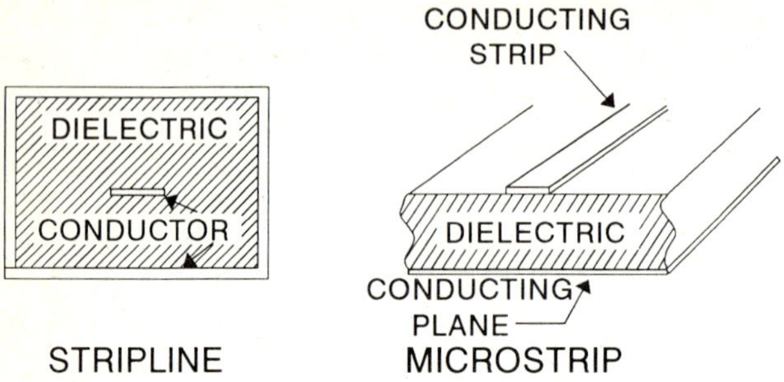

FIGURE 1–1. Planar Microwave Waveguides.

range. The second type is the fully integrated circuit and is called the *monolithic microwave/millimeter wave integrated circuit (MMIC).* Microstrip lines have been used in both of these circuit types.

The microstrip line is not readily adapted for mounting components shunted to the ground plane. Thus, energizing the microstrip from an active element usually requires connecting a conductor through the dielectric to the ground plane. These holes in the dielectric are called *through holes* or *via holes.* Similarly, making a short to ground on the transmission line requires a conductor connecting the microstrip through the dielectric to the ground plane. These connections are called *via connections.* At the higher range of microwave frequencies, the conductor through the hole acts as an inductor. In addition to this disadvantage, the propagation in the microstrip line is subject to an energy leakage resulting in a loss of signal.

The dual of the microstrip line was proposed in 1968 by S. B. Cohn, partly to overcome the necessity of through holes.[3] This consists of a slot in a conductive coating on a dielectric slab as shown in Figure 1–2A. This configuration is termed a *slotline.* A component such as a diode or a resistor can be mounted between the two sides of the slot with ease. If the dielectric constant is sufficiently large, then the field is concentrated in the region of the slot and the dielectric material helps to reduce radiation from the slot.[4]

An effort to overcome the problem associated with energizing waveguides from nonreciprocal magnetic devices that have a circularly polarized magnetic field led to the development of the *coplanar waveguide (CPW)* shown in Figure 1–2B.[5] The principal propagation mode in CPW has an elliptical polarization allowing it to be excited by such magnetic devices. Another favorable feature of this guide is that there are three conducting surfaces, all in the same plane, to which components can be mounted. These are the two ground planes and the center conductor. The *coplanar strip line* is the complement of this guide. It consists of two parallel metallic strips mounted on a dielectric slab as shown in Figure 1–2C.

All of the above types of transmission lines can be used without being enclosed by a metallic grounded shield and can be fabricated on a plane dielectric surface. This leads to fairly simple circuit construction.

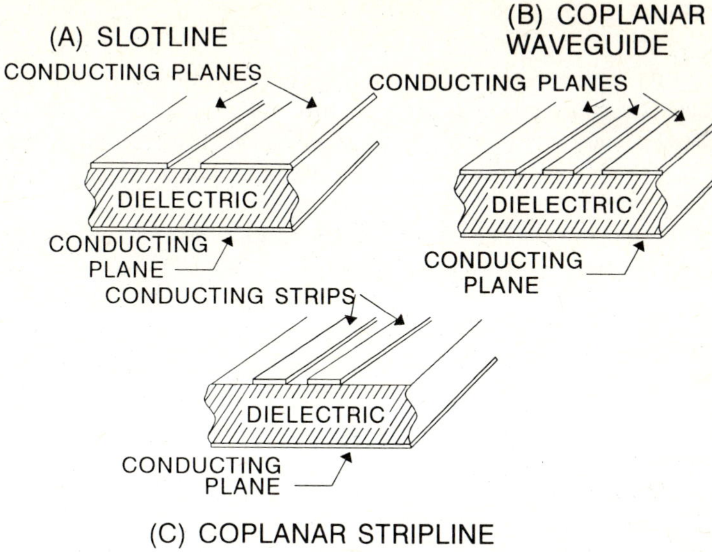

FIGURE 1–2. Printed Circuit Waveguides.

The strip line is a metallic strip embedded in a uniform dielectric and enclosed inside a grounded conductor. The *finline* is also a waveguide which is enclosed in a grounded conductor. The finline is shown in Figure 1–3. The finline originally was proposed for use at frequencies above 30 GHz as a way to reduce radiation losses, decrease dispersion, and increase structure sizes to allow the use of the components available at that time.[6]

As the technology progressed toward smaller integrated circuits, interest arose in lines fabricated on semiconducting materials. The *metal-insulator-semiconductor (MIS)* transmission line is such a line. It can consist of structures like a microstrip line, a coplanar waveguide, or coplanar strips on an insulating layer that is backed by a semiconducting layer. These were first discussed in 1967 by Guckel, Brennan, and Palócz.[7] Theoretical and experimental results were published by Hasegawa, Furukawa, and Yanai in 1971.[8]

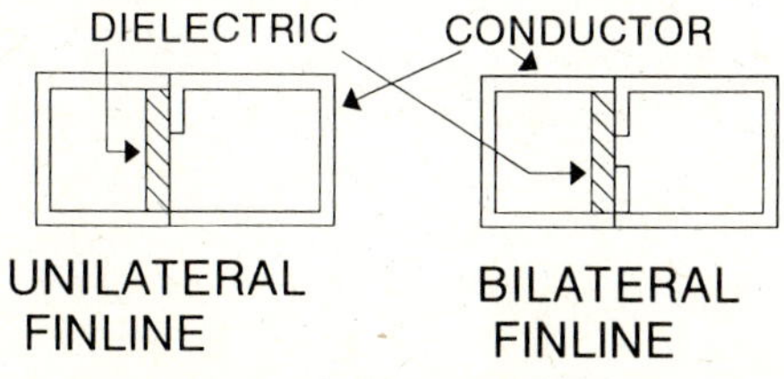

FIGURE 1–3. Finlines.

1.1 MICROWAVE CIRCUITS

The term "microwave circuit" is often used to refer to widely diverse configurations. For instance, the configuration of a microstrip connecting transistors, resistors, and capacitors is called a microwave circuit if it operates at microwave frequencies. If the microstrip lines connecting the circuit elements have lengths that are an appreciable fraction of a wavelength of the microwaves, then the microstrip lines act as transmission lines and all the theory concerning waves propagating on transmission lines pertains. Devices having conventional rectangular cross section waveguides and containing couplers, attenuators, and other distributed elements are sometimes referred to as microwave circuits. Certainly the primary analysis of the operation of such waveguides must be in terms of a solution of the vector wave equation for the electromagnetic fields. The basic reason for classifying these as microwave circuits is that they can be analyzed to a first approximation as transmission lines connecting lumped elements. This simplifies the description and design of the circuits containing these devices.

When analyzing these devices, it is convenient to use voltage and current variables in the same way as one might when analyzing low frequency circuits. However, one cannot always identify or measure the voltage or current in the actual device. In the case of a conventional waveguide of rectangular cross section, one cannot identify a voltage that can be measured with a voltmeter, for example. We shall define a voltage variable for such a case and then apply the usual formulas for describing how a voltage propagates down a transmission line.

We use the term *cylindrical waveguide,* or waveguide, to refer to any "cylindrical" device capable of wave propagation down the axis along which a cross section of the device does not change. For instance, the rectangular cross section guide mentioned above is a "cylindrical" waveguide. Similarly, a pair of parallel wires is a cylindrical waveguide with propagation down the direction of the wires. In each case, the z axis of a cylindrical guide is chosen to be the direction of propagation of the wave down the guide and the geometry of a cross section does not change along the z direction.

1.2 LOSSLESS TRANSMISSION LINE PROPAGATION

Although the propagation of electromagnetic waves along transmission lines is characterized most exactly by Maxwell's equations, frequently the simplest analysis is performed by using an equivalent circuit for the line. In that case, a short element of the line from the point at z meters to the point at $z + \Delta z$ meters can be characterized as shown in Figure 1–4.

Using Kirchhoff's laws, we have the following expressions.

$$v = L\,\Delta z\,\frac{\partial i}{\partial t} + v + \Delta v$$

$$i = C\,\Delta z\,\frac{\partial (v + \Delta v)}{\partial t} + i + \Delta i$$

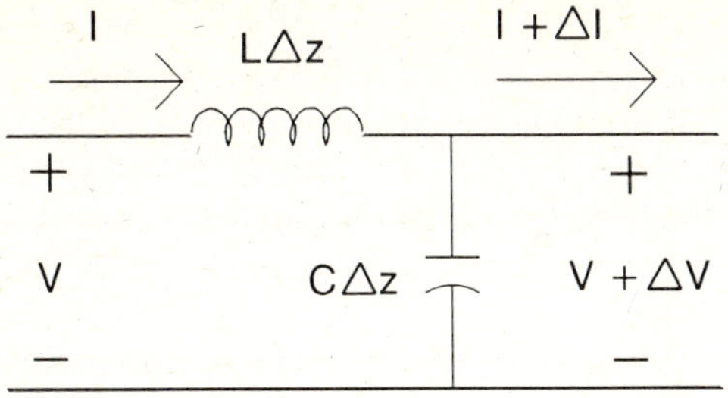

FIGURE 1–4. Incremental Length of Transmission Line.

If we use the expressions

$$\lim_{\Delta z \to 0} \Delta v = \frac{\partial v}{\partial z} dz \text{ and } \lim_{\Delta z \to 0} \Delta i = \frac{\partial i}{\partial z} dz$$

with the above expressions we can arrive at the differential Eqs. (1–1) and (1–2).

$$\partial v/\partial z = -L\, \partial i/\partial t \qquad (1\text{–}1)$$

$$\partial i/\partial z = -C\, \partial v/\partial t \qquad (1\text{–}2)$$

A simultaneous solution of Eqs. (1–1) and (1–2) yields the *wave equation* $\partial^2 v/\partial z^2 = LC\, \partial^2 v/\partial t^2$ that has solutions describing the voltage on the line. These voltage waveforms progress along the line in the z direction while maintaining their shape. A way of expressing this property is to write the voltage v as a function of the variable $\tau = t - z/V_\ell$, where V_ℓ is the speed of the wave. Frequently, the variable τ is called the *retarded time* variable. Functions that have the variables t and z appearing additively as above are solutions to the wave equation as shown by D'Alembert. In this form, the function $v(\tau)$ can describe any reasonable voltage waveform such that when the time variable t is increased; the distance variable scaled by the velocity z/V_ℓ must be increased by a like amount so that $v(\tau)$ has the same value. For instance, a step increase in the voltage on a line may be caused by connecting a one-volt battery to the line at the time $t = 0$ and at the location $z = 0$. Such a waveform is described by the expression $v = U(\tau)$, where $U(\tau)$ is the unit step function. We observe that the voltage at a location $z = 10$ meters has to be zero at the time $t = 0$ seconds, because the wave has not had time to propagate to that point. If the propagation velocity, V_ℓ, *is* 3×10^8 m/s then the voltage measured at a point 10 meters down the line from where the battery voltage was applied will take a one volt step increase at the time $t = \frac{1}{30}$ μs. This is the time required for the wave to propagate to that point on the line.

Because the positive range of z is not the only direction that waves can progress along the line, we also can use $\tau = t + z/V_\ell$ as the variable in the voltage waveform. This describes a wave traveling in the opposite direction.

Usually the analysis of microwave circuits is made with the assumption that the signal has a sinusoidal waveform. In this case, the time variation can be described with the exponential $e^{j\omega t}$. Because we are discussing solutions of the wave equation in which the variables t and z must appear in a linear combination, as in D'Alembert's solution, the

z dependence must be described by the exponential $e^{\pm j\beta z}$, where the two choices of signs correspond to the two different directions of wave travel. Assume that a transmission line is energized at one end with a $\frac{1}{2}$ V signal at 1 GHz. Then the wave on the line is expressed by 0.5 exp $(j(\omega t - \beta z))$. The factor ω is $2\pi 10^9$. This expression could also be written as 0.5 exp $[j\omega(t - z/V_\ell)]$. We see that this has the form required of a solution to the wave equation where the time and distance arguments appear additively. We also note that the quantity β, which is called the *phase constant,* is related to the velocity V_ℓ by $\beta = \omega/V_\ell$. If we recall from the study of physics that the frequency f, the wavelength λ, and the velocity V_ℓ, are related by $f\lambda = V_\ell$, then we have Eq. (1–3). If we substitute the wave function exp $[j(\omega t - \beta z)]$ into the wave equation, then we can derive the second equality in Eq. (1–3).

$$\beta = 2\pi/\lambda = \omega\sqrt{LC} \tag{1–3}$$

Using the factor $e^{j\omega t}$ as in the above discussion presumes that signal has existed for all time. In that case, the notion of the leading edge of a signal progressing down a transmission line, as discussed in the case of the unit step function above, no longer applies and we now have a change in the phase angle along the line as time increases. The sinusoid will appear to move down the line as time increases for a positively traveling wave. We measure the wave progression by observing the time interval required for a particular phase angle to move from one location, z_1, on the line to another location z_2. Thus, we have the angle $\theta_1 = \omega(t_1 - z_1/V_\ell)$ equal to the angle $\theta_2 = \omega(t_2 - z_2/V_\ell)$. Setting these two expressions equal to each other and solving for V_ℓ, results in the expression: $V_\ell = (z_2 - z_1)/(t_2 - t_1) = \Delta z/\Delta t$. This velocity with which the phase front progresses along the line is called the *phase velocity* and is denoted V_p. The phase velocity is given by Eq. (1–4).

$$V_p = \omega/\beta \tag{1–4}$$

The differential Eqs. (1–1) and (1–2) can be solved for a wave equation in terms of the current variable i. This equation is $\partial^2 i/\partial z^2 = LC\,\partial^2 i/\partial t^2$. From this we can infer that all the comments made concerning the solutions for voltage apply equally well to solutions made for current.

1.3 WAVE IMPEDANCE

We can substitute the expressions for the voltage and current $v(t) = V$ exp $[j(\omega t - \beta z)]$ and $i(t) = I$ exp $[j(\omega t - \beta z)]$, where V and I are possibly complex quantities, into Eq. (1–1) and also substitute for β from Eq. (1–3). The resulting expression can be solved for Eq. (1–5), where Z_0 is called the *characteristic impedance* of the line.

$$Z_0 = V/I = \sqrt{L/C} \tag{1–5}$$

If we follow the same steps using the wave functions for the wave traveling in the opposite direction as shown in Figure 1–5 and use a positive sign in the exponent, then the result will be $V_r/I_r = -Z_0$. The negative sign occurs because the positive reference direction for the current has been chosen the same as in Figure 1–4. The reason for this is that the

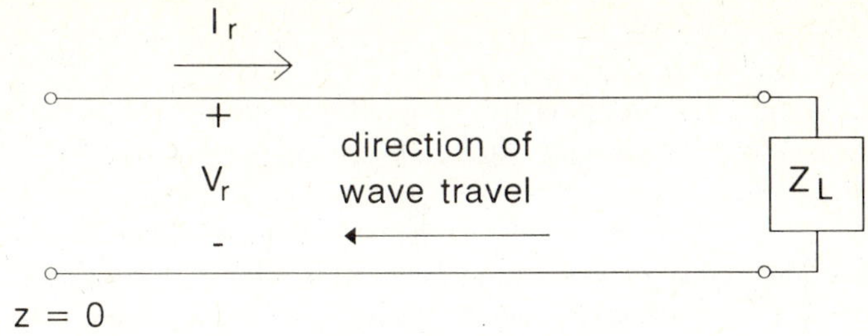

FIGURE 1–5. Oppositely Directed Wave on a Transmission Line.

wave equation was derived with this assumed positive reference direction. It is important to remember this change in sign.

1.4 TRANSMISSION LINE LOSSES

Power losses occur in waveguides in the form of copper losses, i^2R, and in the form of power dissipation in dielectrics described by the *loss tangent,* tan (δ), of the dielectric material. The angle δ is the complement of the angle between the electric field in the dielectric material and the displacement current through it. To the first order, these losses can be modeled by a series resistor and a shunt conductance inserted into the circuit of Figure 1–4; see problem 1–4. The effect of these added elements is to change the differential equation from the lossless wave equation to the equation

$$\partial^2 v/\partial z^2 = LC\, \partial^2 v/\partial t^2 + (LG + RC)\, \partial v/\partial t + RGv.$$

In the case when the voltage is sinusoidal in time, where we can use $e^{j\omega t}$, this becomes

$$\partial^2 v/\partial z^2 = (A + jB)v, \quad A = RG - \omega^2 LC, \quad B = \omega(RC + LG).$$

In this case, the solutions are of the form $v = V_0 e^{\gamma z + j\omega t}$ where the propagation constant is complex as in $\gamma = \alpha + j\beta = \pm\sqrt{A + jB} = \pm\sqrt[4]{A^2 + B^2}\, \sqrt{e^{j2\theta}}$. The term α is called the *attenuation constant* and β is the phase constant. From this, we define the angle θ as $\theta = (1/2)$ arctan (B/A) and we have the formula $\alpha = \pm\sqrt[4]{A^2 + B^2} \cos(\theta)$. The negative sign is to be used for the waves in the positive z direction and the positive sign for those in the opposite direction. The factor $e^{\alpha z}$ describes the decrease in the wave amplitude with an increase in distance.

Example: As an example, we take a coaxial cable. A commonly used coax is one that has the designation of RG58C/U. This letter-number combination refers to the joint Army-Navy (JAN) specification for coaxial cables. This cable is flexible with a center conductor surrounded by a polyethylene dielectric, which in turn is encased in a braided wire shield. The entire cable is encased in a plastic covering. The capacitance of the cable is nominally 93.5 pF/m. This cable is usable up to at least 1 GHz at which frequency it has about 1 dB/m attenuation. This cable is designed to have a characteristic impedance of 50 Ω. Using Eq. (1–5) we can calculate that the inductance per meter is about 0.2 μH/m.

1.5 WAVE REFLECTIONS

An instrument that uses propagating waveforms, similar to one of those described above, is the *time delay reflectometer (TDR)*. This instrument energizes a line with an approximate step function of voltage. The voltage waveform, $U(t - z/V)$, progresses down the line until it reaches a point at $z = L$ where the line has a discontinuity such as a short circuit. At that point, another voltage waveform $-U(t + z/V - 2L/V)$, called a *reflected wave,* is generated to satisfy the condition that the voltage across the short circuit must be zero at all times. This second waveform progresses backward along the line to the origination point. The total elapsed time from when the original wave was applied to the time when the reflected wave arrived back at the TDR gives a measure of twice the distance to the short circuit.

In general, a reflected wave is caused by any change in the characteristic impedance of the line. If the line is terminated by any impedance other than an impedance equal to the characteristic impedance, then a reflected wave is caused at that point. Consider a coaxial cable connected to another cable with a different characteristic impedance. A reflection occurs at the point of connection. Similarly, if the cable is terminated in a resistor with a resistance different from Z_0 then a reflection occurs at the termination.

A transmission line energized with a sinusoid $V_i \exp[j(\omega t - \beta z)]$, called the *incident wave,* and terminated in an impedance Z_L other than Z_0 will have a reflected wave that must have a wave function of the form $V_r \exp[j(\omega t + \beta z + \theta)]$ where the magnitude and phase angle $V_r e^{j\theta}$ are determined from the *boundary conditions* at the termination. For this discussion it will be convenient to drop the time variation and work just with the phasors $V_i e^{-j\beta z}$ and $V_r e^{j(\beta z + \theta)}$. It is conventional to choose the distance parameter to be zero at the load and a negative value of L meters at the input to the transmission line.

We wish to determine the phasor $V_r e^{j\theta}$ in terms of the known quantities. We assume that the line impedance and the load impedance are known. We also assume that $V_i e^{+j\beta L}$ is known, because this is the phasor representing the incident wave at the generator end of the line. The boundary conditions in this case are that the voltage and current just to the left of the load must be equal to the voltage and current at the load (refer to Figure 1–6 for this discussion of boundary conditions). However, the voltage and current on the load must satisfy Ohm's law as in $V_L = I_L Z_L$. At the location of the load, ($z = 0$), the line voltage is given as the sum of the voltages of the incident wave and the reflected wave. The line voltage at the load is given as $V_i + V_r e^{j\theta}$. The second boundary condition is that the line current at $z = 0$ must be equal to the load current. These boundary conditions yield the equations $V_i + V_r e^{j\theta} = V_L$ and $I_i + I_r e^{j\theta} = I_L$. When we solve these equations along with the relations $V_L = Z_L I_L$, $V_i = I_i Z_0$, and $V_r = -I_r Z_0$, we get the important formula for the *reflection coefficient* Γ shown in Eq. (1–6).

$$\Gamma = \frac{V_r\, e^{j\theta}}{V_i} = \frac{Z_L - Z_0}{Z_L + Z_0} \tag{1–6}$$

The definition of the reflection coefficient is the reflected wave divided by the incident wave as measured at the load. This formula for Γ in terms of the impedances is worth remembering.

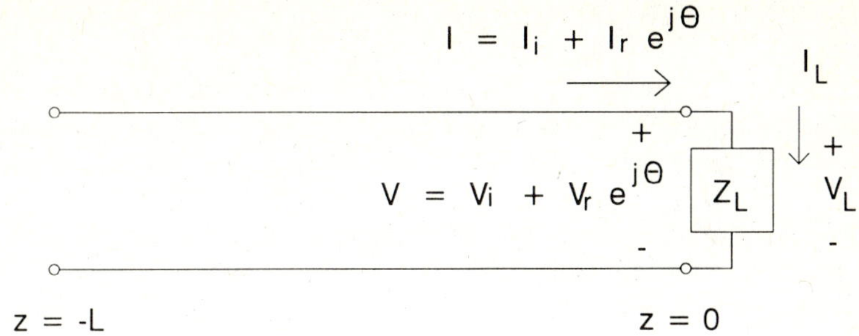

FIGURE 1–6. Boundary Conditions at a Load on a Line.

Example: A 50 Ω cable is terminated in a 10 Ω resistor. The reflection coefficient calculated from Eq. (1–6) is –0.6667. If the incident wave is a 10 V signal whose phase is 30°, then the reflected wave set up at the termination is –6.667∠30° or 6.667∠210°. The signal developed across the resistor is the phasor sum of the incident and reflected waves. It is 3.333 V and the power dissipated in the resistor is 1.111 W. The incident and reflected wave phasors rotate oppositely as functions of distance along the cable due to the different sign in the phase factor, $e^{\pm j\beta z}$. The voltage at any point along the line is a phasor sum of the voltage due to the incident wave and the voltage due to the reflected wave. This sum may be expressed as $V_{line} = 10 \exp(-j(\beta z - \pi/6)) - 6.667 \exp(j(\beta z + \pi/6))$ and $|V_{line}| = |e^{j(\pi/6 - \beta z)}|\,|10 - 6.667e^{j2\beta z}| = |10 - 6.667e^{j2\beta z}| = 10|1 + \Gamma e^{j2\beta z}|$. The line voltage magnitude is 3.333 where $z = 0$ and is 16.667 at the points on the line where $2\beta z$ is an odd multiple of $-\pi$ radians. A plot of $|V_{line}|$ versus z is shown in Figure 1–7. We say that a *standing wave* has been set up on the cable and the distance between minima of $|V_{line}|$ is $\frac{1}{2}$ of a wavelength. Note that the waveform in Figure 1–7 is not a sinusoid. The minima are narrower than the maxima.

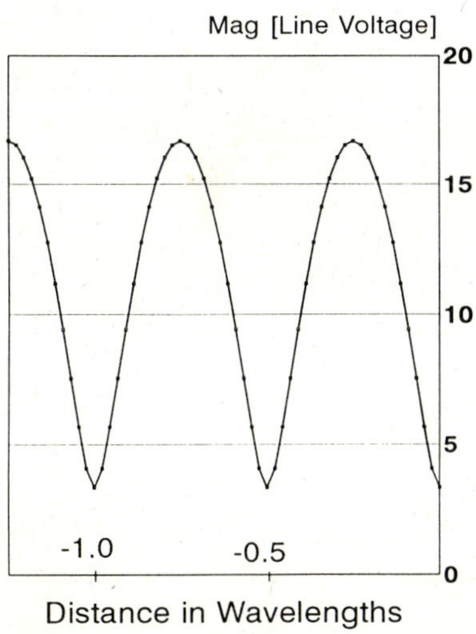

FIGURE 1–7. Standing Wave of Voltage on a Line.

1.6 SWR MEASUREMENTS

In the previous example, the standing waveform was deduced from knowledge of the reflection coefficient. A more common situation occurs when measurements of the standing wave on a line are made and the reflection coefficient is deduced from these measurements. Commonly, one can measure the points along the line where the voltage (or electric field) magnitude is at a minimum. The minima locations can usually be determined more accurately than the locations of the maxima. Furthermore, one can measure the ratio between the maximum line voltage magnitude and the minimum line voltage magnitude. This ratio is called the *standing wave ratio (SWR)* and is denoted by S as $S = \max|V_{\text{line}}|/\min|V_{\text{line}}|$. We determined above that the line voltage on a line terminated with a load having a reflection coefficient Γ was given by the equivalent of Eq. (1–7) where the notation $V_{\text{inc}}(0)$ stands for the incident voltage wave phasor at the location where the reflection occurs ($z = 0$).

$$|V_{\text{line}}| = |V_{\text{inc}}(0)e^{-j\beta z}|\,|1 + \Gamma e^{j2\beta z}| \tag{1–7}$$

From this, we see that the max $|V_{\text{line}}|$, min $|V_{\text{line}}|$, and S are given by

$$\max|V_{\text{line}}| = |V_{\text{inc}}(0)|(1 + |\Gamma|)$$

$$\min|V_{\text{line}}| = |V_{\text{inc}}(0)|(1 - |\Gamma|)$$

$$S = \frac{1+|\Gamma|}{1-|\Gamma|}$$

If we solve the last formula for the magnitude of the reflection coefficient $|\Gamma|$, we have Eq. (1–8).

$$|\Gamma| = \frac{S-1}{S+1} \tag{1–8}$$

To find the angle of the reflection coefficient, we must establish a *reference plane,* a location along the line to which we can refer our measurements. To establish the reference plane, we can measure the location z_s at which a minimum occurs in the line voltage when the load is replaced with a short circuit. This can be compared to the location z_ℓ where the line voltage is minimum when the load terminates the line. The difference between these two locations is related to the phase angle of Γ. In the case of the short circuit for which Γ_s is -1, we see from Eq. (1–7) that the condition $2\beta z_s = -2m\pi$ must hold for $|V_{\text{line}}|$ to be at a minimum. Then for the case when the short circuit is replaced by a load such that $\Gamma_\ell = |\Gamma_\ell|e^{j\theta}$, we see from Eq. (1–7) that $2\beta z_\ell + \theta = -(2m + 1)\pi$ again for $|V_{\text{line}}|$ to be at a minimum. We use the difference between these two relations to arrive at Eq. (1–9) for the phase angle. Note that the numbers used in this formula for z_s and z_ℓ must be negative.

$$\theta = \pi + 2\beta(z_s - z_\ell) \tag{1–9}$$

Equations (1–8) and (1–9) are used in the classical method of determining the reflection coefficient and the impedance of an unknown load. The modern method of measuring reflection coefficients is to use network analyzers. Despite their automated circuitry, they require a calibration procedure that usually involves establishing a reference plane by connecting a short circuit to their terminals. Once the complex value of Γ is determined, then Eq. (1–6) can be solved to obtain the important formula for calculating the load impedance Z_L given as Eq. (1–10). This formula is worth remembering.

$$Z_L = Z_0 \frac{1+\Gamma}{1-\Gamma} \tag{1–10}$$

Example: When connecting microwave antennas to transmitters, an adapter frequently is used to couple from a coaxial cable to the antenna. The change from one transmission line to another must be done as carefully as possible to minimize the reflections at the adapter. Any reflection at that point reduces the usable signal and thus the range of the transmitter. To check how well the adapter is working, we can measure the reflection coefficient and use Eq. (1–8) and (1–9). Suppose that such measurements were made for an adapter at 2 GHz resulting in the following data with adapter connecting to a matched load: V_{max} = 11.71 volts; V_{min} = 8.29 volts voltage minima occur at –10 and –17.5 cm. with short circuit in place of the adapter: voltage minima occur at –15.83, –23.33 and –30.83 cm. Note that the distance between minima is $\frac{1}{2}$ wavelength. The reason for this is that the incident and reflected waves have opposite phase rotation with distance away from the short circuit. Thus, they go from in phase to 180° out of phase in $\frac{1}{4}$ wavelength. This allows us to calculate λ = 15 cm. Further, note that the distance z_s does not correspond to an integer number of half wavelengths at 2 GHz. This implies that the distance measurements were not referenced to the point of connection. This will not cause an error as long as the same reference point is used for both measurements, because only the difference between these measurements is used. Dividing V_{max} by V_{min}, we have the SWR equal to 1.413. From $\beta = 2\pi/\lambda$, we have β = 0.4189 rad/cm. Then, we can calculate the reflection coefficient from Eq. (1–8) and (1–9) to be $\Gamma = 0.171\angle{-99.85°}$. If the incident wave is 10 volts, then the reflected wave at the adapter is 1.71 volts, and it has an angle relative to the incident wave of –99.85°. Carrying the calculation further, we see that the admittance looking into the adapter is $Y_L = 1/Z_L = 0.02 + j0.00694\ S$, as calculated using Eq. (1–10) and assuming that the cable has $Z_0 = 50\ \Omega$. If the antenna has an input impedance of 50 Ω, then this corresponds to a parallel combination of a 50 Ω resistor and a capacitor of 0.5 pF. The adapter can be modeled as a lumped capacitance of 0.5 pF in parallel with the antenna as shown in Figure 1–8.

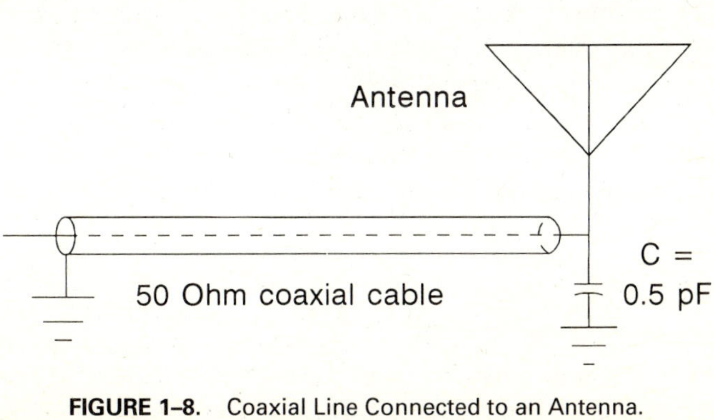

FIGURE 1–8. Coaxial Line Connected to an Antenna.

1.7 TRANSMISSION LINE IMPEDANCE

Corresponding to the standing waves in the voltage on a line, we also have standing waves in the current. The incident and reflected voltage waves are coupled to incident and reflected current wave functions as described by the relations $V_i = Z_0 I_i$ and $V_r = -Z_0 I_r$. The total current on the line at any point is given by the equation $I_{\text{line}} = (V_i/Z_0)\exp[j(\omega t - \beta z)] - (V_r e^{j\theta}/Z_0)\exp[j(\omega t + \beta z)]$. From these, we have the impedance seen at any point on the line given by Eq. (1–11).

$$Z_{\text{line}} = V_{\text{line}} / I_{\text{line}} = Z_0 \frac{1+\Gamma\, e^{j2\beta z}}{1-\Gamma\, e^{j2\beta z}} \tag{1–11}$$

This equation is an expression for the impedance that would be seen by a generator connected to that point on the line. This expression is quite important and forms the basis for a graphical method for determining the action of transmission lines which will be presented in chapter 3.

1.8 RECTANGULAR WAVEGUIDE PROPAGATION

Propagation of microwaves on the waveguides shown in section 1–0 is often described in terms of the equivalent transmission line voltage and current and such properties as the characteristic impedance and the *dispersion.* Although we desire to have a circuit representation of waveguides, these properties are usually determined through numerical solutions of Maxwell's equations. The results of such solutions are frequently given in tabular or graphical form. These solutions and their results are not well suited to giving an understanding of the propagation.

One of the simplest waveguides to treat analytically is one that has a rectangular cross section as shown in Figure 1–9. The results of such a treatment are functional expressions for impedances and other characteristics. We analyze this first to give the reader a feeling for these parameters that you cannot get when using computer solutions of the wave equation.

The electromagnetic fields in a waveguide of rectangular cross section have interactions described by Maxwell's equations, Eq. (1–12), when we assume that there is no free charge or true current interior to the guide (excluding the guide walls).

$$\nabla \times \mathbf{E} = -\,\partial\mathbf{B}/\partial t\,, \qquad \nabla \times \mathbf{H} = \partial\mathbf{D}/\partial t\,, \qquad \nabla\cdot\mathbf{B} = 0, \qquad \nabla\cdot\mathbf{D} = 0 \tag{1–12}$$

To derive a wave equation, we can take the curl of $\nabla \times \mathbf{H}$ and cross substitute from the equation for the curl of **E.** Using the vector identity $\nabla \times \nabla \times \mathbf{A} = \nabla(\nabla\cdot\mathbf{A}) - \nabla^2\mathbf{A}$, we can derive the wave equation

$$\nabla^2\mathbf{H} = \mu\varepsilon\,\partial^2\mathbf{H}/\partial t^2\,, \qquad \text{alternately } \nabla^2\mathbf{E} = \mu\varepsilon\,\partial^2\mathbf{E}/\partial t^2.$$

(For a more detailed discussion of the wave equation and its solutions one can refer to any electromagnetic field theory text book). For this case, this equation can be expressed

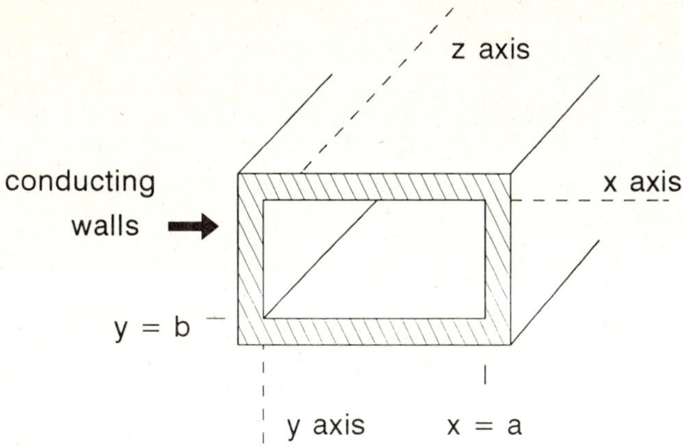

FIGURE 1–9. Cross Sectional View of a Rectangular Waveguide.

as three separate equations, one for each component of the magnetic field. If we use a time variation of $e^{j\omega t}$ in this wave equation, the result is the *Helmholtz equation* Eq. (1–13).

$$\frac{\partial^2 H_z}{\partial x^2} + \frac{\partial^2 H_z}{\partial y^2} + \frac{\partial^2 H_z}{\partial z^2} = -\omega^2 \mu\varepsilon\, H_z \tag{1–13}$$

For the case of propagation down the waveguide, the z variation must be exponential as in exp $[j(\omega t \pm \beta z)]$, where the negative sign again implies propagation in the positive z direction. If we assume that the x and y variation are in the form of exp $[j(k_y x + k_y y)]$ and substitute this into Eq. (1–13) the result is $-k_x^2 H_z - k_y^2 H_z - \beta^2 H_z = -\omega^2\mu\varepsilon H_z$. This simplifies to the *characteristic equation* (or *eigen equation*) of Eq. (1–14).

$$\beta^2 = \omega^2\mu\varepsilon - k_x{}^2 - k_y{}^2, \tag{1–14}$$

From the characteristic equation we infer that β can be calculated from knowledge of the *separation constants* k_x and k_y.

From a knowledge of the two components E_z and H_z, we can find the remaining field components which comprise this mode of oscillation. To show this, we write some of the terms from Maxwell's equations as

$$-j\omega H_x - j\beta E_y = \partial E_z/\partial y$$
$$-j\omega\mu H_y + j\beta E_x = -\partial E_z/\partial x$$
$$j\omega\varepsilon E_x - j\beta H_y = \partial H_z/\partial y$$
$$j\omega\varepsilon E_y + j\beta H_x = -\,\partial H_z/\partial x.$$

In these equations, we have assumed that the time variation is $e^{j\omega t}$, the z variation is $e^{-j\beta z}$ and that there is no true current density in the interior of the waveguide (not including the guide walls). This set of 4 equations contains E_x, E_y, H_x, and H_y that we shall treat as dependent variables. These equations can be solved to express these components in terms of the E_z and H_z components. These solutions are shown in Eq. (1–15).

$$\left.\begin{aligned}
H_x &= \frac{j}{k^2}\left(\omega\varepsilon\frac{\partial E_z}{\partial y} - \beta\frac{\partial H_z}{\partial x}\right)\\
H_y &= \frac{j}{k^2}\left(-\omega\varepsilon\frac{\partial E_z}{\partial x} - \beta\frac{\partial H_z}{\partial y}\right)\\
E_x &= \frac{j}{k^2}\left(-\beta\frac{\partial E_z}{\partial x} - \omega\mu\frac{\partial H_z}{\partial y}\right)\\
E_y &= \frac{j}{k^2}\left(-\beta\frac{\partial E_z}{\partial y} + \omega\mu\frac{\partial H_z}{\partial x}\right)
\end{aligned}\right\} \quad \text{where } k^2 = \omega^2\mu\varepsilon - \beta^2 \qquad (1\text{–}15)$$

The reason for this formulation is that we can determine all components by solving the wave equation for H_z and again for E_z. Then by superposing these solutions with the use of Eq. (1–15), we can obtain the solution which fits the pertinent boundary conditions.

1.9 PROPAGATION MODES

From the physics of the situation, we realize that any propagating wave must be excited from one end or the other of the waveguide. We will assume that the guide is excited by a field at the location $z = 0$. In order for this expression to fit all physically possible excitations at this boundary, we use two dimensional Fourier series expressions for both E_z and H_z as in Eq. (1–16).

$$H_z\,(x,y,0) = \Sigma_{m,n}\,[a_m \cos\,(k_{xm}x) + \\ + b_m \sin\,(k_{xm}x)]\,[c_n \cos\,(k_{yn}y) + d_n \sin\,(k_{yn}y)]e^{j\omega t} \qquad (1\text{–}16)$$

To fit the boundary conditions $E_y = (j\omega\mu/k^2)\ \partial H_z/\partial x = 0$ at $x = 0$, a and $E_x = -(j\omega\mu/k^2)\ \partial H_z/\partial y$ at $y = 0$, b the constants k_{xm} and k_{yn} must be equal to $m\pi/a$ and $n\pi/b$ respectively and the constants b_m and d_n must be zero.

We classify the possible solutions according to the possible *modes* of oscillation. Each term in the sum represents a mode. So that a particular term in Eq. (1–16) $a_m c_n$ cos $(k_{xm}x)$ cos $(k_{yn}y)$ exp $[j(\omega t - \beta_{mn}z)]$ represents one possible mode of oscillation in the guide. The value of β_{mn} can be found by substituting the values of k_{xm} and k_{yn} into the characteristic equation shown in Eq. (1 17).

$$\beta_{mn} = \sqrt{\omega^2\mu\varepsilon - (m\pi/a)^2 - (n\pi/b)^2} \qquad (1\text{–}17)$$

The solution obtained with the H_z component has no E_z component and is referred to as the *transverse electric (TE)* mode. Similarly, the modes for E_z will have no H_z component and they are referred to as *transverse magnetic (TM)* modes. Thus the *m-n* mode shown above would be classified as a TE_{mn} mode.

1.10 CUT-OFF FREQUENCY

The β of Eq. (1–17) becomes imaginary when the argument of the radical becomes negative for a given m-n mode. This defines the *cut-off frequency* f_c of the mode. At frequencies higher than f_c, waves will propagate in that mode and below it the energy attenuates. Solving for the cut-off frequency, we have the expression $f_c = (C/2\pi)\sqrt{(m\pi/a)^2 + (n\pi/b)^2}$. In this equation, we have $C = 1/\sqrt{\mu\varepsilon}$ as the phase velocity of a plane wave in the medium. For the case when $\beta = -j|\beta|$, the wave function has the dependence $e^{j\omega t}e^{-|\beta|z}$. We use α for $|\beta|$ and the factor $e^{-\alpha z}$ represents a wave attenuation in the z direction. In this case, the quantity α is called an *attenuation coefficient.* The "wave" having this z dependence no longer propagates in the truest sense because it does not change phase with distance and it rapidly decreases to zero. We sometimes refer to this type of "wave" as an *evanescent wave.*

The lowest cut-off frequency occurs for the TE_{10} mode when $m = 1$ and n = 0. From the boundary conditions at $x = a$ and $y = b$, we can show that a TM_{10} mode cannot exist in a waveguide. Then, the next lowest cut-off frequency depends upon the dimensions a and b of the guide. However, in the frequency range between the cut-off frequency for the TE_{10} mode and the next cut-off frequency, there is only one mode that can propagate: the TE_{10} mode. For this reason, the TE_{10} mode is called the *dominant mode.* This is the frequency range in which the guide normally is used, and the waveguide dimensions are chosen to make the frequency range for this mode reasonably wide. Here, we concentrate on the formulas concerning propagation in this mode.

1.11 DOMINANT MODE PROPAGATION

To discuss dominant mode propagation in a waveguide, we choose $m = 1$ and $n = 0$ in Eq. (1–16). So H_z in the TE_{10} mode is $H_z = H_0 \cos(\pi x/a)e^{j(\omega t-\beta z)}$. For this mode, we solve Eq. (1–17) for β and the *guide wavelength* $\lambda_g = 2\pi/\beta$. The result is Eq. (1–18) where f_c is the cut-off frequency for the TE_{10} mode.

$$\beta = \omega\sqrt{\mu\varepsilon}\sqrt{1-(f_c/f)^2}, \quad \lambda_g = \frac{2\pi}{\omega\sqrt{\mu\varepsilon}\sqrt{1-(f_c/f)^2}} \tag{1–18}$$

From the expression for H_z substituted into Eq. (1–15), we can obtain Eq. (1–19) for E_y and H_x in which we have used $E_0 = j\omega\mu a H_0/\pi$ and the *guide impedance* $Z_g = \omega\mu/\beta$. First, we note that the transverse electric field and magnetic field components differ by a negative sign and a factor Z_g.

$$\begin{aligned} E_y &= E_0\,[\sin(\pi x/a)]e^{j(\omega t - \beta z)} \\ H_x &= -(E_0/Z_g)\,[\sin(\pi x/a)]e^{j(\omega t-\beta z)} \end{aligned} \tag{1–19}$$

From Eq. (1–18) and $Z_g = \omega\mu/\beta$, we get Eq. (1–20). The negative sign in Eq. (1–19) occurs because the positive direction for the y axis was chosen downward in Figure 1–9 in order to have a right-handed coordinate system.

$$Z_g = \frac{\sqrt{\mu/\varepsilon}}{\sqrt{1-(f_c/f)^2}} \tag{1–20}$$

We see that the direction of power flow as given by $\mathbf{E} \times \mathbf{H}$ is the direction into the waveguide.

Example: Let us consider a waveguide suitable for operation in a satellite. The frequency band designated as the K band extends from 18 to 27 GHz, and the frequencies from 18.1 to 21.2 GHz have been designated as the down link frequencies for commercial satellites. Assuming that the system being designed is to operate at 20 GHz, we select a waveguide with a JAN designation of RG-53/U. This guide has outside dimensions of 0.5 by 0.25 inches and a wall thickness of 0.04 inches. Using Eq. (1–17) we can calculate a TE_{10} cut-off frequency of 14.06 GHz. The recommended frequency range for operating this guide is from 18 to 26 GHz. Using a frequency of 20 GHz in Eq. (1–20), we can calculate an impedance for the dominant mode of 530 Ω. If we have a maximum electric field at the center of the guide of $\frac{1}{2}$ *V/m,* then the transverse magnetic field has a maximum value of $E_{max}/Z_g = 0.0094$ *A/m.* The time-averaged, power density at the center of the guide where the fields are maximum is given by Poynting's vector and is $E_{max}H_{max} =$ 4.7 mW/m^2.

1.12 TRANSMISSION LINE DESCRIPTION OF A GUIDE

Our aim is to describe the propagation in the waveguide using transmission line theory. To do this, we desire to define a voltage and current which can be used in these calculations. As was mentioned before, we cannot measure a voltage using a voltmeter anywhere on the guide. However, we shall define the voltage by making it proportional to the electric field in the center of the guide as in $V = KE_0 \exp[j(\omega t - \beta z)]$ through the proportionality constant K. Similarly we define a current by $I = KE_0 \exp[j(\omega t - \beta z)]/Z_g$. The current is proportional to the H_x field at the center of the guide with the same proportionality constant so that $I = V/Z_g$. The average power transmitted down the guide can be calculated by integrating Poynting's vector over a cross section of the guide: $P = (\frac{1}{2}) \oint \mathbf{E} \times \mathbf{H}^* \cdot d\mathbf{S}$. This becomes

$$P = (1/2)\int_0^a\int_0^b (E_0^2/Z_g)\sin^2(\pi x/a)\, dy\, dx = abE_0^2/4Z_g$$

The counterpart in the transmission line is power flow down the line as given by $(\frac{1}{2})VI^* = K^2E_0^2/2Z_0$. Requiring these to be equal and setting the transmission line impedance, Z_0, to be equal to Z_g allows us to find that the constant must be $K = \sqrt{ab/2}$.

If we are to treat such a waveguide with transmission line theory then we must find the equivalent circuits for different types of discontinuities in waveguides. One case is that of a thin *iris,* forming a window in the guide as shown in Figure 1–10. The window in the guide is smaller than the opening down the guide and poses an obstruction to the passage of an electromagnetic wave down the guide. In order to determine the action of this portion of the waveguide, we must solve the electromagnetic field boundary value problem involving propagating waves and evanescent waves on both sides of the window that are caused by the requirement that the component of the electric field tangential to the conductor (shaded portion) be zero. These evanescent fields consist of electric and magnetic energy oscillating synchronously in time with the applied wave. There is also a reflected wave and a transmitted wave at this boundary. The net effect can be modeled to the first approximation as a capacitor shunting the equivalent transmission line. The capacitive susceptance required for the model is $B = (Y_0 4b/\lambda_g) \ln [\operatorname{cosec} (\pi d/2b)]$.[9] The reflected voltage wave occurring on a transmission line shunted by such a capacitor will be related to the reflected electromagnetic wave caused by the iris in the waveguide in the same way that the incident wave on the transmission line is related to the incident wave in the guide.

Obstructions in waveguides such as windows, coupling probes, etc., are discussed more completely in a book by Elliott.[10] Methods of numerically solving for the fields in the vicinity of a waveguide iris were given by Lee, Jones, and Campbell,[11] Masterman and Clarricoats,[12] and Wexler.[13] A discussion of other types of waveguides, such as coaxial and ridged waveguides as well as a more complete discussion of rectangular waveguides, are given by Elliott.[14] Microstrip is also a waveguide that is analyzed primarily by solving for the electromagnetic fields. Because of its importance to microstrip circuit analysis, we defer this topic to chapter 2.

Example: A waveguide with outside dimensions of 1 inch by $\frac{1}{2}$ inch and a wall thickness of 0.05 inches (inside dimensions of 2.286 by 1.016 cm) has the JAN designation of RG-52/U and a dominant mode cut-off frequency of $f_c = 6.56$ GHz. The useful frequency range of this guide in the TE_{10} mode is from 8.2 to 12.4 GHz. A window inserted into this guide with dimensions of 2.286 cm by 0.35 cm will cause a reflected

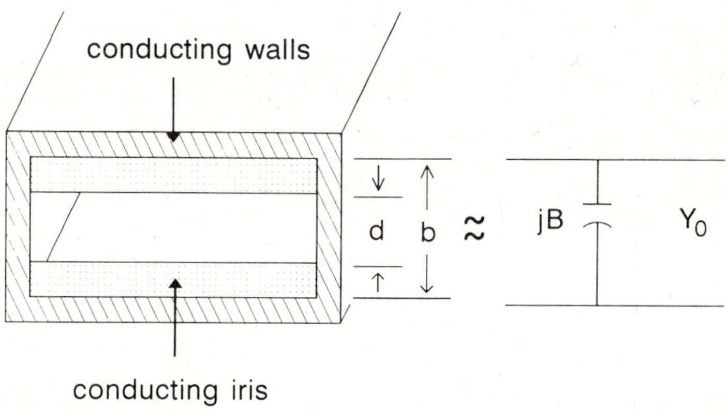

FIGURE 1–10. Waveguide Window and Its Approximate Equivalent Circuit. The shaded portion is a thin conductor partially obscuring the passage of the waves down the guide.

wave. If the guide is excited with an incident wave at 10 GHz, then we can calculate a guide wavelength using Eq. (1–18) of $\lambda_g = 3.97$ cm. We convert the guide dimensions to centimeters and calculate the equivalent normalized admittance of the window to be $b_c = B/Y_0 = 0.678$. Combining this in parallel with the normalized waveguide admittance of $1 + j0$, we have a total admittance at the window of $y_L = 1 + j0.678$. Then, we can convert Eq. (1–6) to the form $\Gamma = (1 - y_L)/(1 + y_L)$ and calculate a reflection coefficient of $\Gamma = 0.3211\angle{-108.7°}$. Because the waveguide characteristic impedance changes with frequency (see Eq. (1–20)), we cannot use this reflection coefficient at other frequencies.

1.13 CYLINDRICAL WAVEGUIDES

Propagation along any structure that has a cross section to the z direction that does not vary with z can be described mathematically in an elegant way by decomposing the operator ∇, or *nabla,* into two parts. One part is ∇_t which operates in the plane transverse to z. The other part, which acts along the z direction, is $\hat{\mathbf{k}}\,\partial/\partial z$, where $\hat{\mathbf{k}}$ is a unit vector in the z direction. Now, Maxwell's curl equation for **H** becomes Eq. (1–21).

$$\boldsymbol{\nabla}_t \times \mathbf{H} + \hat{\mathbf{k}} \times \partial\mathbf{H}/\partial z = j\omega\varepsilon\mathbf{E} \tag{1–21}$$

Expressing **H** as the sum of a transverse part $\mathbf{H}_t$ and a longitudinal part $\hat{\mathbf{k}}H_z$, we obtain Eq. (1–22) where we have used only the transverse component of **E**, dropped terms that are zero, and replaced $\partial/\partial z$ by $-j\beta$.

$$j\omega\varepsilon\mathbf{E}_t = \boldsymbol{\nabla}_t \times \hat{\mathbf{k}}H_z - j\beta\hat{\mathbf{k}} \times \mathbf{H}_t \tag{1–22}$$

Following the same procedure for Maxwell's curl equation for **E**, we obtain Eq. (1–23).

$$-j\omega\mu\mathbf{H}_t = \boldsymbol{\nabla}_t \times \hat{\mathbf{k}}E_z - j\beta\hat{\mathbf{k}} \times \mathbf{E}_t \tag{1–23}$$

These two equations can be solved simultaneously using $\boldsymbol{\nabla}_t \times (\hat{\mathbf{k}}H_z) = -\hat{\mathbf{k}} \times (\boldsymbol{\nabla}_t H_z)$ and the vector identity $\mathbf{A} \times \mathbf{B} \times \mathbf{C} = \mathbf{B}(\mathbf{A} \cdot \mathbf{C}) - \mathbf{C}(\mathbf{A} \cdot \mathbf{B})$ then substituting k^2 for $\omega^2\mu\varepsilon - \beta^2$. The result is Eq. (1–24), which is equivalent to Eq. (1–15). If we can determine the functional form of the operator $\boldsymbol{\nabla}_t$ for the geometry being considered, we can derive the transverse fields in the cylindrical case from the E_z and H_z field components.

$$\left.\begin{aligned} \mathbf{E}_t &= \frac{j}{k^2}\left(-\omega\mu\boldsymbol{\nabla}_t \times (\hat{\mathbf{k}}H_z) - \beta\boldsymbol{\nabla}_t E_z\right) \\ \mathbf{H}_t &= \frac{j}{k^2}\left(\omega\varepsilon\boldsymbol{\nabla}_t \times (\hat{\mathbf{k}}E_z) - \beta\boldsymbol{\nabla}_t H_z\right) \end{aligned}\right\} \tag{1–24}$$

For cylindrical guides, there are TE modes, TM modes, TEM modes, and hybrid modes (which contain both E_z and H_z components). For the TE and TM modes, we can show from Eq. (1–24) that the transverse magnetic and electric fields are perpendicular and that guide impedance given by E_t/H_t is $\beta/\omega\varepsilon$ for the TM modes and $-\omega\mu/\beta$ for the TE modes (see problem 1–22). The TEM mode can not exist in a cylindrical guide having only a single conducting surface. However, in those cylindrical guides that do support a TEM mode (such as a coaxial cable), we can derive the relation $\omega^2\mu\varepsilon = \beta^2$, which then leads to a phase velocity of $1/\sqrt{\mu\varepsilon}$. In the case of a TEM mode, we can derive the relation $E_t/H_t = \sqrt{\mu/\varepsilon}$.

One of the most common of these guides employing hybrid modes is the multimode optical fiber. Another common type of guide that uses hybrid modes is the guide with a nonuniform dielectric medium. We are most interested in a particular guide with a nonuniform dielectric called microstrip. Collin treats guides with nonuniform dielectrics.[15] The fields having no H_z are called *longitudinal section electric (LSE)* modes, and those having no E_z are called *longitudinal section magnetic (LSM)* modes.

1.14 GROUP VELOCITY AND DISPERSION

Two velocities are used to describe the progression of a wave: the phase velocity and the group velocity. The phase velocity is the rate at which z must change in order that the exponent in the function exp $[j(\omega t - \beta z)]$ remains constant. From this, we derived Eq. (1–4) ($V_p = \omega/\beta$). The variation of β with ω can be determined from the characteristic equation. The characteristic equation for waves in a rectangular waveguide, Eq. (1–17), shows that sometimes β is a nonlinear function of ω. Consequently, the phase velocity also can be a nonlinear function of frequency in the general case.

Usually, the signals that are transported in waveguides are modulated sinusoids that can be described as a wave packet having energy distributed over a small range of frequencies. In this case, the phase velocity of the frequency components of the signal differs over the frequency band of the wave packet. It becomes desirable to define a second velocity function to describe the propagation of the entire wave packet. This velocity is called the *group velocity* and is designated V_g.

The expression for the group velocity can be derived from the consideration of the Fourier transform of the wave packet. We restrict this discussion to modulated carriers that have essentially all of their frequency components contained in a narrow band centered upon the carrier frequency. The excitation at the sending end is expressed as usual: $v(t)\cos(\omega_c t)$. The Fourier transform of this is $(\frac{1}{2})[V_\omega(\omega - \omega_c) + V_\omega(\omega + \omega_c)]$, where V_ω is the Fourier transform of the modulation $v(t)$. The solutions of the Helmholtz equation for this excitation are propagating waves given by $(\frac{1}{2})[V_\omega(\omega - \omega_c)e^{-j\beta(\omega)z} + V_\omega(\omega + \omega_c)e^{j\beta(\omega)z}]$. In this expression, we have used $\beta(\omega)$ for β to emphasize the fact that β varies with frequency.

The received waveform at the point $z = L$ is found by taking the inverse Fourier transform of the excitation expression as shown in Eq. (1–25).

$$\text{Received signal} = \frac{1}{4\pi}\left\{\int_{\omega_c-\delta}^{\omega_c+\delta} V_\omega(\omega-\omega_c)e^{j(\omega t-\beta L)}\,d\omega + \right.$$
$$\left. + \int_{-\omega_c-\delta}^{-\omega_c+\delta} V_\omega(\omega+\omega_c)e^{j(\omega t+\beta L)}\,d\omega\right\} \tag{1–25}$$

For this equation, we shall assume the spectrum of the received wave is limited to a band of 2δ radians per second centered about ω_c and $-\omega_c$. The following approximation describes the classic *narrow-band* case.

Because the phase "constant" β is dependent upon the variable of integration, these integrals cannot be evaluated easily. Evaluation of the integrals in an approximate form can be done by first expanding β in the Taylor's series expansion about $+\omega_c$ and about $-\omega_c$ as

$$\beta = \beta(\omega_c) + (\omega - \omega_c)\frac{d\beta}{d\omega}\bigg|_{\omega=\omega_c} + \frac{(\omega-\omega_c)^2}{2}\frac{d^2\beta}{d\omega^2}\bigg|_{\omega=\omega_c} + \ldots \text{ for } \omega > 0 \text{ and}$$

$$\beta = \beta(-\omega_c) + (\omega + \omega_c)\frac{d\beta}{d\omega}\bigg|_{\omega=-\omega_c} + \frac{(\omega+\omega_c)^2}{2}\frac{d^2\beta}{d\omega^2}\bigg|_{\omega=-\omega_c} + \ldots \text{ for } \omega < 0.$$

The characteristic Eq. (1–14), shows β to be an even function of ω so we have β_0 and β_0' as $\beta_0 = \beta(\omega_c) = \beta(-\omega_c)$ and $\beta_0' = d\beta/d\omega\,|_{\omega_c} = -\,d\beta/d\omega\,|_{-\omega_c}$. Using these in the Taylor's series expansions, we discard all but the first two terms of those expansions. This is a first order approximation and if it holds, we say that the signals are "narrow band." Then, we can substitute these first order approximate expansions into Eq. (1–25) resulting in the following expression.

$$\frac{1}{4\pi}\left\{\int_{\omega_c-\delta}^{\omega_c+\delta} V_\omega(\omega-\omega_c)\exp\left[j\omega(t-\beta_0'L) + j\omega_c\beta_0'L - j\beta_0 L\right] d\omega + \right.$$
$$\left. + \int_{-\omega_c-\delta_c}^{-\omega_c+\delta} V_\omega(\omega+\omega_c)\exp\left[j\omega(t-\beta_0'L) - j\omega_c\beta_0'L + j\beta_0 L\right] d\omega\right\}$$

The term $t - \beta_0'L$ is the same form as the retarded time variable that we designated as τ in the discussion on transmission line propagation. We shall use the variable again in the expression ($\tau = t - \beta_0'L$). We can factor several terms out of the integrands that are not dependent upon ω with the result

$$\frac{\exp\left[j(\omega_c\beta_0'L - \beta_0 L)\right]}{2}\left\{\frac{1}{2\pi}\int_{\omega_c-\delta}^{\omega_c+\delta} V_\omega(\omega-\omega_c)e^{j\omega\tau}\,d\omega\right\} +$$
$$+\frac{\exp\left[-j(\omega_c\beta_0'L - \beta_0 L)\right]}{2}\left\{\frac{1}{2\pi}\int_{-\omega_c-\delta}^{-\omega_c+\delta} V_\omega(\omega+\omega_c)e^{j\omega\tau}\,d\omega\right\}.$$

The terms in the curly brackets { } are recognizable as inverse Fourier transforms which result in the terms $v(\tau)\exp\left[j\omega_c(t-\beta_0'L)\right]$ and $v(\tau)\exp\left[-j\omega_c(t-\beta_0'L)\right]$. When these are substituted into the previous expression for the received signal, the result is Eq. (1–26).

$$\text{Received signal} = v(\tau)\cos(\omega_c t - \beta_0 L) \tag{1–26}$$

The modulation on the received signal is $v(\tau)$, which is the same function as the modulation on the original signal with the exception that it is delayed in time. The time difference is $t - \tau = \beta_0'L$. The velocity to be inferred from this expression is the group velocity

$V_g = L/(t - \tau) = 1/\beta_0'$. Then, the general expression for the group velocity is given by Eq. (1–27).

$$V_g = \frac{1}{\left.\dfrac{d\beta}{d\omega}\right|_{\omega = \omega_c}} \tag{1–27}$$

Equation (1–26) still retains the argument of the cosine function such that the velocity of the phase front is the phase velocity $V_p = \omega_c/\beta_0$.

If the waveform is not narrow-band, the first two terms of the Taylor's series expansion for β will not accurately describe the phase constant. In that case, the different "frequency components" of the wave progress at different rates and arrive at the receiving end at different times creating a wave at the output that is markedly different from that which was sent. An interesting example of this phenomenon is whistlers. Pulses of electromagnetic energy created by lightning flashes in the northern hemisphere propagate outward from the earth along the geomagnetic field lines. These lines are in the form of a dipole and they come back to the earth in the southern hemisphere. The signals travel along these lines toward the southern hemisphere and reflect there (where the magnetic field lines converge strongly enough) to propagate back along the geomagnetic field lines to their point of origin. At that point they are no longer in the shape of a pulse. In fact, they are detected as a descending tone best described as a whistle. These pulses are said to have suffered *dispersion.*

Another type of dispersion suffered by waves in some cases is called *modal dispersion.* In this case, there is more than one mode of propagation where the modes have significantly different propagation times. The propagation for any one of the modes produces a nondispersed pulse but, the delay time for the different modes is sufficiently different so that the total output is a sum of many partially overlapping pulses.

Example: An example of this is propagation in a rectangular wave guide at frequencies above the cut-off frequency for the higher order modes. Suppose that we have an RG-52/U waveguide with external dimensions of 2.54 by 1.27 cm, a wall thickness of 1.27 mm, and a length of $\frac{1}{2}$ m. We consider the case of a pulse modulated carrier at 15 GHz. The pulse width is 30 ns. The cut-off frequencies for the waveguide can be calculated to be 6.56 GHz for the TE_{10} mode, 13.12, and 16.16 GHz for the TE_{20} and the TE_{11} modes respectively. We see that two modes will propagate at 15 GHz. If the signal excites both modes, then the signal received at the end of the guide is a composite of signals transmitted by the TE_{10} and the TE_{20} modes. To calculate the group velocity for each mode, we first determine $\partial\beta/\partial\omega$ by differentiating the characteristic equation (see Prob. 1–21). Then Eq. (1–27) results in the expression $V_g = (\sqrt{1 - (f_c/f)^2})/(\sqrt{\mu\varepsilon})$. From this, we calculate group velocities of 2.70×10^8 m/s for the TE_{10} mode and 1.45×10^8 m/s for the TE_{20} mode. The delay time for the signals arriving by these two modes is then 1.85 and 3.44 ns respectively. So the received signal is a composite of signals arriving by these two modes. Each of the individual signals is not dispersed, but the combination is said to be dispersed. The least effect would be a widening of the received pulse by the difference between these two delay times: about 1.6 ns.

PROBLEMS

1–1. By using the chain rule of differentiation: $\left.\frac{\partial\{v(u)\}}{\partial z}\right|_{u=t-z/C} = \frac{dv}{du}\frac{\partial u}{\partial z}$, show that any function $v(t - z/C)$ satisfies the wave equation. In this relation, C is the speed of light. This is D'Alembert's solution of the wave equation.

1–2. Show that the product LC is $\mu\varepsilon$ for a coaxial transmission line operating in a TEM mode where the **E** and **H** are the same as in the static case and where μ and ε are the permeability and permittivity of the material between the inner and outer conductor of the cable.

1–3. Find the wavelength λ_g of a 1 GHz TEM wave on an RG-58C/U coaxial cable.

1–4. A transmission line that has both line and dielectric losses has the circuit representation shown below (see Figure 1–11). Derive the wave equation for this line.

1–5. Find Γ at $z = 0$ for both circuits below (see Figure 1–12).

1–6. The waveform on the face of the oscilloscope on a TDR is shown in Figure 1–13. It represents the voltage measured at the input of a transmission line. Find the load resistance R_L on the transmission line if Z_0 is 50 Ω. Also, find the apparent distance to the load.

1–7. The incident and reflected waves on an RG-58C/U cable are given by $E_i = 10\angle 50°$ and $E_r e^{j\theta} = 5\angle 95°$ measured at the load. Find the reflection coefficient Γ and the load impedance Z_L.

1–8. Derive Eq. (1–10).

1–9. A transmitter operating at 1 GHz feeds power into a 50 Ω lossless transmission line terminated in a 10 Ω antenna.

(a) If 10 watts of power are fed into the antenna how much power is there in the incident wave and in the reflected wave on the transmission line?

(b) What is the impedance seen by the transmitter if the transmission line is $\frac{1}{2}$ wavelength long?

1–10. A quarter wavelength long line is sometimes referred to as a *quarter wave transformer.* From Eq. (1–11), show that the line input impedance, the load impedance, and the characteristic line impedance are related by $Z_{in} Z_L = Z_0^2$ for a quarter wavelength long line.

1–11. If the transmitter and antenna of problem 1–9 were connected by a quarter wave transformer of problem 1–10, what must be the transmitter output impedance for the transmitter to be matched to maximum power output?

1–12. Find the characteristic impedance Z_0 to a TEM wave of a coaxial cable with dielectric mate-

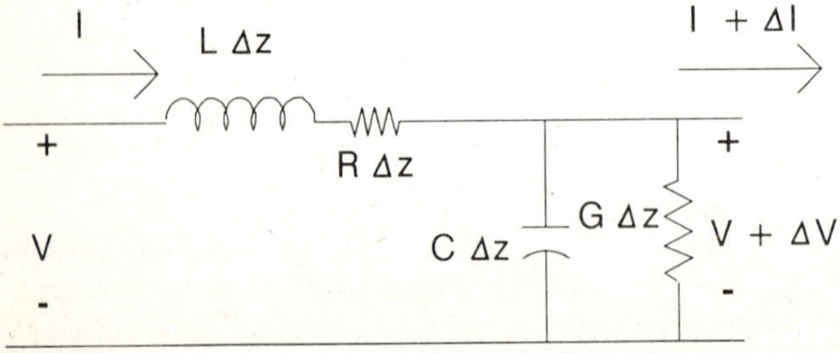

FIGURE 1–11. Circuit for Problem 1–4.

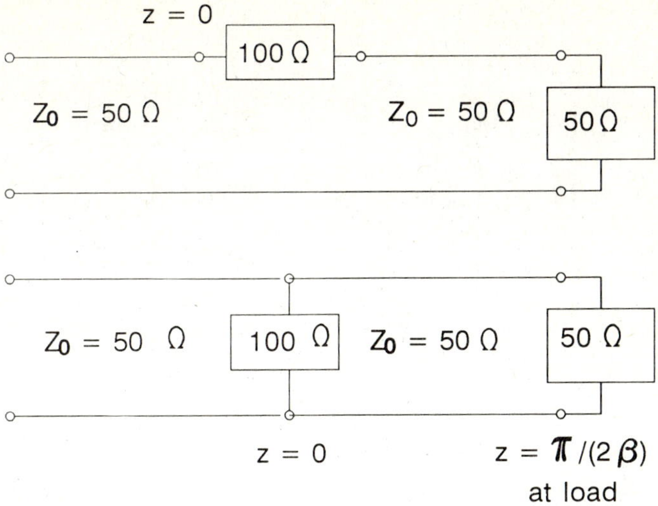

FIGURE 1–12. Circuit for Problem 1–5.

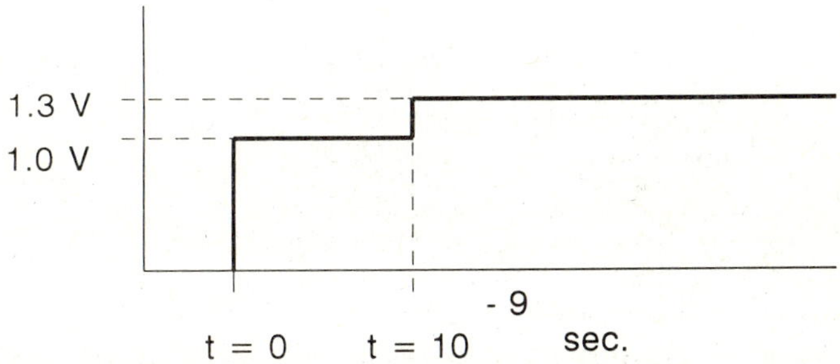

FIGURE 1–13. Waveform for Problem 1–6.

rial between the conductors having $\varepsilon_r = 9$ and $\mu_r = 1$ and having the ratio of outer to inner conductors of 10. Assume that the **E** and **H** fields are the same as for a static field.

1–13. A 50 Ω TEM line terminated in a load impedance Z_L has a standing wave ratio equal to 2.555. At 5.5 cm from the load, the line voltage is minimum. When a short circuit replaces the load, the line voltage minima occur at 4, 8, and 12 cm from the short. Determine Z_L.

1–14. Determine the characteristic impedance of an RG-52/U waveguide at 7 GHz and 10 GHz. What happens to this impedance at cut-off?

1–15. An RG-52/U guide carries $\frac{1}{2}$ watt of power at 10 GHz to a matched load.

(a) Find the maximum electric field E_0 in the guide.

(b) What are the voltage and current on the equivalent transmission line?

1–16. An RG-52/U waveguide is operated at 12 GHz and feeds an antenna whose equivalent admittance is $Y_a = 0.00222 + j0.002$ S. Find the power transmitted to the antenna if the incident wave has a maximum electric field of $E_0 = 1{,}670.7$V/m.

1–17. Find the cutoff frequencies in the TE_{11}, TE_{01}, and TM_{11} modes for an RG-52/U waveguide with the outside dimensions of 1 inch by $\frac{1}{2}$ inches and conducting wall thickness of 0.05 inches.

1–18. Find the dimensions of a symmetrical capacitive iris that has a susceptance, $B = +0.002\ S$ as measured at 12 GHz in an RG-52/U waveguide.

1–19. Derive Eq. (1–15).

1–20. Show that the boundary condition $E_{tan} = 0$ at conducting walls will not allow a TM_{10} mode in a rectangular waveguide.

1–21. Derive the formula for the group velocity in a rectangular waveguide and show that $V_g V_p = C^2$, where $C = 1/\sqrt{\mu\varepsilon}$ is the speed of light in the medium interior to the waveguide.

1–22. From Eq. (1–24) show that the transverse electric field and the transverse magnetic field are perpendicular in either the *TE* or the *TM* modes. Also, show from Eq. (1–24) that the ratio of the transverse electric field to the transverse magnetic field is $\omega\mu/\beta$ for a *TM* mode and $\beta/\omega\varepsilon$ for a *TE* mode in a lossless cylindrical guide.

ENDNOTES

1. Robert M. Barrett, "Microwave Printed Circuits—The Early Years," *IEEE Trans. Microwave Theory Tech., MTT-32, 9* (Sept. 1984), (983–984).
2. Harlan Howe, Jr., "Microwave Integrated Circuits—An Historical Perspective," *IEEE Trans. Microwave Theory Tech., MTT-32, 9* (Sept. 1984), 991.
3. Seymour B. Cohn, "Slot-Line—An Alternative Transmission Medium for Integrated Circuits," *1968 G-MTT International Microwave Symp. Dig.* (104–109) (New York: The Institute of Electrical and Electronic Engineers, 1968).
4. Seymour B. Cohn, "Slot Line on a Dielectric Substrate," *IEEE Trans. Microwave Theory Tech., MTT-17, 10* (Oct. 1969), 768–778.
5. Cheng P. Wen, "Coplanar Waveguide: A Surface Strip Transmission Line Suitable for Nonreciprocal Gyromagnetic Device Application," *IEEE Trans. Microwave Theory Tech., MTT-17, 12* (Dec. 1969), 1087–1090.
6. Paul J. Meier, "Two New Integrated-Circuit Media with Special Advantages at Millimeter Wavelengths," *1972 IEEE-GMTT International Microwave Symp. Dig.* (221–223) (New York: The Institute of Electrical and Electronic Engineers, Inc., 1972).
7. Henry Guckel, Pierce A. Brennan, and István Palócz, "A Parallel-Plate Waveguide Approach to Microminiaturized, Planar Transmission Lines for Integrated Circuits," *IEEE Trans. Microwave Theory Tech., MTT-15, 8* (Aug. 1967), 468–476.
8. Hideki Hasegawa, Mieko Furukawa, and Hisayoshi Yanai, "Properties of Microstrip Line on Si-SiO$_2$ System," *IEEE Trans. Microwave Theory Tech., MTT-19, 11* (Nov. 1971), 869–881.
9. K. F. Sander and G. A. L. Reed, *Transmission and Propagation of Electromagnetic Waves* (290) (London: Cambridge University Press, 1986).
10. Robert S. Elliott, *An Introduction to Guided Waves and Microwave Circuits,* Chap. 12. (Englewood Cliffs, NJ: Prentice Hall, 1993).
11. Shung Wu Lee, William R. Jones, and James J. Campbell, "Convergence of Numerical Solutions of Iris-Type Discontinuity Problems," *Numerical Methods for Passive Microwave and Millimeter Wave Structures,* (326–334) ed. Roberto Sorrentino. (New York: IEEE Press, 1989).

12. P. H. Masterman and P. J. B. Clarricoats, "Computer Field-Matching Solution of Waveguide Transverse Discontinuities," *Numerical Methods,* ed. Roberto Sorrentino.
13. Alvin Wexler, "Solution of Modal Discontinuities by Modal Analysis," *Numerical Methods,* ed. Roberto Sorrentino.
14. Elliott, *Guided Waves . . .*, Chaps. 4, 5.
15. Robert E. Collin, *A Field Theory of Guided Waves,* (2nd ed.), Chap. 6, (New York: IEEE Press, 1991).

Chapter

2

Microstrip Lines

2.0 PLANAR WAVEGUIDES

Various planar microwave waveguides were shown in Figures 1–1 and 1–2. Other variations have been devised using wave propagation along dielectric boundaries. These lines all have different ranges of applicability. The electrical characteristics used for comparison are: (1) range of impedance, (2) loss per unit length (including ohmic and radiation losses), (3) dispersion characteristics, and (4) power handling capability. Comparisons on these bases are given by Gupta, Garg, and Bahl and Edwards.[1,2] Usually, the preferred waveguide in the frequency range from about 1 GHz to a few tens of GHz is microstrip. In this frequency range, impedances on the order of 10 to 100 Ω are realizable and dispersion curves are available if the design tolerances require them. However, equipment using microstrip operating at much higher frequencies (75 to 110 GHz) has been reported.[3] Losses, dispersion, and multimode propagation become important factors at higher frequencies. At these frequencies, finlines and CPW have been made with lower loss and better dispersion characteristics than microstrip.[4,5]

In order to design circuits using planar waveguides at these higher frequencies, one must have design data available concerning characteristic impedance, dispersion, scattering at junctions between lines, effective capacitance or inductance of line configurations, and so on, for these frequencies. The principal way to obtain reliable data on these characteristics has been to solve for the electromagnetic fields propagating along the guide. These fields must satisfy Maxwell's equations and the boundary conditions at the conducting boundaries and at the edges of the surrounding dielectric medium. Because of the complexity of these problems, their solutions usually require numerical computation. In many cases, these computations require large amounts of computer time. However, methods have been devised for making the calculations more efficient. Some of these methods are discussed in this chapter.

Design computations for microwave circuits usually are done with the aid of *computer aided design (CAD)* software. The computations made using this software often rely on empirical formulas determined by fitting curves to the results from the full wave solutions previously mentioned for the electromagnetic fields. The empirical formulas are put in terms of the voltage, current, characteristic impedance for a transmission line that is supposedly equivalent to the microstrip line, and lumped circuit element values equivalent to the effects of line discontinuities. Because of the dispersion present in microstrip lines, considerable effort has been expended in determining empirical formulas for the phase constant β and for an effective (relative) dielectric constant. Some of these results are discussed in this chapter.

2.1 NON-TEM PROPAGATION

With the exception of the stripline, some of the waveguides mentioned in section 1.0 are incapable of TEM wave propagation. Because the dielectric medium around the conductor in the strip line is homogeneous, the strip line will support a TEM mode. The phase velocity and characteristic impedance of the TEM mode do not vary with frequency if the permittivity and permeability are not frequency dependent. However, the microstrip fields are not symmetrical and are not purely TEM, so the phase velocity in it is frequency dependent. For these reasons, the stripline enjoyed an early popularity however, the printed-circuit-like nature of the microstrip line, which allows integrated circuit manufacturing techniques, soon made microstrip the more favored waveguide.[6]

One can use the following argument to determine that a TEM wave cannot propagate on a microstrip line.[7] We refer to Figure 2–1 for the coordinate axes location. We assume that there is a TEM wave propagating with the propagation factor $e^{-\gamma z}$ and then we show that this wave cannot exist. With the TEM assumption, we have E_z and H_z both equal to zero. We can solve Maxwell's equation $\nabla \times \mathbf{H} = j\omega\varepsilon\mathbf{E}$ to get Eq. (2–1).

$$E_x = (\gamma / j\omega\varepsilon)H_y \tag{2–1}$$

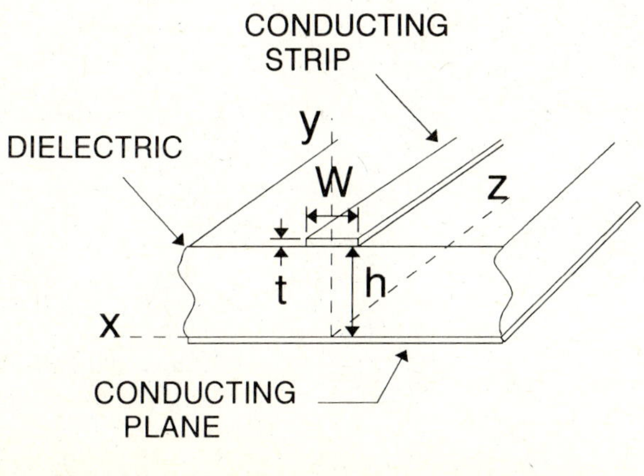

FIGURE 2–1. Microstrip Geometry.

The condition on the electric fields at the dielectric boundary $y = h$ as shown in Figure 2–1 is $E_{\tan}(x, h-, z) = E_{\tan}(x, h+, z)$. However, the tangential component of **E** is E_x so, we have $E_{\tan} = E_x = (\gamma/j\omega\varepsilon) H_y$. Substituting this into the boundary component of the condition above, we arrive at Eq. (2–2).

$$(\gamma/j\omega\varepsilon_0\varepsilon_r)H_y(x, h-, z) = (\gamma/j\omega\varepsilon_0)H_y(x, h+, z) \tag{2–2}$$

The normal magnetic field at the dielectric boundary $y = h$ has the boundary condition $B_{\text{norm}}(x, h-, z) = B_{\text{norm}}(x, h+, z)$. This becomes $H_y(x, h-, z) = H_y(x, h+, z)$ when the permeability of the dielectric is unity. Substituting this into Eq. (2–2), we get $(1 - 1/\varepsilon_r) H_y(x, h, z) = 0$. Because the dielectric constant ε_r is greater than unity, the first factor cannot be zero, so H_y must be zero for this to be true. However, this implies either there is no wave propagating on the line or the TEM assumption is false. The waves that propagate along this structure must not be TEM.

2.2 LINE IMPEDANCE DEFINITIONS

The fact that the microstrip fields must have a longitudinal component (z component) introduces a complication in defining the line impedance.[8] To understand the nature of the problem, we first consider a dc voltage applied to the line and a dc current along it. The line integral of the electric field from the ground plane to the strip and the line integral of the magnetic field taken around the strip are related to the line voltage V and line current I as in Eq. (2–3). Now, consider the case of microstrip.

$$\int \mathbf{E} \cdot d\mathbf{l} = V, \qquad \oint \mathbf{H} \cdot d\mathbf{l} = I \tag{2–3}$$

As shown in Figure 2–2, the line integral from the ground plane to the strip can be taken over two different paths a and b. These paths can be connected through the conductors resulting in the line integral $\oint \mathbf{E} \cdot d\mathbf{l}$ around a closed path such that the path lies in a plane transverse to the z axis. For the voltage to be defined uniquely, this line integral must have a zero value. But, this line integral can be expressed as in Eq. (2–4) using the integral form of Maxwell's equations.

$$\oint \mathbf{E} \cdot d\mathbf{l} = \int (\partial \mathbf{B}/\partial t) \cdot d\mathbf{S} \tag{2–4}$$

The line integral is related to a surface integral of the time derivative of the magnetic field. The surface is that enclosed between the two paths. From Figure 2–2, we see that the longitudinal component of **B** is perpendicular to the surface of the surface integral and contributes a nonzero value for this surface integral. From Eq. (2–4) we infer that the line integral around the closed path also must not be zero. This leads to the conclusion that the line integral along path a is not the same as that along path b. A similar argument can be given for the line integral $\oint \mathbf{H} \cdot d\mathbf{l} = \int \mathbf{J} \cdot d\mathbf{S} + \int (\partial \mathbf{D}/\partial t) \cdot d\mathbf{S}$ in which the line integral is related to the current as in Eq. (2–3). The consequence is that we cannot uniquely relate a voltage or current to the **E** and **H** fields respectively by Eq. (2–3) except at zero frequency. Thus, the "characteristic impedance" as a ratio of V/I is not unique.

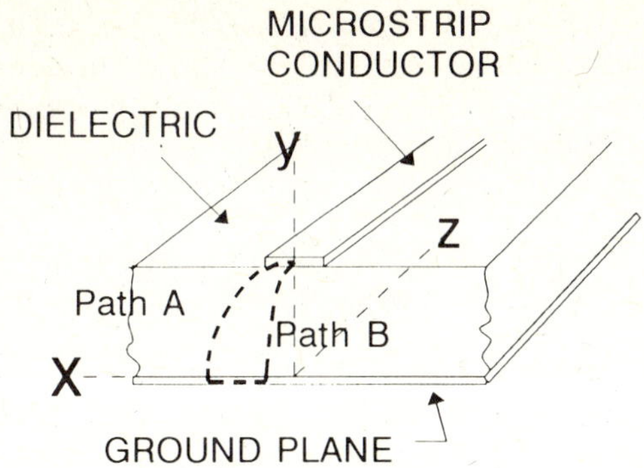

FIGURE 2–2. Line Integral Paths for Voltage Calculation.

For the case where there is only a single hybrid mode, the choice of voltage and current definitions in terms of the **E** and **H** fields can be made so that the transmission line model of microstrip is valid. The key to this is that they must be chosen so that the complex power, given by $P + jQ = VI^*/2$, also agrees with the power transfer down the microstrip as calculated by $P + jQ = (\frac{1}{2}) \int \mathbf{E} \times \mathbf{H}^* \cdot d\mathbf{S}$.[9] In the case where there are many modes propagating along microstrip, the characteristic impedance cannot be defined uniquely. The practical consequence of this matter lies in the choice of formulas to be used for computing the characteristic impedance of a dispersive microstrip line. One of the ways in which it has been defined is according to the reflection measured when a TEM line is connected to it. Another way is by the ratio $V^2/2P$, where V is the peak voltage at the strip center and P is the power transfer down the line. A third definition is $P/2I^2$, where I is defined as the surface integral of the longitudinal current density over the cross section of the microstrip conductor: $I = \int\int J_z \, dx \, dy$. The power-current definition of characteristic impedance is that most frequently adopted.

2.3 SUBSTRATE MATERIALS

Some of the more common materials on which microstrip circuits are fabricated are alumina, RT/ duroid® microwave laminates, gallium arsenide, and sapphire. A discussion of such materials is given by Edwards[10] and Hoffman.[11] Table 2–1 lists dielectric constants for these materials. Sapphire has an anisotropic nature. It is crystalline and has different

TABLE 2–1 Dielectric Constants.

Material	Dielectric Constant
alumina	8–10
RT/duroid® microwave laminates	2.23–2.33
GaAs	12.8
sapphire	9.4/11.6*

*anisotropic dielectric

dielectric properties along different axes of the material. Although this makes the design more difficult, sapphire has the advantages that it is very pure and the dielectric constant is accurately known and achieved in production. Gallium arsenide is generally used for higher frequencies.

2.4 QUASI-STATIC APPROXIMATION

The propagation along a microstrip line is governed by the Helmholtz equation (Eq. (2–5) or Eq. (1–13)), in which E_j can be any of the E_x, E_y, or electric field components.

$$\partial^2 E_j/\partial x^2 + \partial^2 E_j/\partial y^2 + \partial^2 E_j/\partial z^2 = -\omega^2 \mu\varepsilon\, E_j \tag{2–5}$$

The z variation is assumed to be $e^{\gamma z}$ so that the second derivative with respect to z for a lossless line becomes multiplication by $\gamma^2 = -(2\pi/\lambda)^2$, where λ is the guide wavelength. The guide wavelength λ and the radian frequency ω bear an inverse relation; for example see Eq. (1–18). So, the third and fourth terms in Eq. (2–5) decrease with decreasing frequency. When ω is sufficiently small, the last two terms are negligible with respect to the first two terms. In that case, Eq. (2–5) can be approximated by the 2-dimensional Laplace's equation in Eq. (2–6).

$$\partial^2 E_j/\partial x^2 + \partial^2 E_j/\partial y^2 = 0 \tag{2–6}$$

This is the *quasi-static* approximation. In this approximation, the field becomes purely transverse. The field becomes TEM in the dc limit. The propagation mode asymptotic to this is referred to as a *quasi-TEM* mode. In the lower frequency range (a few GHz), the fields must be asymptotic to those at zero frequency. For this frequency range, the longitudinal components are usually ignored and the quasi-static approximation is used. The accuracy of this approximation improves as the ratio of the wavelength to the thickness of the dielectric increases.

2.5 QUASI-STATIC LINE PARAMETERS

Equations have been determined for designing microstrip lines in the quasi-static case. These design equations are used for determining the characteristic impedance as a function of the line width, dielectric thickness, and dielectric constant. Equations for the inverse computation also are available. These equations result from solutions for the quasi-static fields. One method for finding these fields is by the use of *conformal transformations* in which a spatial transformation for relating the microstrip geometry to a parallel plane geometry is used.[12,13] Other methods used are a numerical solution of Eq. (2–6), a numerical solution of the integral equation obtained from Eq. (2–6), and a variational method using the Fourier transform.[14,15] The design equations are then determined empirically by fitting equations to the numerical data from these solutions.

The line parameters such as the characteristic impedance and wavelength along the line are related to an *effective dielectric constant* ε_{eff}. We define ε_{eff} such that the phase velocity V_p and the guide wavelength λ are given by Eq. (2–7), where λ_0 is the free space wavelength and C is the speed of light in free space.

$$V_p = \frac{1}{\sqrt{\mu_0 \varepsilon_0 \varepsilon_{eff}}} = \frac{C}{\sqrt{\varepsilon_{eff}}}, \quad \lambda = \frac{\lambda_0}{\sqrt{\varepsilon_{eff}}} \tag{2–7}$$

Equations (2–8) and (2–9) are design equations for obtaining Z_0 in the case where the thickness of the conductor t is negligible. These equations fit the numerical quasi-static solutions within 0.2% for $\varepsilon_r \leq 128$ and $0.01 \leq W/h \leq 100$.[16]

$$\left. \begin{aligned} Z_0 &= \frac{59.9585}{\sqrt{\varepsilon_{eff}}} \ln\left\{ \frac{h}{W}\left[6 + (2\pi - 6)\, e^{-A}\right] + \sqrt{1 + (2h/W)^2} \right\} \\ &\text{where } A = (30.666\ h/W)^{0.7528} \end{aligned} \right\} \tag{2–8}$$

$$\left. \begin{aligned} \varepsilon_{eff} &= \frac{\varepsilon_r + 1}{2} + \frac{\varepsilon_r - 1}{2}(1 + 10\ h/W)^{-ab} \\ &\text{where } b = 0.564\left(\frac{\varepsilon_r - 0.9}{\varepsilon_r + 3}\right)^{0.053} \text{ and} \\ a &= 1 + (1/49)\ln\left(\frac{(W/h)^4 + (W/52h)^2}{(W/h)^4 + 0.432}\right) + (1/18.7)\ln\left(1 + (W/18.1h)^3\right) \end{aligned} \right\} \tag{2–9}$$

The inverse problem of determining the width W of the microstrip conductor and the thickness h of the dielectric can be solved by using Eq. (2–10) and Eq. (2–11).[17] These equations were selected to give the best result for the case where ε_r is in the range of 8 to 12. The formulas in Eq. (2–10) and Eq. (2–11) have a slight disagreement at the crossover point $Z_0 = 44 - 2\varepsilon_r$.

for $Z_0 \leq 44 - 2\varepsilon_r$:

$$\left. \begin{aligned} \frac{W}{h} &= \frac{2}{\pi}\left\{ B - 1 - \ln(2B - 1) + \frac{\varepsilon_r - 1}{2\varepsilon_r}\left[\ln(B - 1) + 0.293 - \frac{0.517}{\varepsilon_r}\right]\right\} \\ &\text{where } B = \frac{377\,\pi}{2\,Z_0\sqrt{\varepsilon_r}} \end{aligned} \right\} \tag{2–10}$$

for $Z_0 \geq 44 - 2\varepsilon_r$:

$$\left. \begin{aligned} \frac{W}{h} &= \frac{8\,e^{\eta}}{e^{2\eta} - 2}, \\ &\text{where } \eta = \sqrt{\frac{\varepsilon_r + 1}{2}\frac{Z_0}{60} + \frac{\varepsilon_r - 1}{\varepsilon_r + 1}\left(0.226 + \frac{0.12}{\varepsilon_r}\right)} \end{aligned} \right\} \tag{2–11}$$

At $\varepsilon_r = 9$, the difference between the two is less than 0.4%, while at $\varepsilon_r = 3$ the disagreement is about 2%. More accurate design equations are available for reducing this disagreement.[18]

Example: Suppose we desire to fabricate a 75 Ω line on alumina with a thickness of 0.635 mm and a dielectric constant ε_r of 9. Because Z_0 is greater than $44 - 2\varepsilon_r$, we use the formula in Eq. (2–11) as:

$$\eta = \sqrt{\frac{9+1}{2}\frac{75}{60} + \frac{8}{10}\left(0.226 + \frac{0.12}{9}\right)} = 2.987$$

$$W/h = \frac{8\,e^{2.987}}{e^{5.974} - 2} = 0.4058, \quad \text{or } W = 0.4058 \times 0.635 = 0.2577 \text{ mm.}$$

The resulting line must be about 0.26 mm wide. To cross check these formulas, we can use the parameters $W/h = 0.4058$ and $\varepsilon_r = 9$ in Eq. (2–9). Performing this calculation as shown below, we obtain $\varepsilon_{eff} = 5.739$ and $Z_0 = 74.7\ \Omega$.

$$a = 1 + (1/49)\ln\left(\frac{(0.4058)^4 + (0.4058/52)^2}{(0.4058)^4 + 0.432}\right) + (1/18.7)\ln\left(1 + (0.4058/18.1)^3\right)$$

$$a = 0.9423$$

$$b = 0.564(8.1/12)^{0.053} = 0.5524$$

$$\varepsilon_{eff} = 10/2 + 4(1 + 10/0.4058)^{-0.5206} = 5.739$$

$$A = \left(\frac{30.666}{0.4058}\right)^{0.7528} = 25.943$$

$$Z_0 = \frac{59.9585}{\sqrt{5.739}}\ln\left(\frac{6 + (2\pi - 6)\,e^{-25.943}}{0.4058} + \sqrt{1 + (2/0.4058)^2}\right) = 74.745$$

The formulas previously given apply to the case in which the microstrip is not enclosed inside a conducting box. Corrections must be made when packaging a microstrip circuit in that manner.[19]

2.6 MICROSTRIP OPEN CIRCUITS AND GAPS

The subject of microstrip discontinuities has been covered in detail by Hoffman.[20] Only a relatively brief collection of published results will be presented here.

The electromagnetic fields at the end of a long straight microstrip line terminated with a rectangular end, as shown in Figure 2–3, exhibit reflection, radiation, mode conversion, and a concentration of reactive energy that make circuit modeling difficult at the higher frequency range. The concentration of reactive energy creates a capacitive effect. This capacitive effect and reflection have been calculated using the quasi-static calcula-

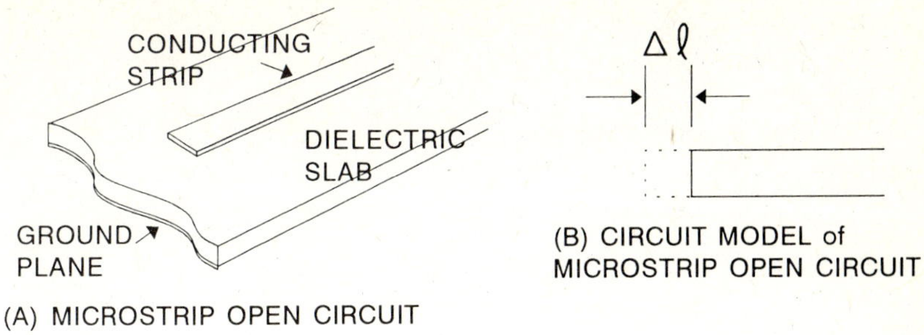

FIGURE 2–3. $\Delta\ell$ is the extension which models the capacitive effect of the microstrip open circuit.

tions, which ignore radiation and inductive effects. These results have been used to model the open circuit as a pure reflection on a transmission line with a length extended by $\Delta\ell$ as shown in Figure 2–3. In this figure, the circuit model of the microstrip line is shown as a rectangular box. The extension is shown as dotted lines. Hammerstad introduced the empirical formula in Eq. (2–12) to account for this effect.[21]

$$\frac{\Delta\ell}{h} = 0.412\frac{\varepsilon_{eff}+0.3}{\varepsilon_{eff}-0.258}\times\frac{(W/h)+0.262}{(W/h)+0.813} \tag{2–12}$$

The ε_{eff} here can be calculated as in Eq. (2–9). A more accurate, but more cumbersome, formula for this has been given.[22] That formula is reported to be better than 2.5% at 1 GHz in the ranges: $0.01 \le W/h \le 100$ and $\varepsilon_r \le 50$.

A gap in a microstrip line such as is shown in Figure 2–4 is modeled at the lower microwave frequencies as a pi network of capacitors as shown. An integral equation derived from Eq. (2–6) was solved numerically to determine quasi-static values for these capacitances as functions of microstrip width-to-substrate-thickness ratio (W/h).[23] These data were presented in graphical form. Measurements have been made which agree with these quasi-static calculations up to 17 GHz on 0.66 mm thick alumina substrates.[24] However, these calculations do not predict the frequency dependence of the capacitive effect.

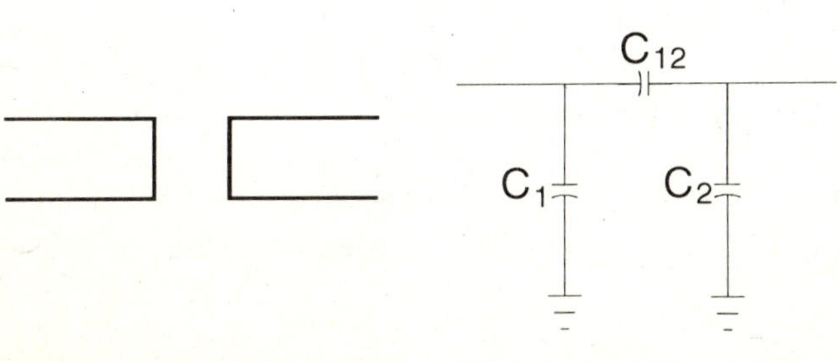

FIGURE 2–4. Microstrip Gap Discontinuity and Equivalent Circuit.

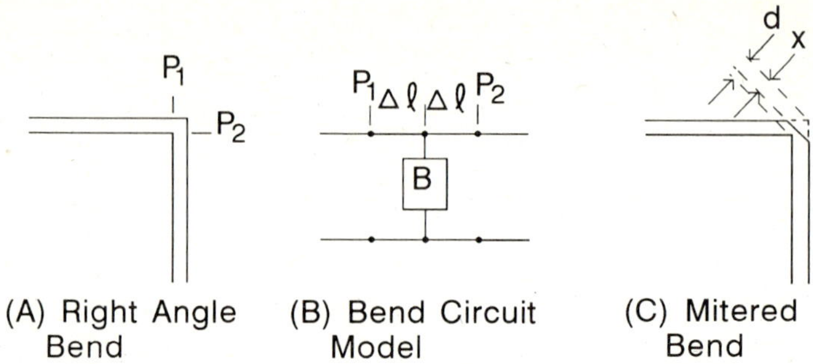

FIGURE 2–5. Microstrip Bends and Their Equivalent Circuits.

2.7 MICROSTRIP CORNERS

A right-angled bend, such as is shown in Figure 2–5, perturbs the wave traveling down a microstrip line causing reflections. The effects of this bend can be modeled for the quasi-static case by the circuit shown in Figure 2–5. Graphs made from measurements detailing these effects have been reported.[25] These effects can be minimized somewhat by a beveled joint as shown in Figure 2–5. This is called a *mitered corner.* The percent ratio of x to d is defined to be M: $M = 100(x/d)$. For the 45 degree miter, the optimum M has been shown to be given by Eq. (2–13) for dielectric constants in the range 2.5 to 25, $W/h \geq 0.25$ and $h = 5$ mm.[26]

$$M = 52 + 65 \exp(-1.35W/h) \tag{2–13}$$

This expression agrees with measurements within ± 4 and results in a maximum VSWR of 1.1 below 2 GHz. By suitable rescaling, Eq. (2–13) may be applied at higher frequencies to substrates with smaller h, such as alumina with a thickness of 0.635 mm.

Aside from the capacitive effect at corners, there is a crowding of the current near to the corner that causes an inductive effect modeled as shown in Figure 2–5. Quasi-static calculations have been made to predict the magnitude of this effect.[27,28]

2.8 STEP CHANGES IN WIDTH

As one would expect, a step change in the width of a microstrip line will not only cause a change in the impedance of the line but will also result in evanescent waves at this change that cause reactive energy storage near to the step. The equivalent circuit used to model these effects is shown in Figure 2–6. The components in this equivalent circuit have been determined from quasi-static calculations.[29,30,31,32] Data showing these circuit components are given in Figure 2–7. These data do not apply at the higher microwave frequencies. The discontinuity capacitances graphed in Figure 2–7 are normalized to the product of the line capacitance $C_{\infty w2}$ and the substrate thickness h. We have

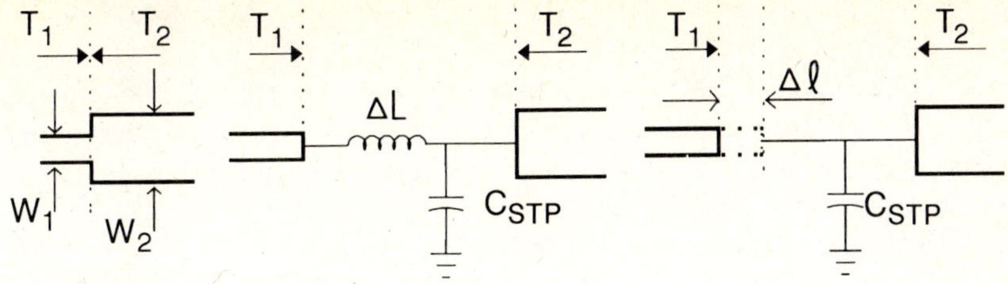

FIGURE 2–6. Step Discontinuity Approximate Equivalent Circuit. The reference planes T_1 and T_2 show where the equivalent circuit fits into the microstrip line configuration.

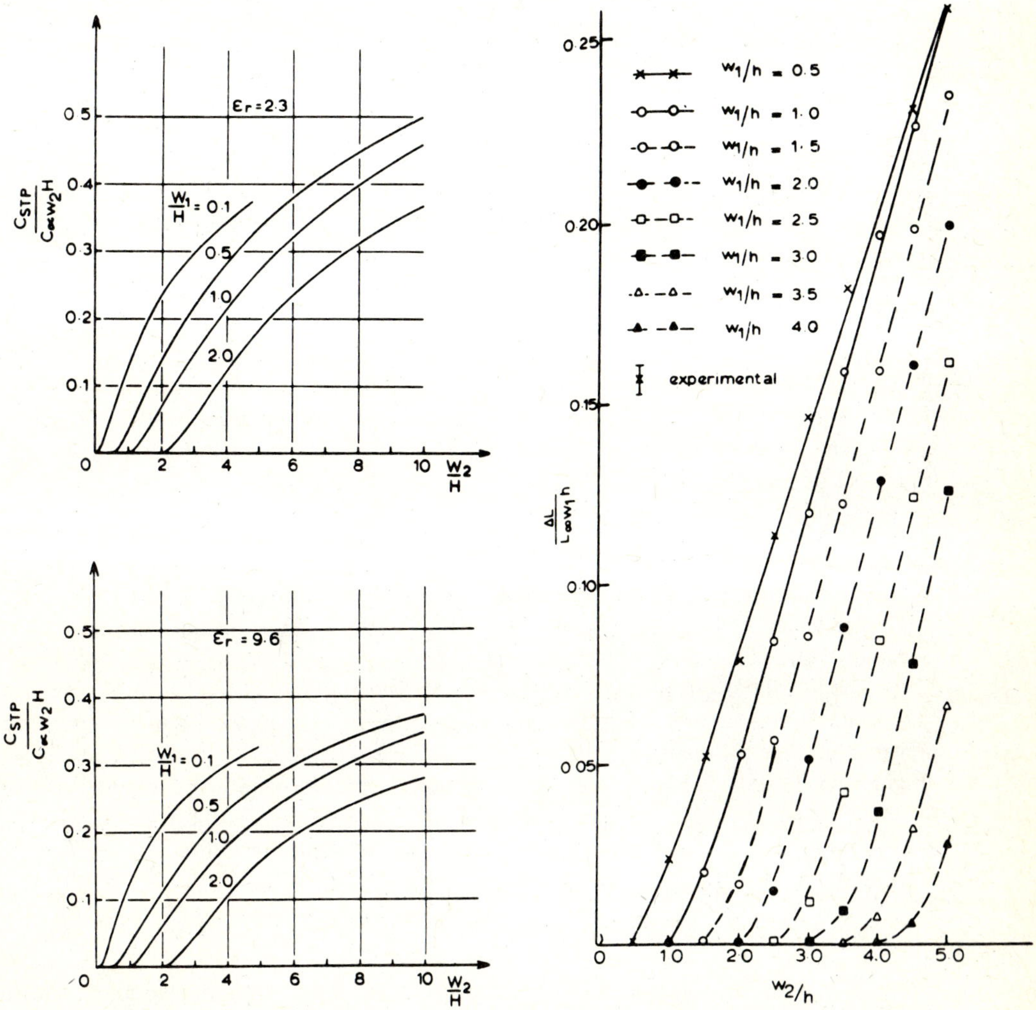

FIGURE 2–7. Microstrip Step Discontinuity Circuit Parameters. The left-hand charts are for the step capacitance normalized to the product of the line capacitance $C_{\infty W2}$ and h.[31] The right-hand chart is for the step inductance normalized to the product of $L_{\infty W1}$ and h.[32] © 1977 IEEE and © 1976 IEEE. Reprinted with the permission of the IEEE.

$Z_0 = \sqrt{L_\infty / C_\infty}$ and $V_p = 1/\sqrt{L_\infty C_\infty}$ from which we obtain the line capacitance $C_\infty = 1/(Z_0 V_p)$. Similarly, the discontinuity inductances shown in these figures are normalized to $L_{\infty w1}$ times h. The line to the left of the step in Figure 2–6 has the inductance $L_{\infty w1}$ equal to Z_0/V_p.

Example: Step Change in Line Width. To determine the equivalent circuit of a step change in line width, we take a case where the microstrip line changes from 0.33 mm to 1.32 mm, where 0.33 mm is W_1 in Figure 2–6 and 1.32 mm is W_2. The line is on a substrate with $h = 0.66$ mm and a dielectric constant of $\varepsilon_r = 9.6$. Using $W_2/h = 2.0$ and $\varepsilon_r = 9.6$ in Eq. (2–9), we get $\varepsilon_{eff} = 6.897$. Using these in Eq. (2–8), we have $Z_{02} = 33.9\ \Omega$. Then the line capacitance for the wide line is $C_{\infty W2} = 1/(Z_0 V_p) = (\sqrt{6.89707})/(33.897 \times 3 \times 10^8) = 258.2$ pF/m. From Figure 2–7B, we scale $C_{STP}/(C_{\infty W2} h) = 0.215$. Then, we find $C_{STP} = 36.65$ fF. To determine the inductance ΔL, we calculate $\varepsilon_{eff} = 6.157$ and $Z_{01} = 67.181$ from Eq. (2–9) and (2–8). We scale $\Delta L/(L_{\infty W1} h) = 0.083$ from Figure 2–7C. Then, we can find $L_{\infty W1} = Z_{01}/V_p = 67.181\ (\sqrt{6.157})/3 \times 10^8 = 0.5557\ \mu H$/m. The inductance ΔL is then $\Delta L = 0.083 \times 0.5557 \times 10^{-6} \times 0.66 \times 10^{-3} = 30.44$ pH. Instead of using an inductance in our circuit model, we can use a line extension as shown in Figure 2–6 where the line extension is $\Delta\ell = \Delta L / L_{\infty W1} = 30.44 \times 10^{-12} / 5.557 \times 10^{-7} = 0.0548$ mm.

2.9 DISPERSION ANALYSIS

The quasi-static analysis formulas of the previous sections are valid approximations only so high in frequency. Hoffman has offered an empirical formula, Eq. (2–14), which gives a very conservative upper frequency limit, $f_{g,stat}$ in GHz, above which the quasi-static formulas are no longer accurate.[33]

$$f_{g,stat} = 0.04\ Z_0/(h\sqrt{\varepsilon_{eff}}) \qquad (2\text{–}14)$$

In this formula h is the substrate thickness in mm and one is to use the quasi-static Z_0 and ε_{eff} as calculated with Eqs. (2–8) and (2–9). The relevance of this formula depends on the accuracy required for a particular application.

The initial designs of microstrip circuits are generally made using a TEM transmission line approximate description of the microstrip lines. This approximation uses the transmission line formulas such as: $Z_0 = \sqrt{L/C}$, $V_p = 1/\sqrt{LC} = f\lambda_0/\sqrt{\varepsilon_{eff}}$, where L and C are the inductance and capacitance per unit length of the transmission line that simulates the microstrip line, λ_0 is the free space wavelength of the wave, f is the frequency of the wave, and ε_{eff} is the appropriate (relative) dielectric constant. From these, one can derive the expressions for L and C:

$$L = (Z_0/f\lambda_0)\sqrt{\varepsilon_{eff}}, \quad C = \sqrt{\varepsilon_{eff}}/(Z_0 f\lambda_0)$$

To use a TEM transmission line description of a microstrip line, we require some reasonably accurate way of determining the "effective dielectric constant" ε_{eff} and the "characteristic impedance" Z_0 for the microstrip line at frequencies higher than those for which we can use quasi-static formulas. Full wave solutions have been made for the electromagnetic fields on a microstrip line as functions of frequency, substrate dielectric constant, and the *W/h* ratio to determine these parameters at these higher frequencies. The data from these solutions were fitted to empirical formulas. Equation (2–15) is one formula that has been proposed for the effective dielectric constant.[34,35] This effective dielectric constant depends on the frequency *f* through the function *P(f)* and on the effective dielectric constant at zero frequency $\varepsilon_{eff}(0)$.

$$\varepsilon_{eff} = \varepsilon_r - \frac{\varepsilon_r - \varepsilon_{eff}(0)}{1 + P(f)} \qquad (2\text{–}15)$$

There have been several versions of *P(f)*, one of which is shown as Eq. (2–16). This formula was given by Kobayashi and has been reprinted with the permission of the IEEE.[36] (© 1988 IEEE).

$$P(f) = \left(f/f_{50}\right)^m \quad \text{where:}$$

$$f_{50} = \frac{f_{K,TMO}}{0.75 + (0.75 - (0.332/\varepsilon_r^{1.73}))W/h}$$

$$f_{K,TMO} = \frac{C \tan^{-1}\left(\varepsilon_r \sqrt{\dfrac{\varepsilon_{eff}(0) - 1}{\varepsilon_r - \varepsilon_{eff}(0)}}\right)}{2\pi h \sqrt{\varepsilon_r - \varepsilon_{eff}(0)}}$$

$$m = m_0\, m_c$$

$$m_0 = 1 + \frac{1}{1 + \sqrt{W/h}} + 0.32\left(\frac{1}{1 + \sqrt{W/h}}\right)^3 \qquad (2\text{–}16)$$

$$m_c = \begin{cases} 1 + \dfrac{1.4}{1 + W/h}\left(0.15 - \dfrac{0.235}{\exp(0.45 f/f_{50})}\right), & W/h \le 0.7 \\ 1 \quad , & W/h \ge 0.7 \end{cases}$$

W = microstrip conductor width, h = dielectric thickness,

f = frequency, ε_r = relative dielectric constant,

C = velocity of light in free space

This formula is reported to fit the full wave solutions within 0.6% for the ranges: $0.1 \le W/h \le 10$, $1 \le \varepsilon_r \le 128$ and any h/λ_0 where λ_0 is *C/f*. This last specification implies that this expression fits full wave solutions to frequencies well above 100 GHz: However, its practical application lies in the range where only the lowest order (quasi-TEM) mode propagates.[37] Equations (2–15) and (2–16) have been compared to measured dielectric constants for a range of substrates line widths and frequencies.[38] Except for low impedance lines and high dielectric substrates, these equations gave valid results over a wide range of substrates, impedances, and frequencies. A BASIC program for these equations is given in the Chapter appendix.

Example: Using data measured by Edwards and Owens for a sapphire substrate we can check the accuracy of the dispersion formulas Eqs. (2–15) and (2–16).[39] One of their measurements was of the effective dielectric constant at 14.7 GHz for a line 0.642 mm in width on a sapphire substrate with h = 0.510 mm and ε_r = 11.18. Because of the anisotropy of sapphire, this is an equivalent relative dielectric constant. To calculate the quasi-static effective dielectric, we use Eq. (2–9) and get a = 0.99678, b = 0.55447, and a quasi-static effective dielectric constant

$$\varepsilon_{eff}(0) = \frac{11.18+1}{2} +$$

$$+\frac{11.18-1}{2}(1+10\times 0.510/0.642)^{-0.99678\times 0.5547} = 7.6064$$

The effective dielectric constant at 14.7 GHz can be calculated using Eq. (2–15) and Eq. (2–16). Using Eq. (2–16), one can calculate the factor $f_{K,TMO}$ to be 74.54 × 10^9. Then f_{50} is 44.166 × 10^9, m_0 is 1.505, m_c is 1, and $P(f)$ is 0.1910.

From Eq. (2–15), we get the effective dielectric constant

$$\varepsilon_{eff} = 11.18 - \frac{11.18-7.6064}{1+0.1910} = 8.180.$$

We compare this to an ε_{eff} of about 8.29 at a frequency of 14.7 GHz obtained by scaling from Edwards and Owens Figure 5. This is within about 1.3% of the value calculated here. This is within the accuracy of our scaling of data from Edwards and Owens Figure 5.

2.10 MICROSTRIP CHARACTERISTIC IMPEDANCE

As well as the requirement for an effective dielectric constant, we need a "characteristic impedance" to use with our transmission line description of the propagation along a microstrip line. Empirical formulas for this parameter have been fitted to the results from full wave solutions for the propagating fields on the microstrip lines.

$$Z_0(f) = Z_0\,(0)\,(R_{13}/R_{14})^{R_{17}} \qquad (2\text{–}17)$$

The formula, Eq. (2–17), has been proposed for computing the characteristic impedance.[40] The characteristic impedance computed with this formula is the power-current version given in section 2–2. In this formula, the quantity Z_0 (0) is the quasi-static characteristic impedance computed by Eq. (2–8) and the parameters R_i are those computed by the expressions in Eq. (2–18).

$$R_1 = 0.0389\,\varepsilon_r^{1.4}$$
$$R_2 = 0.267\,(W/h)^7$$
$$R_3 = 4.766\,\exp\,(-3.228(W/h)^{0.641})$$
$$R_4 = 0.016 + (0.0514\,\varepsilon_r)^{4.524}$$
$$R_5 = (fh/28.843)^{12}$$
$$R_6 = 22.2\,(W/h)^{1.92}$$
$$R_7 = 1.206 - 0.3144\,\exp\,(-R_1)[1 - \exp\,(-R_2)]$$
$$R_8 = 1 + 1.275\{1 - \exp\,[-0.004625\,R_3\,\varepsilon_r^{1.674}\,(fh/18.365)^{2.745}]\}$$
$$R_9 = \frac{5.086\,R_4\,R_5\,(\varepsilon_r - 1)^6 \exp(-R_6)}{(0.3838 + 0.386\,R_4)(1 + 1.2992\,R_5)[1 + 10(\varepsilon_r - 1)^6]} \qquad (2\text{–}18)$$
$$R_{10} = 0.00044\,\varepsilon_r^{2.136} + 0.0184$$
$$R_{11} = (fh/19.47)^6/[1 + 0.0962\,(fh/19.47)^6]$$
$$R_{12} = 1/[1 + 0.00245(W/h)^2]$$
$$R_{13} = 0.9408\,\varepsilon_{eff}^{R_8}(f) - 0.9603$$
$$R_{14} = (0.9408 - R_9)\,\varepsilon_{eff}^{R_8}(0) - 0.9603$$
$$R_{15} = 0.707\,R_{10}\,(fh/12.3)^{1.097}$$
$$R_{16} = 1 + 0.0503\,\varepsilon_r^2\,R_{11}\{1 - \exp\,[-(\frac{W}{15\,h})^6]\}$$
$$R_{17} = R_7\{1 - 1.1241\,(R_{12}/R_{16})\,\exp\,[-0.026\,(W/h)^{1.15656} - R_{15}]\}$$

The $\varepsilon_{eff}(f)$ is the effective dielectric constant as computed by Eq. (2–15) and $\varepsilon_{eff}(0)$ is the quasi-static dielectric constant as computed by Eq. (2–9). The term *fh* is the product of the frequency *f* in GHz and the dielectric thickness *h* in *mm.* A BASIC program for computing with these equations is given in the chapter appendix.

2.11 SYMMETRIC T JUNCTION

A common discontinuity in a microstrip line is the junction where one line terminates at another line, forming a T junction. Each of the microstrip lines might have a different characteristic impedance but both are on the same substrate. Calculations for determining the T-junction equivalent circuit have been made where the microstrip lines were simulated with waveguide models. This model of a microstrip line consists of parallel conducting plates with widths labeled W_{eff} forming the top and the bottom of the guide. The top and bottom are separated by the thickness *h* of the dielectric substrate. The sides are modeled with *magnetic walls,* planes over which the tangential magnetic field is required to be zero. The region interior to these boundaries is considered to be filled with a dielectric equal to that of the substrate. The width of the guide is then given by Eq. (2–19), where *h* is the substrate thickness, Z_0 is the line characteristic impedance given by Eq. (2–17), and ε_{eff} is the effective dielectric constant given by Eq. (2–15).

$$W_{effi} = (h/Z_0)\sqrt{\mu_0/(\varepsilon_0 \varepsilon_{effi})}, \qquad w_{ei} = W_{effi}/\lambda_i \qquad (2\text{–}19)$$

The results obtained with this model of the symmetric T junction of Figure 2–8 correspond to the circuit shown in Figure 2–9. The components of this equivalent circuit can be calculated by the formula given in Eq. (2–20).[41,42,43] The distances d_1 and d_2 in Figure 2–8 are those by which the reference planes are shifted to T_3, T_4, and T_5.

$$\left.\begin{aligned} B_{p2} &= [2\tan(2\pi d_1/\lambda_1) - (n^2 Z_{01}/Z_{02})\tan(2\pi d_2'/\lambda_2)]/Z_{01} \\ n &= n_1 \frac{\cos(2\pi d_2'/\lambda_2)}{\cos(2\pi d_1/\lambda_1)} \\ d_1 &= (\lambda_1/2\pi)\arctan(-X_1/Z_{01}) \\ d_2 &= (W_{eff1}/2) - d_2' \end{aligned}\right\} \quad (2\text{–}20)$$

The constants required in Eq. (2–20) are those that can be computed from Eqs. (2–19) and (2–21). A BASIC program for computing with these equations is listed in the appendix.

$$\left.\begin{aligned} & n_1 = \frac{\sin(\pi w_{e2})}{\pi w_{e2}}, \qquad n_1' = n_1\sqrt{Z_{01}/Z_{02}}, \qquad X_1/Z_{01} = -w_{e2}\left(n_1'\,\pi/4\right)^2 \\ & \text{if } Z_{01}/Z_{02} > 0.5 \text{ then :} \\ & X_2/Z_{02} = -X_1/2Z_{02} + \frac{2w_{e2}}{n_1^2}\left[\ln\left(\frac{1.43\,Z_{02}}{Z_{01}}\right) + 2(w_{e1})^2\right] \\ & \text{if } Z_{01}/Z_{02} \le 0.5 \text{ then :} \\ & \frac{B_t Z_{01}}{2} = 2\,w_{e1}\left[\ln\left(\sqrt{1+\tan^{-2}[\pi Z_{01}/(2Z_{02})]}\right) + \right. \\ & \qquad\qquad \left. + 0.5\,(w_{e1})^2\cos^4\left(\frac{\pi Z_{01}}{2Z_{02}}\right)\right] \\ & X_2/Z_{02} = -X_1/2Z_{02} + \left[\frac{B_t Z_{01}^2}{2\,Z_{02}} + 2w_{e2}\left(\ln(2) + \frac{\pi Z_{01}}{6Z_{02}} + 1.5\,(w_{e1})^2\right)\right]\frac{1}{n_1^2} \\ & d_2' = (\lambda_2/2\pi)\arctan(n_1^2\,X_2/Z_{02}) \end{aligned}\right\} \quad (2\text{–}21)$$

Example: A symmetric T junction is on a 1.5875 mm (62.5 mil) thick substrate with a relative dielectric constant of $\varepsilon_r = 2.3$. The equal arms of the junction are 50 Ω and the right angled-arm is 90 Ω, all at 3 GHz. Using Eq. (2–15) and (2–17) together, we find that the width of 4.75 mm results in 50.1 Ω and a width of 1.71 mm results in 90.1 Ω. We also determine the effective dielectric constants of 1.9689 and 1.8515 for the 50 Ω and the 90 Ω lines respectively. From these dielectric constants, we get 3 GHz wavelengths along these lines as in: $\lambda = 3 \times 10^8/(f\sqrt{\varepsilon_{eff}})$. The wavelengths are 7.122 cm for the 50 Ω line and 7.344 cm for the 90 Ω line. Using Eq. (2–19), we have effective widths W_{eff1} = 8.51 and W_{eff2} = 4.88 mm for the waveguide models of the 50 and 90 Ω lines. Using

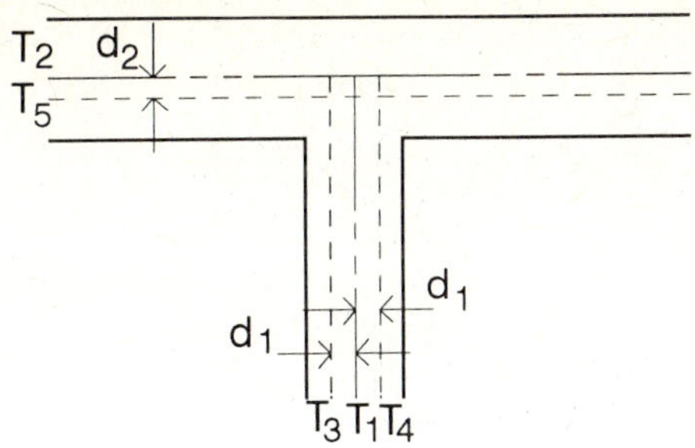

FIGURE 2–8. A Microstrip Symmetric T-Junction. The reference planes shown correspond to the distances given in Eq. (2–21).

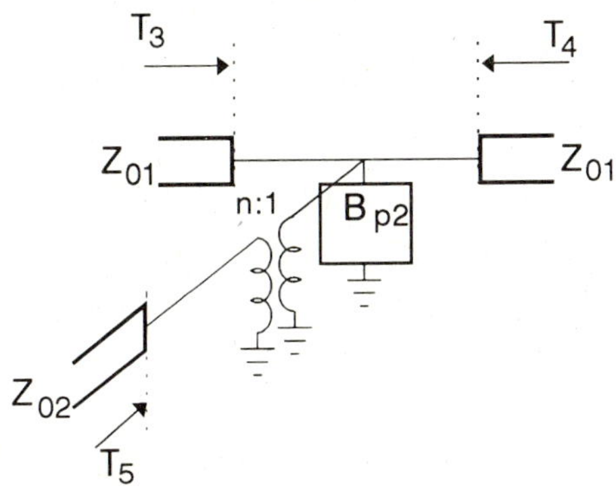

FIGURE 2–9. The Symmetric T-Junction Equivalent Circuit.

these values in Eqs. (2–20) and (2–21) we get reference plane displacements of d_1 = 0.253 mm and d_2 = 2.68 mm. The transformer ratio found from these is n = 0.984 and the shunt susceptance normalized to Y_{01} is $b_{p2} = -0.0284$.

2.12 FULL WAVE ANALYSIS OF MICROSTRIP PROPAGATION

The formulas for Z_0 and ε_{eff}, Eqs. (2–17) and (2–15), have been derived from solutions for the propagating electromagnetic fields that satisfy the boundary conditions on the microstrip. These solutions can be obtained directly by solving the Helmholtz equation found from Maxwell's equations. Alternately, one can define a vector or scalar potential function from which one can determine these fields. These potential functions must also satisfy a Helmholtz equation.

The fact that the resulting electromagnetic fields are the desired solution is assured by a uniqueness theorem. The uniqueness theorem states that there is only one solution that can satisfy both Maxwell's equations in a region (at least slightly lossy) and the conditions at the boundaries of the region.[44] The two pertinent boundary conditions are that the component of the electric field tangential to the boundaries must be continuous across the boundaries and that the discontinuity in the tangential component of the magnetic field across the boundary must be proportional to the surface current density at the boundary. These are expressed as in Eq. (2–22) where $\hat{\mathbf{n}}$ is a unit vector normal to the bounding surface and $\Delta\mathbf{E}$ and $\Delta\mathbf{H}$ are the changes in the fields across the surface.

$$\hat{\mathbf{n}} \times \Delta\mathbf{E} = 0, \qquad \hat{\mathbf{n}} \times \Delta\mathbf{H} = \mathbf{J}_s \tag{2–22}$$

The problem of finding solutions to the Helmholtz equation can be expressed more generally using the notation of a linear operator L operating on the unknown function f, as in $L(f) = g$, where g is equal to a source function.[45] In this simplified explanation we will consider the one-dimensional case where f is a function only of x. This approach uses the theory of *linear vector spaces* where we have a *function space* consisting of all possible solutions f that satisfy $L(f) = g$. This space is called the *domain of L*. This function space is analogous to ordinary three-dimensional space of geometry. This function space is said to be *spanned* by the set of *basis functions* (or *expansion functions*) $\{f_n\}$, if any f in the space can be expressed as a sum satisfying $\lim_{N\to\infty} \int |f(x) - \sum_{n=0}^{N} a_n f_n(x)|^2\, dx = 0$, where the integration is over the interval of x for which the solution is defined. We express this as: $f = \sum a_n f_n$. The basis functions are analogous to unit vectors in vector analysis, and the a_n are to be compared to the components of geometrical vectors. Further, we define an orthogonality relation by saying that two basis functions are orthogonal if $\int f_i f_j\, dx = 0$, where the integration is over the x domain for which the functions f_n are defined. This latter integration is frequently called an *inner product* and is written as: $\langle f_i, f_j \rangle$.

Example: Shielded Microstrip. To show the application of the theory, we discuss propagation on microstrip enclosed in a conducting shield as shown in Figure 2–10 following Krage and Haddad.[46] In this figure, we have assumed that the thickness of the microstrip conductor is negligible so the currents on this conductor are represented as surface currents. All components of the fields can be determined from the E_z and H_z as

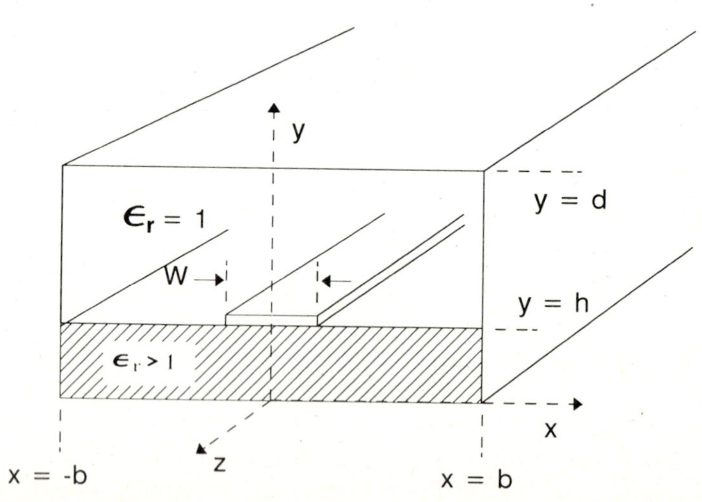

FIGURE 2–10. Shielded Microstrip Geometry. The walls at $y = 0$, $y = d$, $x = -b$, b are assumed to be perfectly conducting planes.

discussed in section 1–13. We can write a general solution for the even/odd mode as shown in Eq. (2–23), where the boundary conditions at $x = \pm b$, $y = 0$, and $y = d$ are satisfied, and where

$$\alpha_n = \begin{Bmatrix} (2n+1)\pi/b \\ 2n\pi/b \end{Bmatrix} \text{ for the } \begin{Bmatrix} \text{even} \\ \text{odd} \end{Bmatrix} \text{ mode and } \kappa_{in} = \sqrt{\gamma^2 + \omega^2\mu_0\varepsilon_i - \alpha_n^2}.$$

In this case, the basis functions are of the form: $\sinh(\kappa_{1n}y + \phi)\cos(\alpha_n x + \theta)$.

$$\left.\begin{aligned} E_{z1} &= \sum_n A_{1n} \frac{\sinh(\kappa_{1n}y)}{\sinh(\kappa_{1n}h)} \begin{Bmatrix} \cos(\alpha_n x) \\ \sin(\alpha_n x) \end{Bmatrix} \\ H_{z1} &= \sum_n B_{1n} \frac{\cosh(\kappa_{1n}y)}{\cosh(\kappa_{1n}h)} \begin{Bmatrix} \sin(\alpha_n x) \\ \cos(\alpha_n x) \end{Bmatrix} \end{aligned}\right\} y \le h$$

$$\left.\begin{aligned} E_{z2} &= \sum_n A_{2n} \frac{\sinh(\kappa_{2n}(d-y))}{\sinh(\kappa_{2n}(d-h))} \begin{Bmatrix} \cos(\alpha_n x) \\ \sin(\alpha_n x) \end{Bmatrix} \\ H_{z2} &= \sum_n B_{2n} \frac{\cosh(\kappa_{2n}(d-y))}{\cosh(\kappa_{2n}(d-h))} \cos \begin{Bmatrix} \sin(\alpha_n x) \\ \cos(\alpha_n x) \end{Bmatrix} \end{aligned}\right\} h \le y \le d \qquad \begin{Bmatrix} \text{even} \\ \text{odd} \end{Bmatrix} \text{mode} \tag{2–23}$$

The function space spanned by these basis functions consists of all possible solutions of the Helmholtz equation, Eq. (2–24).

$$\nabla_t^2 \begin{pmatrix} E_z \\ H_z \end{pmatrix} + k^2 \begin{pmatrix} E_z \\ H_z \end{pmatrix} = 0 \tag{2–24}$$

The linear operator is $L = \nabla_t^2 + k^2$ where $k^2 = \omega^2\mu_0\varepsilon + \gamma^2$. The surface current density $\mathbf{J}_s$ on the microstrip takes the place of g. To solve for the fields surrounding the microstrip, one must find E_x and H_x from E_z and H_z using Eq. (1–15), and require that they meet the boundary conditions at $y = h$. These boundary conditions are: $\hat{\mathbf{j}} \times (\mathbf{H}_{\tan 2} - \mathbf{H}_{\tan 1}) = \mathbf{J}_s$ and $E_{\tan 1} = E_{\tan 2}$. In the last expression, $\hat{\mathbf{j}}$ is a unit vector in the y direction. Using Eq. (1–15) for E_x and H_x from E_z and H_z, and substituting this result into these boundary conditions at $y = h$, one can arrive at a set of relations for the unknown quantities A_{in} and B_{in} in terms of the unknown surface current density $\mathbf{J}_s$. We note that this is not the final solution, because A_{in}, B_{in}, and $\mathbf{J}_s$ are not known.

2.13 LSE AND LSM POTENTIALS

We notice that the solution shown in the example of section 2–12 was put in terms of the E_z and H_z components. An alternate way of obtaining a solution is to use potentials that are directly related to these field components. One particular set of potentials are the LSE and LSM potentials mentioned in section 1–13. Much of the literature on this subject uses these potentials so the discussion here will be in terms of these potentials.

We will assume that these potentials can be factored into two functions. One is $\psi(x, y)$, which is a function of only the coordinates transverse to the propagation direction, and the other is $e^{\gamma z}$. We define two potential fields ψ^h and ψ^e by specifying the curl of the vector fields $\hat{\mathbf{k}}\psi^h e^{\gamma z}$ and $\hat{\mathbf{k}}\psi^e e^{\gamma z}$ as in Eq. (2–25), where $\hat{\mathbf{k}}$ is the unit vector in the z direction.

$$\mathbf{E}^h = -j\omega\mu\nabla \times (\hat{\mathbf{k}}\psi^h e^{\gamma z}), \qquad \mathbf{H}^e = j\omega\varepsilon\nabla \times (\hat{\mathbf{k}}\psi^e e^{\gamma z}) \tag{2–25}$$

As in section 1–13, we recognize that each potential will contribute to part of the total electromagnetic field, so the fields $\mathbf{E}^h$ and $\mathbf{H}^e$ are partial fields where the total fields are $\mathbf{E} = \mathbf{E}^e + \mathbf{E}^h$ and $\mathbf{H} = \mathbf{H}^e + \mathbf{H}^h$. The quantity $\hat{\mathbf{k}}\psi^h e^{\gamma z}$ is usually called the *magnetic Hertzian potential* and given the symbol $\mathbf{\Pi}^h$. Similarly, $\hat{\mathbf{k}}\psi^e e^{\gamma z}$ is called the *electric Hertzian potential* denoted by $\mathbf{\Pi}^e$.

This does not completely define these fields. The Helmholtz theorem states that any vector field is completely specified by its curl and its divergence. To substantiate a choice for the divergence, we use Maxwell's equation $\nabla \times \mathbf{H}^h = j\omega\varepsilon\mathbf{E}^h$ (in the absence of currents) with Eq. (2–25) to get $\nabla \times \mathbf{H}^h = \omega^2\mu\varepsilon\nabla \times (\hat{\mathbf{k}}\psi^h e^{\gamma z})$. From this, we get Eq. (2–26), where ϕ is arbitrary.

$$\mathbf{H}^h = \hat{\mathbf{k}}\omega^2\mu\varepsilon\psi^h e^{\gamma z} + \nabla\phi \tag{2–26}$$

Now, we take the curl of the both sides of the first equation in Eq. (2–25) to get $\nabla \times \mathbf{E}^h = -j\omega\mu\nabla \times \nabla \times (\hat{\mathbf{k}}\psi^h e^{\gamma z})$. Substituting from Maxwell's equation for $\nabla \times \mathbf{E}^h$ and the expression for $\mathbf{H}^h$ given in Eq. (2–26) for $-j\omega\mu\mathbf{H}^h$, we have $\hat{\mathbf{k}}\omega^2\mu\varepsilon\psi^h e^{\gamma z} + \nabla\phi = \nabla \times \nabla \times (\hat{\mathbf{k}}\psi^h e^{\gamma z})$. The right side of this can be expanded using the vector identity, which applies only in rectangular coordinates $\nabla \times \nabla \times (\mathbf{A}) = \nabla(\nabla \cdot \mathbf{A}) - \nabla^2 (\mathbf{A})$ to get Eq. (2–27).

$$\nabla(\nabla \cdot \hat{\mathbf{k}}\psi^h e^{\gamma z}) - \hat{\mathbf{k}}\nabla^2\psi^h e^{\gamma z} = \hat{\mathbf{k}}\omega^2\mu\varepsilon\psi^h e^{\gamma z} + \nabla\phi \tag{2–27}$$

Because ϕ was an arbitrary potential, we are free to specify it as the divergence: $\phi = \nabla \cdot (\hat{\mathbf{k}}\psi^h e^{\gamma z}) = \gamma\psi^h e^{\gamma z}$. Then, Eq. (2–27) reduces to the Helmholtz equation in Eq. (2–28).

$$\nabla^2(\psi^h e^{\gamma z}) + \omega^2\mu\varepsilon\psi^h e^{\gamma z} = 0 \tag{2–28}$$

The z component of $\nabla\phi$ is $\partial^2(\psi^h e^{\gamma z})/\partial z^2 = \gamma^2\psi^h e^{\gamma z}$. When this is substituted into Eq. (2–26), we obtain Eq. (2–29).

$$H_z^h = (\omega^2\mu\varepsilon + \gamma^2)\psi^h e^{\gamma z} = k^2\psi^h e^{\gamma z} = H_z, \text{ where } k^2 = \omega^2\mu\varepsilon + \gamma^2 \tag{2–29}$$

In Eq. (2–29), we see that ψ^h is proportional to H_z^h through a factor containing the frequency and the propagation constant. The $\mathbf{H}^e$ partial field is $\mathbf{H}^e = j\omega\varepsilon\nabla \times (\hat{\mathbf{k}}\psi^e e^{\gamma z})$. Because $\nabla \times (\hat{\mathbf{k}}f)$ is perpendicular to $\hat{\mathbf{k}}$ for any scalar function f, we see that H_z^e must be zero, so we have $H_z = H_z^h$. Using the transverse operator ∇_t and assuming lossless propagation, $\gamma = \pm j\beta$, we can get Eq. (2–30) from Eq. (2–28).

$$\nabla_t^2\psi^h + (\omega^2\mu\varepsilon - \beta^2)\Psi^h = 0 \tag{2–30}$$

Because Maxwell's equations have symmetry with respect to the **E** and **H** vectors, we can start with a potential ψ^e satisfying $\mathbf{H}^e = j\omega\varepsilon\, \nabla \times (\hat{\mathbf{k}}\psi^e e^{\gamma z})$ and use the same steps used above to arrive at Eqs. (2–31) and (2–32).

$$E_z = (\omega^2\mu\varepsilon - \beta^2)\psi^e e^{\gamma z} = k^2\psi^e e^{\gamma z} \qquad (2\text{–}31)$$

$$\nabla_t^2\psi^e + (\omega^2\mu\varepsilon - \beta^2)\psi^e = 0 \qquad (2\text{–}32)$$

Frequently, the process of solving for the electromagnetic fields along any cylindrical guide, including microstrip, reduces to solving a set of equations in several unknown constants and functions as shown in the example of section 2–12. Several methods have been developed for obtaining solutions from relations such as those in the example. They generally require the aid of a computer and do not result in analytical expressions. In the next section, we discuss a solution method widely reported in the technical literature on microstrip propagation. A more complete discussion of methods for obtaining full wave solutions has been compiled by Sorrentino.[47] Here, we are concerned primarily with the methods used for obtaining data on dispersion and equivalent circuit elements for microstrip discontinuities.

2.14 SPECTRAL DOMAIN ANALYSIS

We do not need to find an explicit expression for the fields in order to find the frequency dependence of the phase constant β. We can find a characteristic equation and solve it for β. The earliest of such solutions consumed a lot of computer time because of the way that a characteristic equation for β was determined. More recently, the method called *spectral domain analysis* (SDA) has been used. This method is more efficient in its use of computer time. Several of the more important articles based on the use of SDA have been reprinted.[48]

In this method, one uses a spatial Fourier transform of the Helmholtz equation, Eq. (2–30) or (2–32), to transform one spatial variable into a spectral domain variable α. Then, the partial differential Helmholtz equation becomes an ordinary differential equation. Next, one uses the boundary conditions and Parseval's theorem to obtain a set of homogeneous algebraic equations. For this set of equations to have a nontrivial solution, the determinant of the coefficient matrix must be zero. One sets this determinant to zero to form the characteristic equation. The terms in this coefficient matrix are functions of the β and ω, and they incorporate integrals over the spectral domain variable α. Usually, one requires a computer to determine values of β that satisfy this characteristic equation. The values of β that make the determinant zero at different values of ω then exhibit the mode structure of the microstrip. The β for the lowest order mode yield dispersion data usually given in graphical or tabular form. This method is especially useful for analyzing circuits in which the conducting surfaces are thin and the analysis can be done assuming that their thickness is negligible. This makes it ideal for analyzing microstrip circuits. A review article covering SDA and its applications to microstrip circuits has been written by Jansen.[49]

2.15 DISPERSION RELATION FOR OPEN MICROSTRIP

We discuss the method used by Itoh and Mittra for finding a dispersion relation for open (i.e., unshielded) microstrip in this section.[50] The geometry used for this solution is shown in Figure 2–11. The x axis is taken to be one that is parallel to the dielectric surface and the y axis is perpendicular to the dielectric surface. The surface of the dielectric is at $y = h$. On this surface, the conducting microstrip lies along the z direction and it is W meters in width. The thickness of the conductor is assumed to be negligible. In this discussion, we use $k_1^2 = \omega^2\mu_o\varepsilon_o\varepsilon_{r1}$, $\mathscr{k}_1^2 = k_1^2 - \beta^2$, and $\kappa_1^2 = \mathscr{k}_1^2 - \alpha^2$. Please note carefully these different constants.

Taking the Fourier transform as indicated by Eq. (2–33), we obtain second order ordinary differential equations from Eqs. (2–30) and (2–32) as shown in Eq. (2–34).

$$\int_{-\infty}^{\infty} f(x)e^{j\alpha x}\,dx = \tilde{f}(\alpha) \qquad (2\text{–}33)$$

$$\frac{d^2\tilde{\psi}^h}{dy^2} + (\omega^2\mu\varepsilon - \beta^2 - \alpha^2)\tilde{\psi}^h = 0 \text{ and } \frac{d^2\tilde{\psi}^e}{dy^2} + (\omega^2\mu\varepsilon - \beta^2 - \alpha^2)\tilde{\psi}^e = 0 \qquad (2\text{–}34)$$

These have solutions of the form: sin $(\kappa_1 y)$, cos $(\kappa_1 y)$, where $\kappa_1^2 = \omega^2\mu_0\varepsilon_1 - \beta^2 - \alpha^2$ for $y \le h$ and exp $(-\kappa_2 (y - h))$, where $\kappa_2^2 = \alpha^2 + \beta^2 - \omega^2\mu_0\varepsilon_0$ for $y \ge h$. The dielectric permittivity is $\varepsilon_1 = \varepsilon_0\varepsilon_r$ for the region $0 < y < h$ and $\varepsilon_2 = \varepsilon_0$ for the region $y > h$. These solutions must meet the boundary conditions at $y = 0$, $\tilde{E}_z(\alpha, 0) = 0$, $d\tilde{H}_z(\alpha, y)/dy\,|_{y=0} = 0$, and not exhibit radiation such that $\tilde{E}_z(\alpha, \infty)$ and $\tilde{H}_z(\alpha, \infty)$ go to zero exponentially as y tends to infinity. The solutions for these potentials that meet these boundary conditions are given in Eq. (2–35).

$$\begin{aligned} &\tilde{\psi}_1^e = A\sin(\kappa_1 y), \quad 0<y<h, \qquad \tilde{\psi}_2^e = B\exp(-\kappa_2(y-h)), \quad h<y \\ &\tilde{\psi}_1^h = C\cos(\kappa_1 y), \quad 0<y<h, \qquad \tilde{\psi}_2^h = D\exp(-\kappa_2(y-h)), \quad h<y \end{aligned} \qquad (2\text{–}35)$$

The constants A, B, C, and D are evaluated from the boundary conditions at $y = h$ as given in Eq. (2–36), where

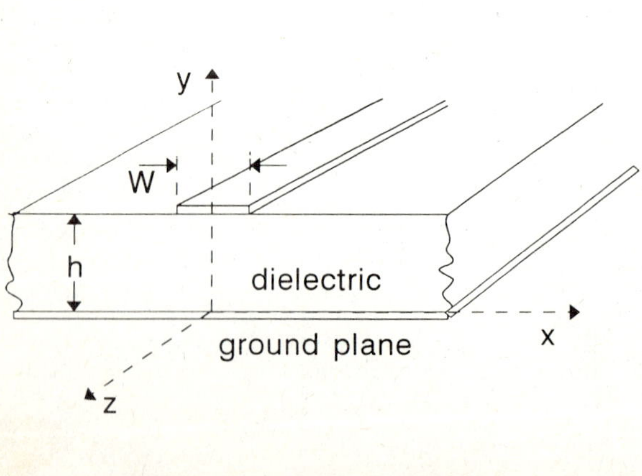

FIGURE 2–11. Microstrip Geometry.

$$
\begin{aligned}
\tilde{E}_{z1}(\alpha, h) &= \tilde{E}_{z2}(\alpha, h) \\
\tilde{E}_{x1}(\alpha, h) &= \tilde{E}_{x2}(\alpha, h) \\
\tilde{H}_{z1}(\alpha, h) - \tilde{H}_{z2}(\alpha, h) &= -\tilde{J}_x(\alpha) \\
\tilde{H}_{x1}(\alpha, h) - \tilde{H}_{x2}(\alpha, h) &= \tilde{J}_z(\alpha)
\end{aligned}
\tag{2–36}
$$

$\tilde{J}_x(\alpha)$ and $\tilde{J}_z(\alpha)$ are the spatial Fourier x and z components of the current density at $y = h$. The fields E_{xi} and H_{xi} are determined from E_{zi} and H_{zi} using Eq. (1–15). Using Eqs. (2–31) and (2–29), these are put in terms of the potential functions and substituted into Eq. (2–36) to get:

$$
\begin{aligned}
&k_2^2\, \tilde{\psi}_1^e(\alpha, h) = k_2^2\, \tilde{\psi}_2^e(\alpha, h) \\
&-\alpha\beta\, \tilde{\psi}_1^e(\alpha, h) - j\omega\mu_0 \frac{\partial \tilde{\psi}_1^h}{\partial y}\Big|_{y=h} = -\alpha\beta\, \tilde{\psi}_2^e(\alpha, h) - j\omega\mu_0 \frac{\partial \tilde{\psi}_2^h}{\partial y}\Big|_{y=h} \\
&k_1^2\, \tilde{\psi}_1^h(\alpha, h) - k_2^2\, \tilde{\psi}_2^h(\alpha, h) = -\tilde{J}_x(\alpha) \\
&j\omega\varepsilon_0\varepsilon_r \frac{\partial \tilde{\psi}_1^e}{\partial y}\Big|_{y=h} - \alpha\beta\, \tilde{\psi}_1^h(\alpha, h) - j\omega\varepsilon_0 \frac{\partial \tilde{\psi}_2^e}{\partial y}\Big|_{y=h} + \alpha\beta\, \tilde{\psi}_2^h(\alpha, h) = \tilde{J}_z(\alpha),
\end{aligned}
$$

where $k_1^2 = \kappa_1^2 + \alpha^2$ and $k_2^2 = -\kappa_2^2 + \alpha^2$. These relations can be solved for A, B, C, and D in terms of the currents $\tilde{J}_x(\alpha)$ and $\tilde{J}_z(\alpha)$. To do this, we solve the first and the third of these relations to get $A = Bk_2^2/(k_1^2 \sin(\kappa_1 h))$ and $C = (Dk_2^2 - \tilde{J}_x(\alpha))/(k_1^2 \cos(\kappa_1 h))$. Then, we use the remaining two relations to express the constants B and D in terms of $\tilde{J}_x(\alpha)$ and $\tilde{J}_z(\alpha)$. The result is Eq. (2–37), where the terms b_{ij} are given in Eq. (2–38).

$$
\left.
\begin{aligned}
B &= \frac{j}{k_1^2 \det}\Big(-[\omega\mu_0 b_{22}\kappa_1 \tan(\kappa_1 h) + j\alpha\beta\, k_1^2 b_{12}]\, \tilde{J}_x(\alpha) + j b_{12} k_1^2 \tilde{J}_z(\alpha)\Big) \\
D &= \frac{1}{k_1^2 \det}\Big([-\alpha\beta\, k_1^2 b_{11} + j b_{21}\omega\mu_0\kappa_1 \tan(\kappa_1 h)]\, \tilde{J}_x(\alpha) + k_1^2\, b_{11} \tilde{J}_z(\alpha)\Big)
\end{aligned}
\right\}
\tag{2–37}
$$

$$
\left.
\begin{aligned}
b_{11} &= -\alpha\beta[k_2^2/k_1^2) - 1] = b_{22} \\
b_{12} &= j\omega\mu_0\kappa_1[(k_2^2/k_1^2)\tan(\kappa_1 h) - \kappa_2/\kappa_1] \\
b_{21} &= j\omega\varepsilon_0\kappa_1[\varepsilon_r(k_2^2/k_1^2)\cot(\kappa_1 h) + \kappa_2/\kappa_1] \\
\det &= b_{11}b_{22} - b_{12}b_{21}
\end{aligned}
\right\}
\tag{2–38}
$$

Then, we express both $\tilde{E}_{z2}(\alpha, h)$ and $\dfrac{\partial H_{z2}(\alpha, y)}{\partial y}\Big|_{y=h}$ in terms of these "constants". Because A, B, C, and D are linear functions of $\tilde{J}_x(\alpha)$ and $\tilde{J}_z(\alpha)$, then $\tilde{E}_{z2}(\alpha, h)$ and $\dfrac{\partial \tilde{H}_{z2}(\alpha, y)}{\partial y}\Big|_{y=h}$ are also linear functions of $\tilde{J}_x(\alpha)$ and $\tilde{J}_z(\alpha)$. We show that dependence in Eq. (2–39).

$$e^{-\gamma z}\tilde{E}_{z2}(\alpha, h) = k_2^2 B = \tilde{G}_{11}(\alpha, \beta, \omega)\tilde{J}_x(\alpha) + \tilde{G}_{12}(\alpha, \beta, \omega)\tilde{J}_z(\alpha)$$

$$e^{-\gamma z}\frac{\partial \tilde{H}_{z2}(\alpha, y)}{\partial y}\Big|_{y=h} = -k_2^2\,\kappa_2 D = \tilde{G}_{21}(\alpha, \beta, \omega)\tilde{J}_x(\alpha) + \tilde{G}_{22}(\alpha, \beta, \omega)\tilde{J}_z(\alpha) \tag{2–39}$$

Explicit relations for the coefficients $\tilde{G}_{ij}$ are given in Eq. (2–40).

$$\tilde{G}_{11} = \frac{k_2^2\left(-j\omega\mu_0\kappa_1 b_{22}\tan(\kappa_1 h) + \alpha\beta b_{12}k_2^2\right)}{k_1^2 \det}$$

$$\tilde{G}_{12} = -k_2^2 b_{12}/\det$$

$$\tilde{G}_{21} = \frac{\kappa_2 k_2^2\left(\alpha\beta b_{11}k_2^2 - jb_{21}\omega\mu_0\kappa_1\tan(\kappa_1 h)\right)}{k_1^2 \det} \tag{2–40}$$

$$\tilde{G}_{22} = -\kappa_2 k_2^2 b_{11}/\det$$

In principle, Eq. (2–39) is the solution of the wave equation for these fields at the interface. As in section 2–12, the trouble with this solution is that we do not know the functions $\tilde{J}_x(\alpha)$ and $\tilde{J}_z(\alpha)$. We do know approximately how the currents $J_x(x)$ and $J_z(x)$ vary with x and we can use this information to select known functions to be the expansion functions, $J_{xn}(x)$ and $J_{zn}(x)$, which describe $J_x(x)$ and $J_z(x)$ as shown in Eq. (2–41).

$$J_x(x) = \Sigma\, a_n J_{xn}(x) \qquad \text{and} \qquad J_z(x) = \Sigma\, b_n J_{zn}(x) \tag{2–41}$$

The relative weightings a_n and b_n then become the unknowns. These expansion functions are chosen so the partial sums implied by Eq. (2–41) converge rapidly to the actual current. Other criteria necessary for efficient computation are given by Jansen.[51] The expansion functions used in Eq. (2–41) are not necessarily orthogonal.

Substituting the Fourier transforms of these expressions for $J_x(x)$ and $J_z(x)$ into Eq. (2–39), we multiply the resulting equations by $\tilde{J}_{xm}(\alpha)$ and by $\tilde{J}_{zm}(\alpha)$ and integrate with respect to α as shown in Eq. (2–42). The choice of multiplying by $\tilde{J}_{xm}$ and $\tilde{J}_{zm}$ in these integrals is known as *Galerkin's method.*[52]

$$\int_{-\infty}^{\infty}\tilde{E}_{z2}(\alpha, h)\tilde{J}_{xm}(\alpha)\,d\alpha = \sum a_n\int_{-\infty}^{\infty}\tilde{G}_{11}\tilde{J}_{xm}\tilde{J}_{xn}\,d\alpha + \sum b_n\int_{-\infty}^{\infty}\tilde{G}_{12}\tilde{J}_{xm}\tilde{J}_{zn}\,d\alpha$$

$$\int_{-\infty}^{\infty}\tilde{E}_{z2}(\alpha, h)\tilde{J}_{zm}(\alpha)\,d\alpha = \sum a_n\int_{-\infty}^{\infty}\tilde{G}_{11}\tilde{J}_{zm}\tilde{J}_{xn}\,d\alpha + \sum b_n\int_{-\infty}^{\infty}\tilde{G}_{12}\tilde{J}_{zm}\tilde{J}_{zn}\,d\alpha$$

$$\int_{-\infty}^{\infty}\frac{\partial \tilde{H}_{z2}(\alpha, y)}{\partial y}\Big|_{y=h}\tilde{J}_{xm}(\alpha)\,d\alpha = \sum a_n\int_{-\infty}^{\infty}\tilde{G}_{21}\tilde{J}_{xm}\tilde{J}_{xn}\,d\alpha + \sum b_n\int_{-\infty}^{\infty}\tilde{G}_{22}\tilde{J}_{xm}\tilde{J}_{zn}\,d\alpha \tag{2–42}$$

$$\int_{-\infty}^{\infty}\frac{\partial \tilde{H}_{z2}(\alpha, y)}{\partial y}\Big|_{y=h}\tilde{J}_{zm}(\alpha)\,d\alpha = \sum a_n\int_{-\infty}^{\infty}\tilde{G}_{21}\tilde{J}_{zm}\tilde{J}_{xn}\,d\alpha + \sum b_n\int_{-\infty}^{\infty}\tilde{G}_{22}\tilde{J}_{zm}\tilde{J}_{zn}\,d\alpha$$

Now we consider the integrals on the left side of Eq. (2–42) that are of the form:

$$\int_{-\infty}^{\infty} \tilde{E}_{z2}(\alpha, h)\tilde{J}_{xm}(\alpha)\, d\alpha \qquad \text{and} \qquad \int_{-\infty}^{\infty} \frac{\partial \tilde{H}_{z2}(\alpha, y)}{\partial y}\Big|_{y=h} \tilde{J}_{xm}(\alpha)\, d\alpha$$

By Parseval's theorem, these are equal to

$$\int_{-\infty}^{\infty} E_{z2}(x, h)J_{xm}(x)\, dx \qquad \text{and} \qquad \int_{-\infty}^{\infty} \frac{\partial H_{z2}(x, y)}{\partial y}\Big|_{y=h} J_{xm}(x)\, dx$$

respectively. Both of these latter integrands are products of a component of the electric field tangential to the interface and evaluated at the interface, multiplied by a current at the interface. However, the tangential components of the electric field are zero at those (conducting) regions of x where the current is not zero and vice versa in the nonconducting regions. Consequently, both of these integrands are zero over the entire range of integration.

So the left-hand sides of the equations in Eq. (2–42) are all zero, and the right-hand sides are shown in Eq. (2–43), where u and v each take on values of z and x.

$$(0) = (K_{nm})\begin{pmatrix} a_1 \\ \vdots \\ a_n \\ b_1 \\ \vdots \\ b_n \end{pmatrix} \quad \text{where } K_{nm} = \int_{-\infty}^{\infty} \tilde{G}_{ij}(\alpha, \beta, \omega)\, \tilde{J}_{un}(\alpha)\, \tilde{J}_{vm}(\alpha)\, d\alpha \tag{2–43}$$

From the theory of linear algebra, we know that this has a nonzero solution only if the determinant of the coefficient matrix is zero. The dispersion relation that we desire is given by det $(K_{nm}) = 0$. Because this expression has integrals not capable of being evaluated analytically, a computer must be used to find those values of β that satisfy it. From these calculations, one can obtain the variation of β with frequency for the different modes of propagation on the microstrip. These modes are hybrid modes involving both E_z and H_z.

Using this method with some modifications, Mirshekar-Syahkal and Davies computed the mode structure for shielded microstrip on a substrate having a dielectric constant of $\varepsilon_r = 8.875$.[53] Their results are shown in Figure 2–12: Three higher order modes exist in their frequency range. At higher frequencies, there will be more higher order modes. The effective dielectric constant of the dominant mode is asymptotic to the quasi-static effective dielectric constant. This mode is usually called the quasi-TEM mode.

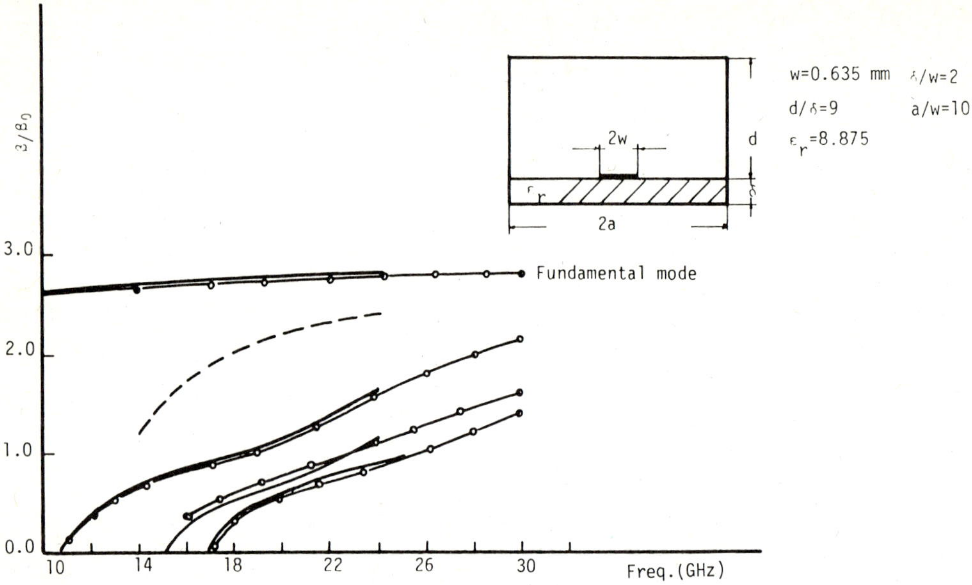

FIGURE 2–12. Mode Structure for Shielded Microstrip. The solid lines with circled points were computed by Mirshekar-Syahkal and Davies.[53] The solid lines and the dashed line were computed by alternate methods and are shown to compare results.[54] © 1979 IEEE. Reprinted with permission of the IEEE.

2.16 SPECTRAL DOMAIN IMPEDANCE ANALYSIS

As shown in the previous section, SDA requires considerable analytical effort to find expressions such as Eq. (2–40) for the coefficients of the current densities at the conductor boundary. An alternate method formalizing this process uses *spectral domain impedances.*[55] Equation (2–39) can be put in the form of Eq. (2–44).

$$\begin{aligned} \tilde{E}_x|_{y=h} &= Z_{xx}\,\tilde{J}_x + Z_{xz}\,\tilde{J}_z \\ \tilde{E}_z|_{y=h} &= Z_{zx}\,\tilde{J}_x + Z_{zz}\,\tilde{J}_z \end{aligned} \qquad (2\text{–}44)$$

The Z_{ij} in this relation are called the spectral domain impedances. We can apply Eq. (2–44) to the shielded microstrip shown in Figure 2–13. The $\tilde{E}_x$ and $\tilde{E}_z$ given in this expression are the spatial Fourier transform of the x and z electric field components at $y = h$. To find the Z_{ij}, we could use the boundary conditions as shown in the previous section. However, a simpler method for finding these functions is to use transmission line analysis. We view the waves as originating at $y = h$ and propagating in the positive and negative y direction toward the conductors at $y = d$ and $y = 0$. These conductors cause reflections of these waves. The transmitted and reflected waves in these regions act similarly to transmitted and reflected waves on transmission lines. The currents $\tilde{J}_x$ and $\tilde{J}_z$ are the sources energizing these two "lines" and the impedance presented to these currents at

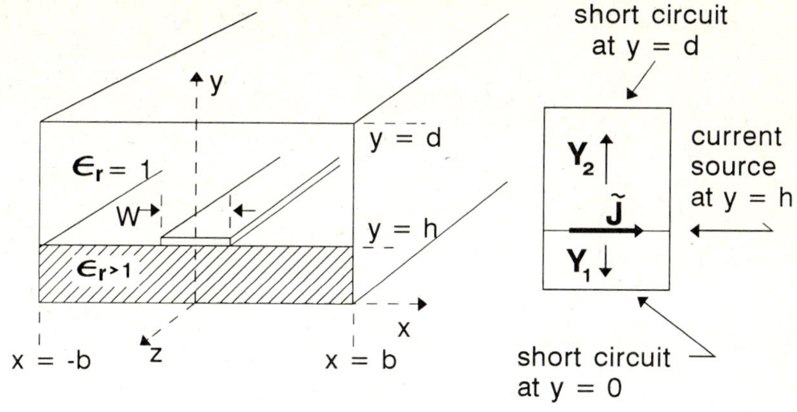

FIGURE 2–13. Shielded Microstrip and Its Representation as Transmission Lines Connected in Parallel. The Y_i are the admittances seen at $y = h$ by the source currents, $\tilde{\mathbf{J}}_s$, on the microstrip conductor. The walls at $y = 0$, $y = d$, $x = b$, and $x = -b$ are conducting surfaces.

$y = h$ are the spectral domain impedances. The $\tilde{E}_x$ and $\tilde{E}_z$ are the responses to these currents. This is shown graphically in Figure 2–13.

To explain this, we start with the two-dimensional inverse Fourier transform of the fields as shown in Eq. (2–45).

$$E_z = (1/4\pi^2)\int\int_{-\infty}^{\infty} \tilde{E}_z \exp\left[-j(\alpha x + \beta z)\right] d\alpha\, d\beta \tag{2–45}$$

Then, we rotate coordinates about the y axis as shown in Eq. (2–46) and Figure 2–14, where $\cos\theta = \alpha/\lambda$ and $\sin\theta = \beta/\lambda$. With that coordinate rotation, we obtain the inverse Fourier integral of Eq. (2–47), where $\lambda = \sqrt{\alpha^2+\beta^2}$.

$$\begin{pmatrix} \cos\theta & \sin\theta \\ -\sin\theta & \cos\theta \end{pmatrix}\begin{pmatrix} x \\ z \end{pmatrix} = \begin{pmatrix} u \\ v \end{pmatrix} \tag{2–46}$$

$$E_z = (1/4\pi^2)\int\int_{-\infty}^{\infty} \tilde{E}_z\, e^{-j\lambda u}\, d\alpha\, d\beta \tag{2–47}$$

All of the field components can be transformed in this way. Equation (2–47) can be viewed as a summation of plane waves $\Delta\mathbf{E} = \tilde{\mathbf{E}}\, e^{-j\lambda u}\, d\alpha\, d\beta$ to form the field $\mathbf{E}$. These incremental field elements, $\Delta\mathbf{E}$, are functions only of the u coordinate and not of the v coordinate. We express the u and v components of $\Delta\mathbf{E}$ and $\Delta\mathbf{H}$ in terms of the TM (to y) mode and the *TE* (to y) mode. For the *TM* mode, ΔH_y must be zero. Substituting this into Maxwell's curl equation, $-j\omega\mu\Delta H_y = (\partial\Delta E_u/\partial v) - (\partial\Delta E_v/\partial u)$, we obtain $(\partial\Delta E_u/\partial v) = (\partial\Delta E_v/\partial u)$. Because of the coordinate transformation of Eq. (2–46), we know that ΔE_u is not a function of v, so $(\partial\Delta E_u/\partial v) = 0$. Thus, we infer that $(\partial\Delta E_v/\partial u) = 0$. Because of the boundary conditions, that also implies that $\Delta E_v = 0$. Looking at the ΔH_u component in

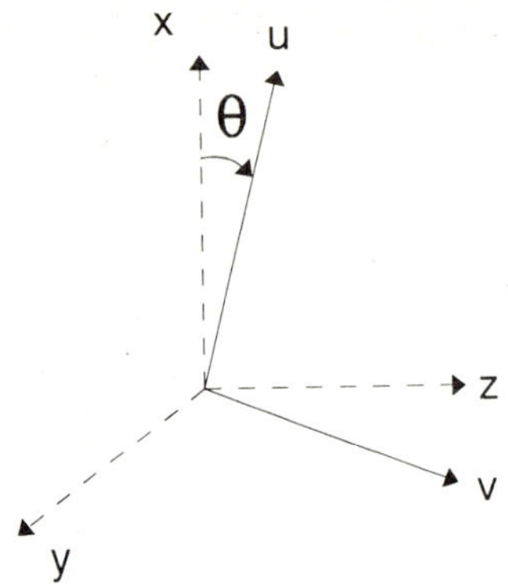

FIGURE 2–14. Coordinate Transformation for Spectral Domain Immittance Method.

Maxwell's curl equation, we have $-j\omega\Delta H_u = (\partial\Delta E_v/\partial y) - (\partial\Delta E_y/\partial v)$. However, we determined above that $\Delta E_v = 0$ and that the components of the field did not depend upon v, so $(\partial\Delta E_y/\partial v)$ is zero. These require that ΔH_u be zero. Thus, the TM mode has field components ΔE_u, ΔE_y, and ΔH_v; which is to say that we have $\tilde{E}_u$, $\tilde{E}_y$, and $\tilde{H}_v$ as the only nonzero field components in the transform domain for this *TM* (to y) mode. Similarly, the *TE* (to y) mode has nonzero components $\tilde{E}_v$, $\tilde{H}_u$, and $\tilde{H}_y$.

The electric fields $\tilde{E}_x$ and $\tilde{E}_y$ are related to $\tilde{E}_u$ and $\tilde{E}_v$ by Eq. (2–48).

$$\begin{pmatrix} \tilde{E}_x \\ \tilde{E}_z \end{pmatrix} = \begin{pmatrix} \cos\theta & -\sin\theta \\ \sin\theta & \cos\theta \end{pmatrix} \begin{pmatrix} \tilde{E}_u \\ \tilde{E}_v \end{pmatrix} \tag{2–48}$$

The relationship between the electric fields and the current densities is Ohm's law, as given in Eq. (2–49).

$$\begin{pmatrix} \tilde{E}_u \\ \tilde{E}_v \end{pmatrix} = \begin{pmatrix} Z^e & 0 \\ 0 & Z^h \end{pmatrix} \begin{pmatrix} \tilde{J}_u \\ \tilde{J}_v \end{pmatrix} \tag{2–49}$$

We also express the current in terms of its components along the u and v axes as shown in Eq. (2–50).

$$\begin{pmatrix} \tilde{J}_u \\ \tilde{J}_v \end{pmatrix} = \begin{pmatrix} \cos\theta & \sin\theta \\ -\sin\theta & \cos\theta \end{pmatrix} \begin{pmatrix} \tilde{J}_x \\ \tilde{J}_z \end{pmatrix} \tag{2–50}$$

Now, we combine Eqs. (2–48), (2–49), and (2–50) along with $\cos\theta = \alpha/\lambda$ and $\sin\theta = \beta/\lambda$ to obtain Eq. (2–51).

$$\begin{pmatrix} \tilde{E}_x \\ \tilde{E}_z \end{pmatrix} = (1/\lambda^2) \begin{pmatrix} Z^e\alpha^2 + Z^h\beta^2, & (Z^e - Z^h)\alpha\beta \\ (Z^e - Z^h)\alpha\beta, & Z^e\beta^2 + Z^h\alpha^2 \end{pmatrix} \begin{pmatrix} \tilde{J}_x \\ \tilde{J}_z \end{pmatrix} \tag{2–51}$$

This is an important relation because we can determine the coefficient matrix elements in terms of the impedances in a relatively simple manner as discussed next.

To find the impedances Z^e and Z^h, we find the wave admittances and invert them as in $Y^e = 1/Z^e$ and $Y^h = 1/Z^h$. These wave admittances are the sum of an admittance encountered by the plane waves propagating in the region for $y > h$, and an admittance encountered by the waves propagating in the region for $y < h$. The *TM* mode admittance in Eq. (2–52) is that at $y = h$ encountered by the wave propagating in the number 1 region with reflection at $y = 0$.

$$Y_1^e = Y_{TM1}\frac{1-\Gamma e^{2\gamma_1 h}}{1+\Gamma e^{2\gamma_1 h}} = \frac{j\omega\varepsilon_0}{\gamma_1}\coth(\gamma_1 h) \tag{2–52}$$

We compute Y_1^e using Eq. (1–11), where Γ is the reflection coefficient of -1 at $y = 0$, γ_1 is the propagation constant in region number 1 equal to $\sqrt{-\omega^2\mu_0\varepsilon_0\varepsilon_r + \alpha^2 + \beta^2}$, and $Y_{TM1} = j\omega\varepsilon_0\varepsilon_r/\gamma_1$ is the intrinsic *TM* mode admittance (see section 1–13 and problem 1–22). The total admittance to the *TM* mode is the sum of that in Eq. (2–52) and the admittance describing the propagation and reflection in the number 2 region. The total *TM* mode admittance is given in Eq. (2–53), where $\gamma_2 = \sqrt{-\omega^2\mu_0\varepsilon_0 + \alpha^2 + \beta^2}$.

$$Y^e = \frac{j\omega\varepsilon_0}{\gamma_2}\coth[\gamma_2(d-h)] + \frac{j\omega\varepsilon_0\varepsilon_r}{\gamma_1}\coth(\gamma_1 h) \tag{2–53}$$

Similarly, the *TE* mode admittance Y^h is given in Eq. (2–54). Substituting from Eqs. (2–53) and (2–54) into Eq. (2–51), we can obtain Z_{xx} as shown in Eq. (2–55).

$$Y^h = \frac{\gamma_2}{j\omega\mu_0}\coth[\gamma_2(d-h)] + \frac{\gamma_1}{j\omega\mu_0}\coth(\gamma_1 h) = 1/(Z^h) \tag{2–54}$$

The remaining impedances can be derived in a similar manner.

$$Z_{xx} = \left(\frac{\alpha^2}{\frac{j\omega\varepsilon_0}{\gamma_2}\coth[\gamma_2(d-h)] + \frac{j\omega\varepsilon_0\varepsilon_r}{\gamma_1}\coth(\gamma_1 h)} + \frac{\beta^2}{\frac{\gamma_2}{j\omega\mu_0}\coth[\gamma_2(d-h)] + \frac{\gamma_1}{j\omega\mu_0}\coth(\gamma_1 h)}\right)\frac{1}{\alpha^2+\beta^2} \tag{2–55}$$

2.17 GREEN'S FUNCTION

The fields obtained in Eq. (2–44) can also be determined by the use of an integral equation in the space domain. The integral equation is formally derived using Green's theorem,[56] which has the form of Eq. (2–56), where $(G_{ij}(\mathbf{r}-\mathbf{r}_0))$ is a 3 by 3 array. Each term in the Green's function is the contribution to the electric field component along the i direction at $\mathbf{r}$ due to a unit current source at $\mathbf{r}_0$ with a current oriented in the j direction. The point indicated by $\mathbf{r}$ is called the *field point* and that indicated by $\mathbf{r}_0$ is called the *source point.* The product indicated by $(G_{ij}) \cdot \mathbf{J}$ is the same as post multiplication of a 3 by 3 matrix by a column matrix. The quantity (G_{ij}) is called the *dyadic Green's function.* Equation (2–56) is the three-dimensional analog of the convolution integral seen when using the Laplace transform.

$$\mathbf{E}(\mathbf{r}) = \int_{-\infty}^{\infty}\int\int \left(G_{ij}(\mathbf{r}-\mathbf{r}_0)\right) \cdot \mathbf{J}(\mathbf{r}_0)\, dx_0\, dy_0\, dz_0 \tag{2–56}$$

Applying Eq. (2–56) to the case of microstrip lines, we have a current $\mathbf{J}(\mathbf{r}_0)$ that is assumed to be two dimensional because the conducting surface of the line is considered to have zero thickness. Then Eq. (2–56) becomes a two-dimensional convolution integral. The points where $\mathbf{J}$ is not zero are for $|x_0| < W/2$. If this integral exists, then we can take the two-dimensional Fourier transform of it as shown in Eq. (2–57).

$$\tilde{\mathbf{E}} = \int_x\int_z e^{-j(\alpha x+\beta z)} \int_{x_0}\int_{z_0} \left(G_{ij}(x-x_0, z-z_0)\right) \cdot \mathbf{J}(x_0, z_0)\, dx_0\, dz_0\, dx\, dz \tag{2–57}$$

From the convolution theorem, we then have the relation: $\tilde{\mathbf{E}} = \tilde{G}_{ij} \cdot \tilde{\mathbf{J}}$. Comparing this with Eq. (2–44), we see that the spectral impedance dyadic (Z_{ij}) is the Fourier transform of the Green's function. For this reason, the spectral impedance dyadic is called the *spectral representation of the Green's function.*

Again, we consider the two-dimensional case of section 2–16. First, we derive Eq. (2–58) from Maxwell's equations.

$$\nabla \times \nabla \times \mathbf{E} - k_0^2\mathbf{E} = j\omega\mu\mathbf{J} \tag{2–58}$$

The right-hand side of this can be expressed as $-j\omega\mu\mathbf{J} = j\omega\mu \int_{-\infty}^{\infty}\int_{-\infty}^{\infty} (I_{ij}) \cdot \mathbf{J}(\mathbf{r}_0)\, \delta(\mathbf{r}-\mathbf{r}_0)\, dx_0\, dy_0$, where $\delta(\mathbf{r}-\mathbf{r}_0) = \delta(x-x_0)\delta(y-y_0)$, and where (I_{ij}) is the two-dimensional unit dyadic: $\left(\begin{smallmatrix}1 & 0\\ 0 & 1\end{smallmatrix}\right)$. We replace the right-hand side as indicated, then substitute the two-dimensional form of Eq. (2–56) into Eq. (2–58) to get Eq. (2–59).

$$\begin{aligned}\int_{-\infty}^{\infty}\int_{-\infty}^{\infty} \left((\nabla\times\nabla\times - k_0^2)G_{ij}(\mathbf{r}-\mathbf{r}_0)\right). \mathbf{J}(\mathbf{r}_0)\, dx_0\, dy_0 &= \\ = \int_{-\infty}^{\infty}\int_{-\infty}^{\infty} \left(-j\omega\mu I_{ij}\delta(\mathbf{r}-\mathbf{r}_0)\right). \mathbf{J}(\mathbf{r}_0)\, dx_0\, dy_0 &\end{aligned} \tag{2–59}$$

In this equation, we have interchanged the integration and the differentiation. This can be done only if the field point is not within the source region where $\mathbf{J}(\mathbf{r}_0)$ is not zero. Now, we equate the bracketed portions of the integrands in Eq. (2–59) to get Eq. (2–60). From this we see that (G_{ij}) is related to $(I_{ij})\delta(\mathbf{r}-\mathbf{r}_0)$ through a wave equation.

$$(\nabla\times\nabla\times - k_0^2)\left(G_{ij}(\mathbf{r}-\mathbf{r}_0)\right) = -j\omega\mu I_{ij}\delta(\mathbf{r}-\mathbf{r}_0) \qquad (2\text{–}60)$$

In addition to the propagation modes determined as in section 2–15, propagation can occur in other modes. One can find the values of β for which the denominator terms in the (Z_{ij}) dyadic are zero. These zeros represent surface wave modes. Conversion of energy from the propagation modes into these modes can occur at points where the microstrip is nonuniform, such as at a step discontinuity. For unshielded microstrip, this energy is lost to the system or part of it can couple back into the circuit at a different point, causing unwanted "cross-talk."

2.18 MILLIMETER WAVE MODELING OF MICROSTRIP LINES

Numerous reports have been presented in the technical literature of computations made using the Green's function and the spectral domain impedances. For instance, Jansen has used these representations to determine integrals from which he computed the microstrip line parameters ε_r, Z_0, and loss for the frequency range from 0 to 16 GHz and for a range of line widths.[57] Figure 2–15 shows the effective dielectric constant of the lowest order hybrid mode as a function of frequency and line width W reported in this reference.

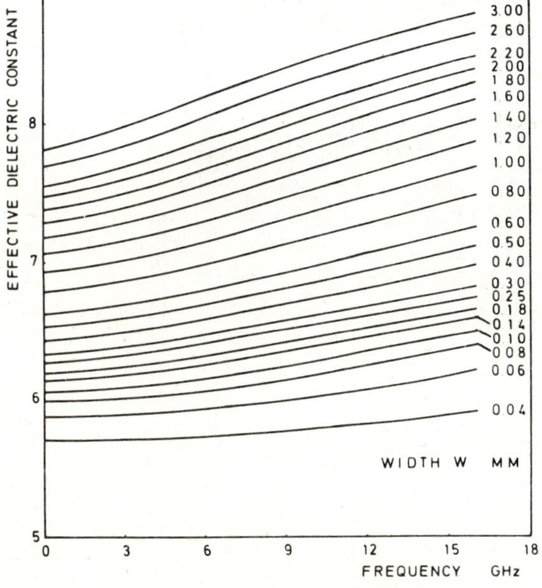

FIGURE 2–15. Effective Dielectric Constant of Microstrip Lines on 0.64 mm Thick Alumina.[57] © 1978 IEEE. Reprinted with permission of the IEEE.

Using spectral domain analysis, Koster and Jansen have computed reflection and transmission coefficients for microstrip step discontinuities; Figures 2–16, 2–17, and 2–18 show some of their results.[58] The graphs are for a substrate with $\varepsilon_r = 10$ and a thickness d of 25 mils. The parameter d/λ corresponds to normalized frequency with λ being the free space wavelength. The value of d/λ equal to 0.04 corresponds to a frequency of nearly 19 GHz. The solid lines are for data computed using SDA and the dashed lines are for data computed using a magnetic wall waveguide model. These authors have found that these transmission and reflection coefficients behave similarly to those for $\varepsilon_r = 2.32$ due to the fact that the lines were designed to have $Z_0 = 50\ \Omega$ in both cases. A different line width is required for the different dielectric constants if the lines each have Z_0 equal to 50 Ω. The effect of the higher order modes is shown in the decrease in $|S_{21}|$ for the wider lines at the higher frequencies.

Another example of the extensive research being reported in the literature is a report of the length extension $\Delta\ell$, shown in Figure 2–19, which characterizes an open end on microstrip on a composite substrate structure at millimeter wave frequencies.[59] The layer thicknesses were 0.15 mm and h was 0.3 mm. These data are the type required when fabricating microstrip lines on semiconductor substrates using a nonconducting dielectric between the semiconductor and the line.

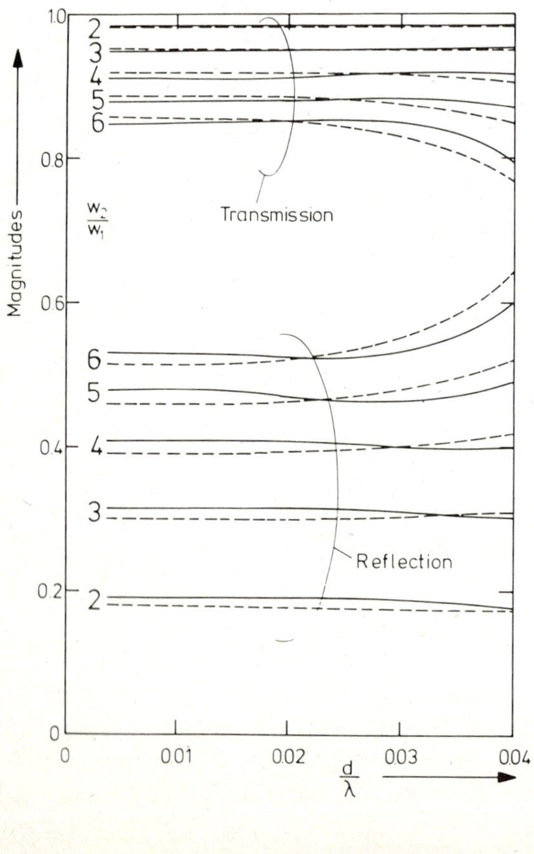

FIGURE 2–16. Reflection and Transmission Coefficient at a Step Discontinuity in a Microstrip Line.[58] © 1986 IEEE. Reprinted with permission of the IEEE.

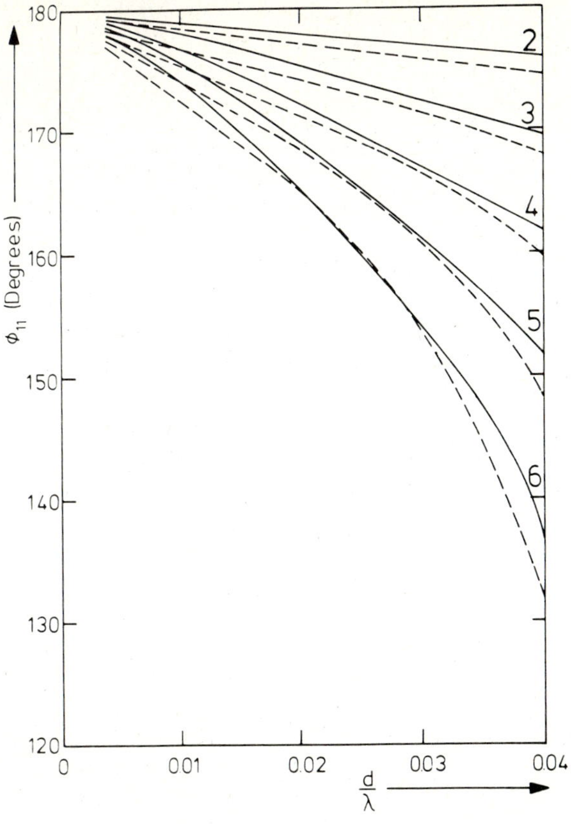

FIGURE 2–17. Reflection Phase Angle at a Step Discontinuity.[58] © 1986 IEEE. Reprinted with permission of the IEEE.

Effective design of integrated millimeter wave circuits poses many problems, such as fabricating such designs on chips is an expensive procedure. The designer would like the chip to have a correct design on the first try. Another problem is that full wave solutions of circuit designs are very consuming of computer time. Thus, it is difficult to check the design prior to implementing it.

Usually, the microstrip circuit design procedure initially involves characterizing the planar guides and their junctions with transmission line analysis techniques. The designer uses such parameters as the characteristic impedance Z_0, the effective dielectric constant ε_{eff}, and the reflection and transmission coefficients at junctions. When possible, discontinuities in microstrip line are modeled to a first order with lumped circuit elements. To finish a design that was started using these characterizations, the designer must have computer software available that allows for more accurate modeling of the circuit. For designs of circuits operating at millimeter wave frequencies, software must necessarily make full wave solutions for the circuits, or it must use a data table with data obtained by a previous full wave solution. The ideal software is capable of a full wave solution for each circuit without requiring excessive computer time. Such software is available presently for planar geometry.

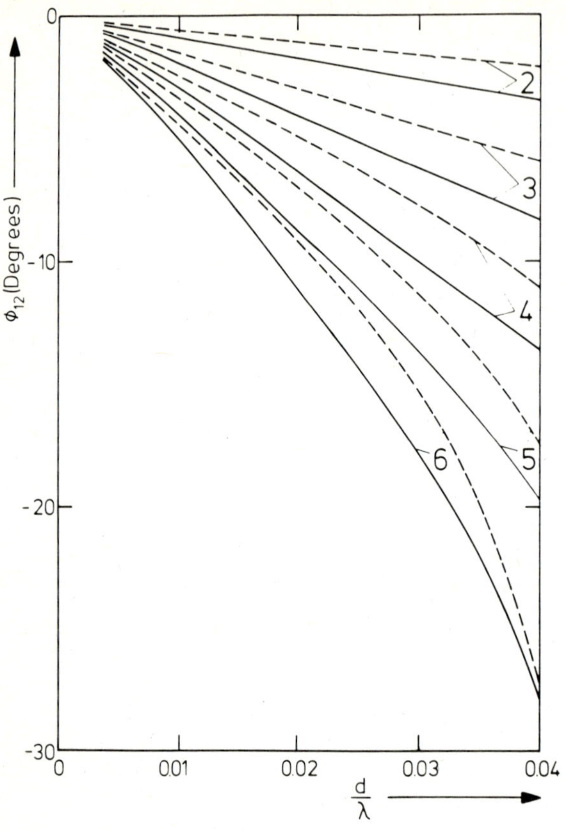

FIGURE 2–18. Transmission Phase Angle at a Step Discontinuity.[58] © 1986 IEEE. Reprinted with permission of the IEEE.

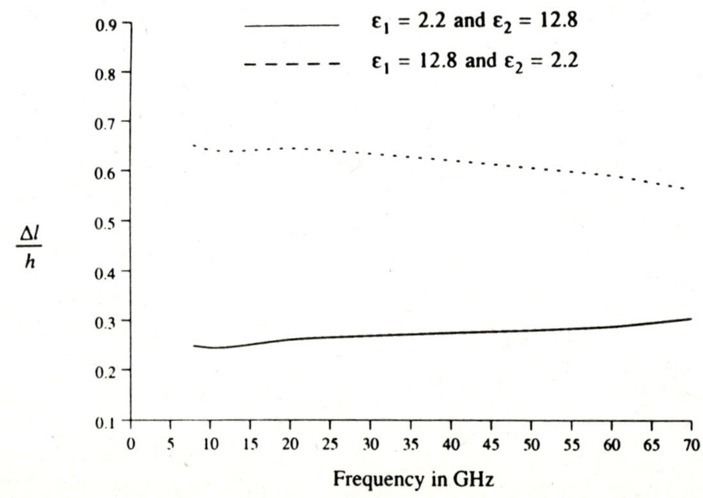

FIGURE 2–19. Length Extension for Modeling an Open Ended Microstrip Line Located on a Composite Substrate with Different Dielectric Layers.[59] © 1989 IEEE. Reprinted with permission of the IEEE.

PROBLEMS

2–1. Calculate the microstrip line width W for a 50 Ω line on a substrate with a thickness of 0.711 mm and a dielectric constant of $\varepsilon_r = 2.32$. Use Eq. (2–11).

2–2. Using the W determined in Prob. 2–1 and the quasi-static formulas of Eq. (2–8), calculate the characteristic impedance of the line in Prob. 2–1.

2–3. Use Eq. (2–11) to determine the line width of a 50 Ω microstrip line on an alumina substrate with a thickness of 0.635 mm and $\varepsilon_r = 9.8$.

2–4. Find the characteristic impedance for a microstrip line having a width of $w = 0.075$ mm on GaAs substrate that is 0.1 mm thick.

2–5. Determine the microstrip line width W for a 20 Ω line on 0.711 mm thick dielectric with $\varepsilon_r = 2.23$.

2–6. Find the length extension $\Delta\ell$ of a 75 Ω, open-ended microstrip line on 0.635 mm alumina with $\varepsilon_r = 8$.

2–7. Calculate M for a 90° mitered corner of a 50 Ω, microstrip line on 0.635 mm thick alumina dielectric substrate with $\varepsilon_r = 9$.

2–8. Determine the quasi-static equivalent circuit parameters ΔL and C_{STP} for a step discontinuity in a microstrip line on a 0.635 mm thick substrate with $\varepsilon_r = 9.6$. The step in line width goes from $W_1 = 0.635$ mm to $W_2 = 1.905$ mm.

2–9. Determine the quasi-static lumped equivalent circuit parameters for the circuit that models the transition between a 93.3 Ω microstrip line to a 50 Ω line. Both lines are on a 0.71 mm substrate with a dielectric constant of $\varepsilon_r = 2.3$. Scale the normalized parameters from Figure 2–7.

2–10. Determine the effective dielectric constant and characteristic impedance of a 0.075 mm wide microstrip line on a 0.1 mm thick GaAs substrate at the frequencies of 2 GHz and 20 GHz.

2–11. For 2 GHz and 20 GHz, determine the characteristic impedance of a 1 mm microstrip line on a 0.711 mm thick substrate with a dielectric constant of $\varepsilon_r = 2.3$.

2–12. Determine the equivalent circuit parameters at 12 GHz for a symmetrical microstrip T junction, where the input and output lines are 50 Ω and the right-angled arm is 75 Ω. All the lines are on a 0.635 mm thick substrate with $\varepsilon_r = 9.8$.

2–13. Show that the total field **H** of section 2–13 is given by $\mathbf{H} = e^{\gamma z}[j\omega\varepsilon_o \nabla_t \times (\hat{\mathbf{k}}\psi^e) + \gamma\nabla_t\psi^h - \hat{\mathbf{k}}\nabla_t^2\psi^h]$.

2–14. Let $\Pi^h = \hat{\mathbf{k}}\psi^h{}_e{}^{\gamma z}$ and $\mathbf{B} = \mu_o\,(\mathbf{H} + \mathbf{M})$ in the discussion of section 2–13. Derive the relation $(\nabla^2 + \omega^2\mu_o\varepsilon_o)\,\Pi^h = -\mathbf{M}$.

2–15. Verify the Y^h is given by Eq. (2–54).

2–16. Determine the spectral impedances for the unshielded microstrip of Figure 2–11.

ENDNOTES

1. K. C. Gupta, Ramesh Garg, and I. J. Bahl, *Microstrip Lines and Slotlines* (297–299) (Dedham, MA: Artech, 1979).
2. Terry Edwards, *Foundations for Microstrip Circuit Design* (2nd ed.), 19–25 (Chichester, England: John Wiley & Sons, 1992).

3. Martin I. Herman, Guey-Liou Lan, James C. Chen, and Cheng-Keng Pao, "Multifunction W-Band MMIC Receiver Technology," *Proc. IEEE, 79, 3* (March 1991), 342–354.
4. Robert W. Jackson, "Considerations in the Use of Coplanar Waveguide For Millimeter-Wave Integrated Circuits," *IEEE Trans. Microwave Theory Tech., MTT-34, 12* (Dec. 1986), 1450–1456.
5. Paul J. Meier, "Integrated Fin-Line Millimeter Components," *IEEE Trans. Microwave Theory Tech., MTT-22, 12* (Dec. 1974), 1209–1216.
6. A. A. Oliner, "Historical Perspective on Microwave Field Theory," *IEEE Trans. Microwave Theory Tech., MTT-32, 9* (Sept. 1984), 1033.
7. K. C. Gupta, Ramesh Garg, and I. J. Bahl, *Microstrip Lines and Slotlines.*
8. William J. Getsinger, "Measurement and Modeling of the Apparent Characteristic Impedance of Microstrip," *IEEE Trans. Microwave Theory Tech., MTT-31, 8* (Aug. 1983), 624–632.
9. John R. Brews, "Characteristic Impedance of Microstrip Lines," *IEEE Trans. Microwave Theory Tech., MTT-35, 1* (Jan. 1987), 30–34.
10. Edwards, *Foundations for Microstrip . . .*, 25–30.
11. Reinmut K. Hoffman, *Handbook of Microwave Integrated Circuits,* trans. by Geoffrey A. Ediss and Nigel J. Keen, English trans. ed. by Harlan H. Howe, Jr. (Norwood, MA: Artech House, 1987), 20–30.
12. Harold A. Wheeler, "Transmission-Line Properties of a Strip on a Dielectric Sheet on a Plane," *IEEE Trans. Microwave Theory Tech., MTT-25, 8* (Aug. 1977), 631–647.
13. Robert E. Collin, *A Field Theory of Guided Waves* (2nd ed.) 259–273 (New York: IEEE Press, 1991).
14. K. C. Gupta, Ramesh Garg, and I. J. Bahl, *Microstrip Lines and Slotlines,* 13–19, (Dedham MA: Artech, 1979).
15. Collin, *A Field Theory of Guided Waves,* 273–285.
16. E. Hammerstad and Ø. Jensen, "Accurate Models for Microstrip Computer-Aided Design," *1980 IEEE MTT-S International Microwave Symp. Dig.,* ed. Leonard M. Schwab, asst. ed. James H. Douglas (New York, NY: The Institute of Electrical and Electronic Engineers (IEEE, 1980), 407–409.
17. R. P. Owens, "Accurate Analytical Determination of Quasi-Static Microstrip Line Parameters," *The Radio and Electronic Engr.,* (46) *7* (July 1976), 360–364.
18. E. H. Fooks and R. A. Zakarevičius, *Microwave Engineering Using Microstrip Circuits,* 56–60, (Sydney: Prentice Hall of Australia Pty Ltd., Prentice Hall, 1990).
19. Steven March, "Microstrip Packaging: Watch the Last Step," *Microwaves,* Dec. 1981, 51–56.
20. Hoffman, *Handbook of Microwave Circuits,* Chap. 10.
21. Hoffman, *Handbook of Microwave Circuits,* 270.
22. M. Kirschning, R. H. Jansen, and N. H. L. Koster, "Accurate Model for Open End Effect of Microstrip Lines," *Electron. Lett., 17, 3* (Feb. 5, 1981), 123–125.
23. Peter Benedek and P. Silvester, "Equivalent Capacitances for Microstrip Gaps and Steps," *IEEE Trans. Microwave Theory Tech., MTT-20, 11* (Nov. 1972), 729–733.
24. A. Gopinath and Chandra Gupta, "Capacitance Parameters of Discontinuities in Microstrip Lines," *IEEE Trans. Microwave Theory Tech., MTT-26, 10* (Oct. 1978), 831–836.
25. Brian Easter, "The Equivalent Circuit of Some Microstrip Discontinuities," *IEEE Trans. Microwave Theory Tech., MTT-23, 8* (Aug. 1975), 655–660.
26. Rene J. P. Douville and David S. James, "Experimental Study of Symmetric Microstrip Bends and Their Compensation," *IEEE Trans. Microwave Theory Tech., MTT-26, 3* (March 1978), 175–182.
27. Alistair F. Thomson and Anand Gopinath, "Calculation of Microstrip Discontinuity Inductances," *IEEE Trans. Microwave Theory Tech., MTT-26, 8* (Aug. 1975), 648–655.

28. Brian M. Neale and A. Gopinath, "Microstrip Discontinuity Inductances," (*IEEE Trans. Microwave Theory Tech., MTT-26, 10* (Oct. 1978), 827–831.

29. Benedek and Silvester, "Equivalent Capacitances . . .".

30. Gopinath and Gupta, "Capacitance Parameters . . .".

31. Chandra Gupta and Anand Gopinath, "Equivalent Circuit Capacitance of Microstrip Step Change in Width," *IEEE Trans. Microwave Theory Tech., MTT-25, 10* (Oct. 1977), 819–822.

32. A. Gopinath, A. G. Thomson, and I. M. Stephenson, "Equivalent Circuit Parameters of Microstrip Step Change in Width and Cross Junctions," *IEEE Trans. Microwave Theory Tech., MTT-24, 3* (March 1976), 142–144.

33. Hoffman, *Handbook of Microwave Integrated Circuits,* 138.

34. William J. Getsinger, "Microstrip Dispersion Model," *IEEE Trans. Microwave Theory Tech., MTT-21, 1* (Jan. 1973), 34–39.

35. M. Kirschning and R. H. Jansen, "Accurate Model for Effective Dielectric Constant of Microstrip with Validity up to Millimetre-Wave Frequencies," *Electron. Lett., 18, 6* (March 18, 1982), 272–273.

36. Masanori Kobayashi, "A Dispersion Formula Satisfying Recent Requirements in Microstrip CAD," *IEEE Trans. Microwave Theory Tech., 36, 8* (Aug. 1988), 1246–1250.

37. Lines operating in the quasi-TEM mode at 100 GHz have been built with minimal dispersion, see K. W. Goossen, H. Roskos, M. C. Nuss, D. W. Kisker, B. Tell, A. E. White, K. T. Short, D. C. Jacobson, and J. M. Poate, "Nearly Dispersionless Microstrip for 100 GHz Pulses Utilizing a Buried Silicide Groundplane", *1991 IEEE MTT-S Int. Microwave Symp. Dig., vol. III,* George L. Heiter, Digest Ed. Inc., (New York: The Institute of Electrical and Electronic Engineers).

38. J. L. Medina, A. Serrano, and F. J. Mendieta, "Microstrip Effective Dielectric Constant Measurements and Test of CAD Models up to 20 GHz", *Microwave Jour., 35, 3* (March 1993), 82–93.

39. Terence C. Edwards and Roger P. Owens, "2-18-GHz Dispersion Measurements on 10–100-Ω Microstrip Lines on Sapphire," *IEEE Trans. Microwave Theory Tech., MTT-24, 8* (Aug. 1976), 506–513.

40. This sequence of formulas was given by Rolf H. Jansen and M. Kirschning, "Arguments and an Accurate Model for the Power-Current Formulation of Microstrip Characteristic Impedance," *Arch. Elektronik u. Übertragungstechn., 37,* 1983, 108–112, and is reprinted with the permission of S. Hirzel Verlag.

41. Parts of Eqs. (2–20) and (2–21) were given by A. G. Franco and A. A. Oliner, "Symmetric Strip Transmission Line Tee Junction", *IRE Trans. Microwave Theory Tech, MTT-10,* (March 1962), 118–123. Permission to reprint was given by IEEE. © 1962 IEEE.

42. Parts of Eqs. (2–20) and (2–21) were given by William H. Leighton, Jr., and Arthur G. Milnes, "Junction Reactance and Dimensional Tolerance Effects on X-Band 3-dB Directional Couplers", *IEEE Trans. Microwave Theory Tech., MTT-19, 10* (Oct. 1971), 818–824. Permission to reprint has been given by IEEE. © 1971 IEEE.

43. Hoffman, *Handbook of Microwave Integrated Circuits,* 285–286.

44. Roger F. Harrington, *Time-Harmonic Electromagnetic Fields,* 100–103, (New York: McGraw-Hill Book Company, 1961).

45. Colin, *Guided Waves,* 61–63.

46. Equation (2–23) was given by Mark K. Krage and George I. Haddad, "Frequency-Dependent Characteristics of Microstrip Transmission Line," *IEEE Trans. Microwave Theory Tech., MTT-20, 10* (Oct. 1972), 678–688. Permission to reprint has been given by the IEEE. © 1972 IEEE.

47. Robert Sorrentino, *Numerical Methods for Passive Microwave and Millimeter Wave Structures,* (New York: IEEE Press, 1989).
48. Tatsuo Itoh, *Planar Transmission Line Structures,* (New York: IEEE Press, 1987).
49. Rolf H. Jansen, "The Spectral-Domain Approach for Microwave Integrated Circuits," *IEEE Trans. Microwave Theory Tech., MTT-33, 10* (Oct. 1985), 1043–1056.
50. The following discussion was adapted from Tatsuo Itoh and Raj Mittra, "Spectral-Domain Approach for Calculating the Dispersion Characteristics of Microstrip Lines," *IEEE Trans. Microwave Theory Tech., MTT-21, 7* (July 1973), 496–99. © 1973 IEEE.
51. Jansen, "Spectral-Domain Approach . . . ," 1053.
52. Roger F. Harrington, "Matrix Methods for Field Problems," *Proc. IEEE, 55, 2* (Feb. 1967), 138.
53. D. Mirshekar-Syahkal and J. Brian Davies, "Accurate Solution of Microstrip and Coplanar Structures for Dispersion and for Dielectric and Conductor Losses," *IEEE Trans. Microwave Theory Tech., MTT-27, 7* (July 1979), 694–699.
54. E. Yamashita and K. Atsuki, "Analysis of Microstrip-Like Transmission Lines by Non-Uniform Discretization of Integral Equations", *IEEE Trans. Microwave Theory Tech., MTT-24, 4* (April 1976), 195–200.
55. Tatsuo Itoh, "Spectral Domain Immitance Approach for Dispersion Characteristics of Generalized Printed Transmission Lines," *IEEE Trans. Microwave Theory Tech., MTT-28, 7* (July 1980), 733–736.
56. Collin, *Guided Waves,* 92–94.
57. Rolf H. Jansen, "High-Speed Computation of Single and Coupled Microstrip Parameters Including Dispersion, High-Order Modes, Loss and Finite Strip Thickness," *IEEE Trans. Microwave Theory Tech., MTT-26, 2* (Feb. 1978), 75–82.
58. Norbert H. L. Koster and Rolf H. Jansen, "The Microstrip Step Discontinuity: A Revised Description," *IEEE Trans. Microwave Theory Tech., MTT-34, 2* (Feb. 1986), 213–223.
59. Hung-Yu Yang, Nicolaos G. Alexopoulos, and David R. Jackson, "Microstrip Open-End Discontinuities in a Substrate-Superstrate Structure," *IEEE Trans. Microwave Theory Tech., 37, 10* (Oct. 1989), 1542–1546.

APPENDIX

This is a program for computing the effective dielectric constant of propagation on a microstrip line using Eqs. (2–15) and (2–16), written in BASIC programming language.

```
10 REM This pgm calc.s Eeff for non-static case
20 E=2.718282
30 INPUT "Er=";ER:INPUT"W IN MM";W:PI=3.141593:C=3*10^8:INPUT"H IN MM";H
40 INPUT "FREQUENCY IN GHZ =";F:F=F*10^9:WH=W/H:H1=H/10^3
50 REM THIS SECTION CALCULATES QUASI-STATIC Eeff LABELED EF
60 A1=LOG(1+(WH/18.1)^3)/18.7:A2=LOG(((WH)^4+(WH/52)^2)/((WH)^4+.432))/49
70 A=1+A1+A2:B=.564*((ER-.9)/(ER+3))^.053
80 EF=((ER+1)+(ER-1)*(1+10/WH)^(-A*B))/2
90 REM THIS SECTION CALCULATES F50
100 FK=C*ATN(ER*SQR((EF-1)/(ER-EF)))/(2*PI*H1*SQR(ER-EF))
110 F50=FK/(.75+(.75-(.332/(ER)^1.73))*WH)
```

```
120 REM THIS SECTION CALC.S P(F) LABELED PF
130 IF WH>.7 THEN GOTO 160
140 MC=1+1.4*(.15-.235/E^(.45*F/F50))/(1+WH)
150 GOTO 170
160 MC=1
170 M0=1+1/(1+SQR(WH))+.32*(1/(1+SQR(WH)))^3:M=M0*MC:PF=(F/F50)^M
180 EEFF= ER-(ER-EF)/(1+PF)
190 PRINT "Eeff=";EEFF
200 END
```

This program calculates the characteristic impedance of a microstrip line using Eq. (2–17) and Eq. (2–18) and is written in BASIC programming language.

```
10 REM This pgm calculates a 'dynamic' Z0 from Jansen & Kirschning, AEU, 1983
20 E=2.718282:PI=3.141593:INPUT"REL. DIEL. CONST.=";ER:INPUT"FREQ. IN GHz =";FF
30 INPUT"W in mm =";W:INPUT"h in mm =";H:U=W/H:FH=FF*H:H1=H/1000
40 REM THIS PART CALCULATES QUASI-STATIC Z0=Z0QS & EFFECTIVE ER CONST.=EFQS
50 A1=LOG(1+(U/18.1)^3)/18.7:A2=LOG(((U)^4+(U/52)^2)/((U)^4+.432))/49
60 A=1+A2+A1:B=.564*((ER-.9)/(ER+3))^.053
70 EFQS=((ER+1)+(ER-1)*(1+10/U)^(-A*B))/2:AZ=(30.666/U)^.7528
80 Z0QS=59.9585*LOG((6+(2*PI-6)*E^(-AZ))/U+SQR(1+(2/U)^2))/SQR(EFQS)
90 REM THIS SECTION CALCULATES EFFECTIVE DIEL. CONST. AT FREQ. F
100 C=3*10^8:F=FF*10^9
110 FK=C*ATN(ER*SQR((EFQS-1)/(ER-EFQS)))/(2*PI*H1*SQR(ER-EFQS))
120 F50=FK/(.75+((.75-(.332/(ER)^1.73))*U)):IF U>.7 THEN GOTO 140
130 MC=1+1.4*(.15-.235/E^(.45*F/F50))/(1+U):GOTO 150
140 MC=1
150 M0=1+1/(1+SQR(U))+.32*(1/(1+SQR(U)))^3:M=M0*MC:PF=(F/F50)^M
160 EEFF = ER-(ER-EFQS)/(1+PF)
170 REM THIS PART OF THE PGM CALCULATES DYNAMIC Z0 = Z0
180 R1=.03891*(ER)^1.4:R2=.267*U^7:R3=4.766/E^(3.228*U^.641)
190 R4=.016+(.0514*ER)^4.524:R5=(FH/28.843)^12:R6=22.2*U^1.92
200 R7=1.206-.3144*(1-(1/E)^R2)*(1/E)^R1
210 R8=1+1.275*(1-E^(-.004625*R3*(ER^1.674)*(FH/18.365)^2.745))
219 R8A=(1/E)^R6
220 R9=5.086*R4*R5*(ER-1)^6/((.3838+.386*R4)*(1+1.2992*R5)*(1+10*(ER-1)^6))*R8A
230 R10=.00044*ER^2.136+.0184:R11=(FH/19.47)^6/(1+9.619999E-02*(FH/19.47)^6)
240 R12=1/(1+.00245*U^2):R13=.9408*EEFF^R8-.9603
250 R14=(.9408-R9)*EFQS^R8-.9603:R15=.707*R10*(FH/12.3)^1.097
260 R16=1+.0503*ER^2*R11*(1-E^(-(U/15)^6))
270 R17=R7*(1-1.1241*R12/(R16*E^(.026*FH^1.15656-R15)))
280 Z0=Z0QS*(R13/R14)^R17
290 PRINT "QUASI-STATIC Z0 =";Z0QS;"QUASI-STATIC EFF. DIEL. CONST.=";EFQS
300 PRINT "Eeff is ";EEFF
310 PRINT "CHAR. IMPEDANCE, Z0,=";Z0
320 END
```

This program calculates the symmetric T junction equivalent circuit parameters given in Eq. (2–20) and (2–21). It is written in BASIC programming language.

```
10 REM THIS PGM CALCULATES THE T JUNCT. EQUIV. CKT. PARAMETERS FROM
20 REM HOFFMAN'S BOOK PAGE 285
30 PI=3.141593:E=2.718282:N0=376.991:C=3*10^8:INPUT "diel. const.";ER
40 INPUT "H in mm";H:INPUT "freq. in GHz";F:FH=F*H:F=F*10^9
50 INPUT "input and output line width in mm";W:GOSUB 180
60 EF1=EEFF:Z01=Z0:INPUT "branch line width in mm";W:GOSUB 180
70 EF2=EEFF:Z02=Z0:W1=N0*H1/(Z01*SQR(EF1)):L1=C/(F*SQR(EF1)): L2=C/(F*SQR(EF2))
80 ANGL=PI*W1*Z01/(L1*Z02):N1=(SIN(ANGL))/ANGL:ZZ=Z01/Z02:N11=N1*SQR(ZZ)
85 WL1=W1/L1
90 BTZ=2*WL1*(LOG(SQR(1+1/(TAN(PI*ZZ/2))^2))+.5*WL1^2*(COS(PI*ZZ/2))^4)
100 XZ=-WL1*ZZ*(N11*PI/4)^2:IF ZZ>.5 THEN GOTO 120
110 X2Z=-XZ/2+(BTZ+2*WL1*(LOG(2)+PI*ZZ/6+1.5*WL1^2))/N1^2:GOTO 130
120 X2Z=-XZ/2+2*WL1*(LOG(1.43/ZZ)+2*WL1^2)/N1^2
130 D1=L1*ATN(-XZ)/(2*PI):D2P=L2*(ATN(N1^2*X2Z*ZZ))/(2*PI):D2=W1/2-D2P
140 NN=N1*COS(2*PI*D2P/L2)/COS(2*PI*D1/L1)
150 BP2=(2*TAN(2*PI*D1/L1)-NN^2*ZZ*TAN(2*PI*D2P/L2))/Z01
160 PRINT "n:1=";NN; "Bp2=";BP2; "d1=";D1; "d2=";D2
170 PRINT "Z01=";Z01; "Z02=";Z02; "Weff1=";W1
175 GOTO 410
179 REM this part calc.s the effective dielectric constant, EEFF
180 WH=W/H:H1=H/10^3:A1=LOG(1+(WH/18.1)^3)/18.7
190 A2=LOG(((WH)^4+(WH/52)^2)/((WH)^4+.432))/49:A=1+A1+A2
200 B=.564*((ER-.9)/(ER+3))^.053:EF=((ER+1)+(ER-1)*(1+10/WH)^(-A*B))/2
210 FK=C*ATN(ER*SQR((EF-1)/(ER-EF)))/(2*PI*H1*SQR(ER-EF))
220 F50=FK/(.75+(.75-(.332/(ER)^1.73))*WH):IF WH>.7 THEN GOTO 240
230 MC=1+1.4*(.15-.235/E^(.45*F/F50))/(1+WH):GOTO 250
240 MC=1
250 M0=1+1/(1+SQR(WH))+.32*(1/(1+SQR(WH)))^3:M=M0*MC:PF=(F/F50)^M
260 EEFF=ER-(ER-EF)/(1+PF)
270 REM THIS PART CALC.S CHAR. IMPEDANCE, Z0
280 AZ=(30.666/WH)^.7528:
290 Z0QS=59.9585*LOG((6+(2*PI-6)*E^(-AZ))/WH+SQR(1+(2/WH)^2))/SQR(EF)
300 R1=.03891*(ER)^1.4:R2=.267*WH^7:R3=4.766/E^(3.228*WH^.641)
310 R4=.016+(.0514*ER)^4.524:R5=(FH/28.843)^12:R6=22.2*WH^1.92
320 R7=1.206-.3144*(1-(1/E)^R2)*(1/E)^R1
330 R8=1+1.275*(1-E^(-.004625*R3*(ER^1.674)*(FH/18.365)^2.745)):R8A=(1/E)^R6
340 R9=5.086*R4*R5*(ER-1)^6/((.3838+.386*R4)*(1+1.2992*R5)*(1+10*(ER-1)^6))*R8A
350 R10=.00044*ER^2.136+.0184:R11=(FH/19.47)^6/(1+9.619999E-02*(FH/19.47)^6)
360 R12=1/(1+.00245*WH^2):R13=.9408*EEFF^R8-.9603
370 R14=(.9408-R9)*EF^R8-.9603:R15=.707*R10*(FH/12.3)^1.097
380 R16=1+.0503*ER^2*R11*(1-E^(-(WH/15)^6))
390 R17=R7*(1-1.1241*R12/(R16*E^(.026*FH^1.15656-R15)))
400 Z0=Z0QS*(R13/R14)^R17
405 RETURN
410 END
```

Chapter

3

Microwave Circuit Analysis Methods

3.0 INTRODUCTION

Microwave circuits are *distributed circuits* in the sense that the dimension of the circuit "components" is at least an appreciable fraction of the wavelength of the voltage and current waves. So, we cannot neglect the effect of the phase changes in the waves traveling along leads connecting circuit elements. Thus, the techniques for analyzing these circuits differ somewhat from those used with lumped circuits—they must be capable of describing the action of lumped circuits as well as distributed circuits. The simplest way to do this is to use transmission line equations and traveling waves. Then, at a lumped element, the voltage across the element is the sum of the voltage waves on the line at that point, and the current through it is the sum of the current waves on the line.

Because Maxwell's equations and the transmission line equations derived from them are linear equations, we can use superposition and reciprocity with them as we do with lumped circuit analysis. When working with transmission line equations, every 2 port must be analyzed to account for the reflection at the terminals. This reflection is described by the reflection coefficient of the 2 port. An illustrative example from chapter 1 is a lossless transmission line terminated in a lumped resistor with a resistance not equal to the line characteristic impedance. In that case, we used Ohm's law to relate the current through the load impedance to the voltage across it. Then this relation constituted the boundary condition upon the voltage waves and current waves on the transmission line. From this, we found the reflection coefficient.

3.1 S PARAMETERS

With transmission lines, we use traveling waves normalized to the square root of the characteristic impedance of the line. We define the *wave functions* $a(z)$ to be the normalized forward traveling (incident) wave and $b(z)$ to be the normalized backward traveling (reflected) wave as in Eq. (3–1).

$$\left.\begin{aligned} a &= V_i / \sqrt{Z_0} = I_i \sqrt{Z_0} \\ b &= V_r / \sqrt{Z_0} = -I_r \sqrt{Z_0} \end{aligned}\right\} \tag{3–1}$$

Usually, we will choose $z = 0$ as the point at which the transmission line terminates on an N port. The line voltage at any point on the line is given by $V_{\text{line}} = (a + b)\sqrt{Z_0}$, the line current by $I_{\text{line}} = (a - b)/\sqrt{Z_0}$, and the power passing down the line at any point is given by Eq. (3–2) (see problem 3–1).

$$\text{power} = \mathcal{Re}\,(VI^*) = aa^* - bb^* \tag{3–2}$$

Conversely, we have: $a = (V_{\text{line}} + I_{\text{line}}Z_0)/2\sqrt{Z_0}$ and $b = (V_{\text{line}} - I_{\text{line}}Z_0)/2\sqrt{Z_0}$ (see problem 3–2).

Example: Consider a lossless 50 Ω line terminated in a 10 Ω resistor as in the example of section 1–5 of chapter 1. The incident voltage is $10\angle 30°$ at the load. The reflection coefficient is $-\frac{2}{3}$. So, the line has voltages $V_i = 10e^{-j(\beta z - \pi/6)}$ and $V_r = -6.667e^{j(\beta z + \pi/6)}$ and currents $I_i = 0.2e^{-j(\beta z - \pi/6)}$ and $I_r = 0.133e^{j(\beta z + \pi/6)}$. The wave functions are: $a = V_i/\sqrt{50} = 1.414e^{-j(\beta z - \pi/6)}$ and $b = V_r/\sqrt{50} = -0.943e^{j(\beta z + \pi/6)}$. The load reflection coefficient is given by b/a evaluated at $z = 0$. The power is $|a|^2 - |b|^2 = 2 - 0.8889$ watts.

We can apply these wave functions to the solution of microwave circuits. To do this, we define the *S parameters* of an N port by the relation in Eq. (3–3), where (a) and (b) are column matrices of wave functions on the input lines of the N port evaluated at the terminals of the N port.

$$(b) = (S)\,(a) \tag{3–3}$$

The S matrix describes the relations between these wave functions. The wave function given by a_i is the incident wave into the ith terminal of the n port and the wave function b_i is the reflected wave at that terminal. If we replace all sources to the input lines as shown in Figure 3–1 with matched impedances and drive the ith input line with a source that generates the wave function a_i, then every term in the matrix (a) will be zero except for the ith term. The output from the jth terminal of the n port will be the b_j wave function. Equation (3–4) can be derived from Eq. (3–3), where the reference point of $z = 0$ on each line is at the terminals of the N port.

$$S_{ji} = \left.\frac{b_j(0)}{a_i(0)}\right|_{a_k = 0 \text{ for } k \neq i} \tag{3–4}$$

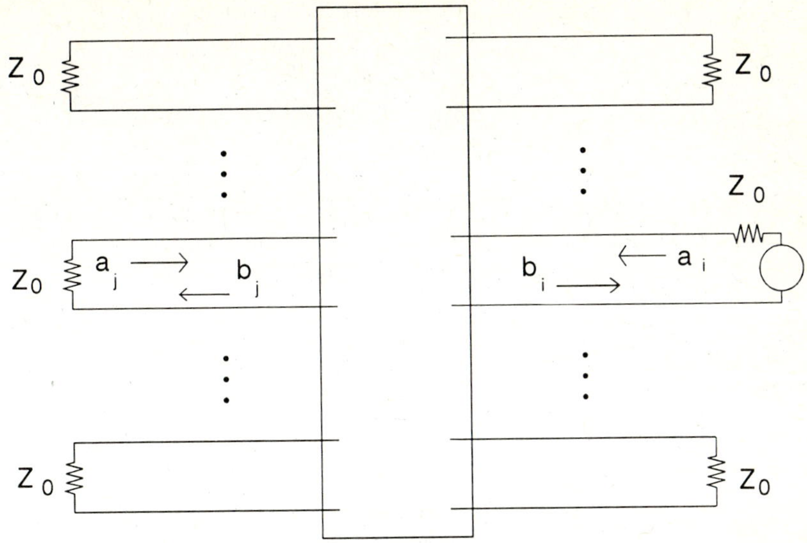

FIGURE 3–1. Graphical Representation of an N port.

S_{ji} is the S parameter which describes the transfer action between an input at the ith port and the output at the jth port. It was previously referred to as the transmission coefficient. In the case of S_{ii}, we have the reflection coefficient at the ith terminal.

Example: Consider transmission lines connected to a 2 port as shown in Figure 3–2. Note that the direction of propagation of a_1 is towards the right, into the 2 port, and the direction of a_2 is towards the left, again into the 2 port. Conversely, b_1 propagates to the left and b_2 to the right. The 2 port consists of a resistor connected from the top line to the bottom line (parallel connection) then the "load" on the input line is the parallel connection of that resistor and the output line impedance as seen at the second port. To determine S_{11}, we use Eq. (3–4). For this relation, we must match the input and output lines with their characteristic impedance and apply an input to port 1 so $Z_g = Z_0$ and $Z_L = Z_0$. To determine the wave functions at the 2 port, we see that the boundary condition on the

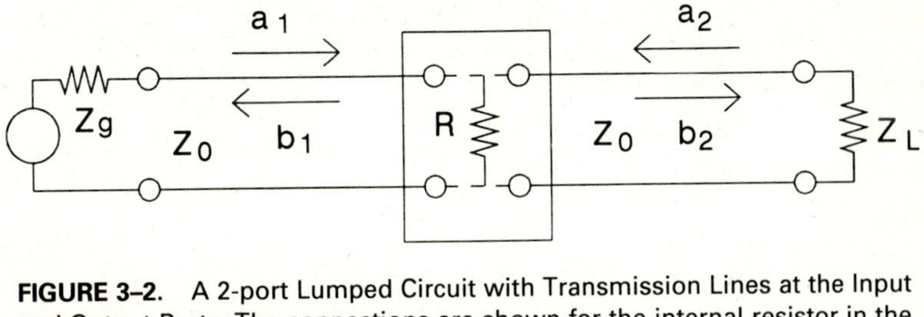

FIGURE 3–2. A 2-port Lumped Circuit with Transmission Lines at the Input and Output Ports. The connections are shown for the internal resistor in the 2 port. Z_0 = 50 and R = 100 Ohms.

line voltage at the resistor is that the voltage is the same at port 2 as at port 1. Because the line voltage is proportional to the sum of the wave functions at these ports, we have

$$[a_1(0) + b_1(0)]\sqrt{Z_0} = [a_2(0) + b_2(0)]\sqrt{Z_0}.$$

The load on port 2 is a matched transmission line so $a_2(0) = 0$. The boundary condition on the voltages becomes $a_1(0) + b_1(0) = b_2(0)$. The voltage and current in the resistor are $V = \sqrt{50}\, b_2(0)$ and $I = [a_1(0) - b_1(0) - b_2(0)]/\sqrt{50}$. Ohm's law for this resistor is $V = 100I$, which becomes $\sqrt{50}\, b_2(0) = 100(a_1(0) - b_1(0) - b_2(0))/\sqrt{50}$. Solving these two relations to eliminate b_2, we get $-5b_1(0) = a_1(0)$. From this we have

$$S_{11} = \frac{b_1(0)}{a_1(0)} = -1/5.$$

To check this result, we note that the parallel combination of the 100 Ω resistor and the 50 Ω (matched) output line is 100/3 Ω. Applying Eq. (1–6) for the reflection coefficient of a 100/3 Ω resistor terminating a 50 Ω line, we verify that the reflection coefficient at the input port is −1/5. Going back to the two equations for $a_1(0)$, $b_1(0)$, and $b_2(0)$, we can eliminate $b_1(0)$ from these two equations and obtain the result

$$S_{21} = \frac{b_2(0)}{a_1(0)} = 4/5.$$

Again, we can verify that this is the transmission coefficient for this case.

Usually, we do not carry along the explicit notation for showing that the wave functions are functions of distance. So, the notation for the output at the ith port when the N port has an input only at the jth port is simply $S_{ij} = b_i/a_j$.

3.2 REFERENCE PLANE SHIFTS

Microwave transistors are usually described in terms of their S parameters. Although they are 3 port devices, the S parameters are given in terms of a particular transistor operation, such as grounded base or grounded emitter, so that the device can be characterized as a 2 port. For example, the Motorola BFR91 transistor has the small signal S parameters for grounded emitter operation at 1 GHz of $S_{11} = 0.3\angle 155°$, $S_{21} = 2.6\angle 60°$, $S_{12} = 0.15\angle 70°$, and $S_{22} = 0.4\angle{-30°}$. A circuit will have connections between the transistor and other parts of the circuit that often must be treated as transmission lines when it is operated at these frequencies. We must account for the traveling wave effects on these lines. Because the S parameters are determined from these waves, the S parameters measured at different points along these lines will have different values. For lossless lines, the change in an S parameter is that in phase. The point on these lines to which the S parameters are referred is the reference plane for the parameters. If we measure the S_{11} parameter (reflection coefficient) at the base of the transistor and then insert a quarter wave, lossless line between this base lead and the point of measurement of the S_{11} parameter, the phase of the measured reflection coefficient would change by $-2\beta L$, corresponding to the phase change by which Γ is multiplied in Eq. (1–11). So, the change in the phase of S_{11} would be $-180°$, resulting in $S_{11} = 0.3\angle -25°$.

To state this phase shift more formally, we assume that the wave functions at the reference plane (the input for the ith port) for the S parameters are a_i and b_i. Then, at some point L_i units along a lossless line away from the ith port, the wave functions are given by $a'_i = a_i \exp(j\beta L_i)$ and $b'_i = b_i \exp(-j\beta L_i)$. Putting this into matrix form we have

$$(b') = \begin{pmatrix} e^{-j\beta L_1} & 0 & \cdots & 0 \\ 0 & e^{-j\beta L_2} & \cdots & 0 \\ \vdots & \vdots & & \vdots \\ 0 & 0 & \cdots & e^{-j\beta L_n} \end{pmatrix} (b) \text{ and}$$

$$(a') = \begin{pmatrix} e^{+j\beta L_1} & 0 & \cdots & 0 \\ 0 & e^{+j\beta L_2} & \cdots & 0 \\ \vdots & \vdots & & \vdots \\ 0 & 0 & \cdots & e^{+j\beta L_n} \end{pmatrix} (a)$$

In these expressions, we have assumed the same phase constant β for the lines feeding into all of the ports. We name this coefficient matrix (E) so that we have $(b') = (E)\,(b)$ and $(a') = (E)^{-1}\,(a)$. Note that the matrices multiplying (b) and (a) are respectively the inverse of each other so that their product is the identity matrix.

Starting with the expression $(b) = (S)(a)$, we premultiply both sides by (E) and substitute (b') for the left side. The result is $(b') = (E)(S)(a)$. Substituting $(E)(a')$ for (a), we get $(b') = (E)(S)(E)(a')$. Comparing this with the expression for the S parameters as measured at the end of these lines $(b') = (S')(a')$, we see that the expression $(S') = (E)(S)(E)$ is the formula for the shifting of the reference planes. This expression can be expanded as Eq. (3–5).

$$(S') = \begin{pmatrix} S_{11}e^{-j\beta(L_1+L_1)} & S_{12}e^{-j\beta(L_1+L_2)} & \cdots & S_{1n}e^{-j\beta(L_1+L_n)} \\ S_{21}e^{-j\beta(L_2+L_1)} & S_{22}e^{-j\beta(L_2+L_2)} & \cdots & S_{2n}e^{-j\beta(L_2+L_n)} \\ \vdots & \vdots & & \vdots \\ S_{n1}e^{-j\beta(L_n+L_1)} & S_{n2}e^{-j\beta(L_n+L_2)} & \cdots & S_{nn}e^{-j\beta(L_n+L_n)} \end{pmatrix} \quad (3\text{–}5)$$

Example: The transistor whose S parameters were given at the start of this section is connected in a circuit where the input and output microstrip leads are 0.2 and 0.3 cm respectively. The microstrip has an effective dielectric constant of 2 at a frequency of 1 GHz. The phase constant for these lines is then $2\pi/\lambda = 2\pi\sqrt{2}/30$ rad/cm. The phase changes for S_{11} and S_{22} are $-\beta(L_1 + L_1) = -0.118$ and $-\beta(L_2 + L_2) = -0.178$ rad respectively while that for S_{12} and S_{21} is -0.148 rad. These changes in phase result in a new set of S parameters: $S_{11} = 0.3\angle 148°$, $S_{21} = 2.6\angle 51.5°$, $S_{12} = 0.15\angle 61.5°$ and $S_{22} = 0.4\angle -40°$ that characterize the device as measured at the ends of the microstrip lines.

3.3 SCATTERING MATRIX OF A PASSIVE LOSSLESS DEVICE

A linear N-port device has a linear relation between the inputs and the outputs at the various ports, expressed in terms of the scattering matrix (S) in Eq. (3–3). If, in addition to being linear, the device is lossless, then we can express the fact that the power input is equal to the power output by summing the net power input at each terminal as in Eq. (3–6).

$$\sum_i (|a_i|^2 - |b_i|^2) = 0 \tag{3–6}$$

This sum can be expressed as $\sum_i (|a_i|^2 - |b_i|^2) = (a^*)^T(a) - (b^*)^T(b) = 0$. When we use Eq. (3–3) in the product $(b^*)^T(b)$, where $(b^*)^T$ is the conjugate of the transpose of (b), this product becomes $(b^*)^T(b) = (a^*)^T(S^*)^T(S)(a)$. Then, we have $(a^*)^T(a) = (a^*)^T(S^*)^T(S)(a)$. So the matrix product $(S^*)^T(S)$ must be the identity matrix. A matrix satisfying this property is called a *Hermitian* matrix. This property of a lossless 2 port is given in Eq. (3–7).

$$(S^*)^T(S) = \begin{pmatrix} 1 & 0 \\ 0 & 1 \end{pmatrix} \tag{3–7}$$

If we examine the individual scalar equations in the matrix equation of Eq. (3–7) by themselves, we can determine the relations that describe a lossless symmetric 2 port. First, we look at the product of the first row of $(S^*)^T$ with the first column of (S): $S_{11}^* S_{11} + S_{21}^* S_{21} = 1$. Reciprocity allows us to set $S_{21} = S_{12}$ (see problem 3–8).[1] From this, we have the relation on the magnitudes given as Eq. (3–8).

$$|S_{12}|^2 = |S_{21}|^2 = 1 - |S_{11}|^2 \tag{3–8}$$

We see that for a lossless 2 port, we can determine $|S_{21}|$ from measurements of the input reflection coefficient $|S_{11}|$. A similar relation holds for $|S_{22}|^2$. If we further require that the device be symmetric, then $S_{11} = S_{22}$. Now, we take the product of the second row of $(S^*)^T$ with the first column of (S) to obtain $S_{12}^* S_{11} + S_{22}^* S_{21} = 0$. Using reciprocity and the assumed symmetry, we have $S_{11}/S_{11}^* = -S_{21}/S_{21}^*$. Letting α be the angle of S_{11} and θ be the angle of S_{21}, this becomes the relation $\alpha - \theta = \pm \pi/2$.

Example: We apply these to the case of an iris in a waveguide considered in section 1–12. In that case, we have a capacitive window with a reflection coefficient of $0.321\angle -108.7°$. From the relation between the angles, we calculate that the angle of S_{12} is 161.3°. The magnitude of S_{12} is $\sqrt{1 - 0.321^2} = 0.947$.

Example: Ideal Directional Couplers. An ideal directional coupler is one where there is no internal loss and the ports are matched so that there is no reflection when a wave is incident upon any port. It should also have perfect directionality, that is, the coupled output should occur for waves traveling in one direction but not for waves traveling in the other direction. The diagrammatic representation of a pair of coupled lines is shown in Figure 3–3. This could represent a pair of coupled lossless waveguides or it could be a pair of coupled lossless transmission lines. We assume that the major amount of power is to pass through from port 1 to port 4 or from port 2 to port 3. Only a small amount of power is to couple from one line to the adjacent line.

We have the following design characteristics:

(1) Matched input and output: $S_{11} = S_{44} = 0$
(2) Directional coupling: $S_{13} = S_{24} = 0$, but neither S_{12} nor S_{34} is zero
(3) Bilateral network: $S_{ij} = S_{ji}$

The S parameter matrix that meets the first two of these criteria is

$$(S) = \begin{pmatrix} 0 & S_{12} & 0 & S_{14} \\ S_{21} & S_{22} & S_{23} & 0 \\ 0 & S_{32} & S_{33} & S_{34} \\ S_{41} & 0 & S_{43} & 0 \end{pmatrix}.$$

If the coupler is lossless, then (S) must be Hermitian, that is $(S^*)^T(S)$ must be equal to the unit matrix.

This Hermitian property becomes

$$\begin{pmatrix} 0 & S_{21}^* & 0 & S_{41}^* \\ S_{21}^* & S_{22}^* & S_{32}^* & 0 \\ 0 & S_{23}^* & S_{33}^* & S_{43}^* \\ S_{14}^* & 0 & S_{34}^* & 0 \end{pmatrix} \cdot \begin{pmatrix} 0 & S_{12} & 0 & S_{14} \\ S_{21} & S_{22} & S_{23} & 0 \\ 0 & S_{32} & S_{33} & S_{34} \\ S_{41} & 0 & S_{43} & 0 \end{pmatrix} = \text{unit matrix.}$$

Multiplying the fourth row of the first matrix by the third column of the second matrix, we have $S_{33}S_{34}^* = 0$, but S_{34}^* must not be zero if we are to have a directional coupler. So, we conclude that S_{33} must be zero. Similarly, from the product of the first row by the second column, we can conclude that S_{22} must be zero. Now, the product of the first row by the third column becomes $S_{21}^*S_{23} + S_{41}^*S_{43} = 0$. Similarly, the product of the fourth row by

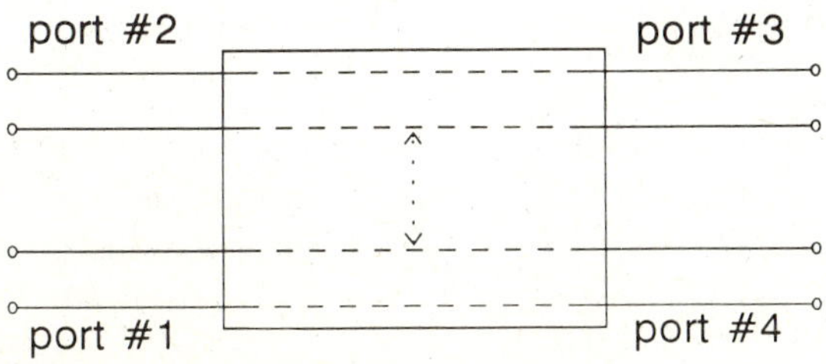

FIGURE 3–3. A 4-port Representation of an Ideal Directional Coupler.

the second column becomes the relation $S_{14}^*S_{12} + S_{34}^*S_{32} = 0$. By combining these two relations and using the requirement $S_{ij} = S_{ji}$, we can derive the relation $|S_{34}| = |S_{12}|$. We let $|S_{12}| = C$. This parameter is the coupling parameter equal to the square root of the ratio of the power out of port 2 to the power into port 1. Similarly, we can derive the relation $|S_{23}| = |S_{14}|$. We call these T. This parameter is the transmission parameter that is equal to the square root of the fraction of the power out of port 4 from power put into port 1. For a 20 dB coupler, this would be 0.99. Then, the scattering matrix is

$$(S) = \begin{pmatrix} 0 & C & 0 & T \\ C & 0 & T & 0 \\ 0 & T & 0 & C \\ T & 0 & C & 0 \end{pmatrix}.$$

Again, we use the fact that (S) is a Hermitian matrix and derive the relation $|T|^2 + |C|^2 = 1$ or $|C| = \sqrt{1 - |T|^2}$. This is a statement of conservation of power. Any power not transmitted to the primary output is coupled to the secondary output. The Hermitian property of the S matrix given above can also be used to obtain the relation $T^*C = -TC^*$. From this, we see that if T is a pure imaginary number, then C must be a real number. If C has a very small magnitude, then we get the approximate scattering matrix

$$(S) \simeq \begin{pmatrix} 0 & C & 0 & j \\ C & 0 & j & 0 \\ 0 & j & 0 & C \\ j & 0 & C & 0 \end{pmatrix}.$$

In this matrix, we have $|S_{14}|^2 = 1$, which corresponds to full transmission of power to port 4 from port 1. S_{23} corresponds to transmission from port 3 to port 2. The fractional coupling is from 1 to 2 or 3 to 4.

Example: 40 dB Directional Coupler. A 40 dB directional coupler couples power in one direction down by 40 dB. That output port has only 0.0001 watt output for every watt put in. Because the (S) matrix relates voltage or current wave functions, the coupling parameter C is 0.01. The transmission parameter is very close to unity in magnitude and the phase angle is 90°, so $T = j$.

3.4 PARAMETER CONVERSIONS

The use of S parameters to describe the properties of N ports has been shown above. However, we know from circuit theory that these N ports also can be described with other parameters such as impedance or admittance. Because all of these different parameters of N ports represent linear relationships between the input and output voltage and current variables, we can convert from a description in terms of one set of parameters to a description in terms of another set of parameters.

To derive the formula for converting from impedances to S parameters, we consider a passive N port with the impedance description as in Eq. (3–9).

$$(V) = (Z)(I) \tag{3–9}$$

In this equation, the matrices (V) and (I) are column matrices of the port voltages and currents. To convert this to an S parameter description, we must express the voltage and current matrices in terms of the matrices (a) and (b) of the wave functions evaluated at the ports. The Figure 3–4 shows these relations. The voltage at the ith port is related to the wave functions as $V_i = a_i\sqrt{Z_{0i}} + b_i\sqrt{Z_{0i}}$, where Z_{0i} is the characteristic impedance of the line connected to the ith port. The matrix relation between the port voltages and the wave functions is given by $(V) = (\sqrt{Z_0})(a) + (\sqrt{Z_0})(b)$, where the matrix $(\sqrt{Z_0})$ is a diagonal matrix with diagonal terms equal to $\sqrt{Z_{0i}}$. Similarly, the current matrix is given by $(I) = (\sqrt{Z_0})^{-1}(a) - (\sqrt{Z_0})^{-1}(b)$ in which the matrix $(\sqrt{Z_0})^{-1}$ is a diagonal matrix with ith diagonal element equal to $1/\sqrt{Z_{0i}}$. From their description, we can verify that $(\sqrt{Z_0})$ *and* $(\sqrt{Z_0})^{-1}$ are inverse matrices.

Substituting the expression for (V) into Eq. (3–9) and multiplying the terms in both sides of that relation by the matrix $(\sqrt{Z_0})^{-1}$, we obtain $(a) + (b) = (\sqrt{Z_0})^{-1}(Z)(\sqrt{Z_0})^{-1}[(a) - (b)]$. Defining the normalized impedance matrix as $(\zeta) = (\sqrt{Z_0})^{-1}(Z)(\sqrt{Z_0})^{-1}$, we can regroup these terms to form the relation $(b) = [(\zeta) + (\mathbf{1})]^{-1}[(\zeta) - (\mathbf{1})](a)$. In this equation, $(\mathbf{1})$ is the unit matrix and the term $[(\zeta) + (\mathbf{1})]^{-1}$ is an inverse of the matrix $[(\zeta) + (\mathbf{1})]$. From the previous relation for $(a) + (b)$ and Eq. (3–3), we see that the S parameter matrix can be calculated from the matrix (ζ) by Eq. (3–10).

$$(S) = [\,(\zeta) + (\mathbf{1})\,]^{-1}[\,(\zeta) - (\mathbf{1})\,] \tag{3–10}$$

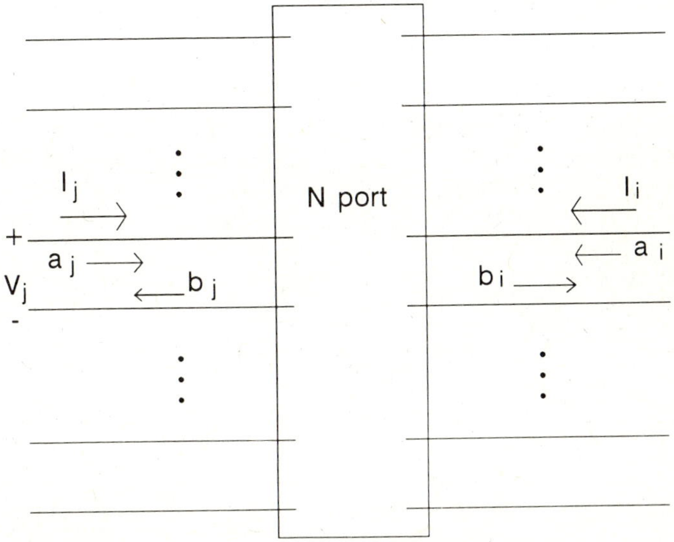

FIGURE 3–4. N-port Voltages, Currents and Wave Functions.

In the case where the characteristic impedances Z_{0i} are all equal, Eq. (3–10) can be written as

$$(S) = ((Z) + (Z_0))^{-1}((Z) - (Z_0)), \tag{3–10}$$

where (Z_0) is

$$(Z_0) = \begin{pmatrix} Z_0 & 0 & \dots & 0 \\ 0 & Z_0 & \dots & 0 \\ \vdots & \vdots & & \vdots \\ 0 & 0 & \dots & Z_0 \end{pmatrix}.$$

We can solve Eq. (3–10) for (ζ) to obtain Eq. (3–11).

$$(\zeta) = ((\mathbf{1}) + (S))((\mathbf{1}) - (S))^{-1} \tag{3–11}$$

Example: Assume that we are given a transistor with the small signal equivalent circuit shown in Figure 3–5 and that the characteristic impedances of the input and output lines are the same. Further, assume that the feedback path with Z_μ and the output capacitance C_0 in this circuit can be ignored. Then, the Z parameter description will be that given in Eq. (3–12).

$$(V) = \begin{pmatrix} Z_{11} & -r_e \\ -r_e(1+g_m Z_\pi) & Z_{22} \end{pmatrix}(I), \qquad \text{where } \begin{Bmatrix} Z_{11} = r_b + Z_\pi + r_e \\ Z_{22} = r_c + r_{ce} + r_e \end{Bmatrix} \tag{3–12}$$

The S parameter matrix can be obtained using Eq. (3–10) as

$$(S) = \begin{pmatrix} Z_{11} + Z_{01} & -r_e \\ -r_e(1+g_m Z_\pi) & Z_{22} + Z_{02} \end{pmatrix}^{-1} \begin{pmatrix} Z_{11} - Z_{01} & -r_e \\ -r_e(1+g_m Z_\pi) & Z_{22} - Z_{02} \end{pmatrix}.$$

By finding the inverse matrix and by expanding this product, we can determine the individual S parameters. The forward transfer parameter S_{21} is

$$S_{21} = -2Z_{01}r_e(1 + g_m Z_\pi)/[(Z_{11} + Z_{01})(Z_{22} + Z_{02}) + r_e^2(1 + g_m Z_\pi)].$$

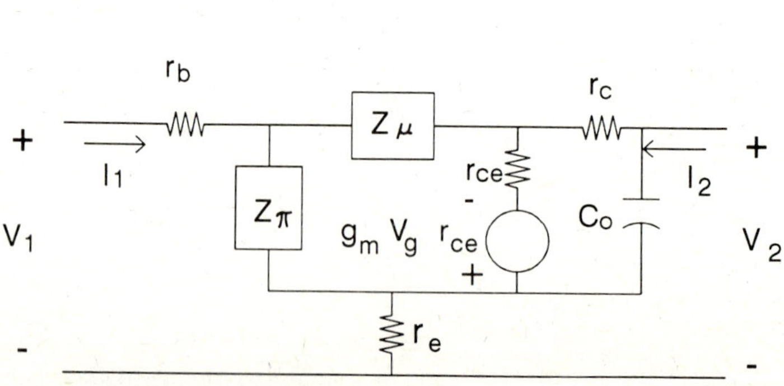

FIGURE 3–5. Small Signal Equivalent Circuit of a Transistor.

Sometimes, we need not use Eqs. (3–10) and (3–11). The usual case is one where we know the input impedance or admittance to a 2 port and we desire to have the S_{11} parameter. In that case, we refer to the fact that S_{11} is the reflection coefficient. So, we can use Eq. (1–6).

3.5 Y PARAMETERS

Formulas for conversion of Y parameters to S parameters and vice versa can be derived in a way similar to the derivation in section 3–4. We start with the matrix relation $(I) = (Y)(V)$ and Eq. (3–1). If we use the normalized admittance matrix $(y) = (\sqrt{Y_0})^{-1}(Y)(\sqrt{Y_0})^{-1}$, then the conversion formulas become those of Eq. (3–13).

$$(y) = \big((\mathbf{1}) + (S)\big)^{-1}\big((\mathbf{1}) - (S)\big), \qquad (S) = \big((\mathbf{1}) + (y)\big)^{-1}\big((\mathbf{1}) - (y)\big) \qquad (3\text{–}13)$$

Notice that these are symmetrical in that interchanging y and S in either equation results in the other equation. This feature is useful when programming these equations on a computer. A BASIC program was written for these conversion equations and is included in the chapter appendix.

Example: Effect of a Shunt Resistor on S Parameters. Frequently, we wish to place a shunt resistor across the input of a transistor as shown in Figure 3–6 to prevent oscillations. The shunt resistor can be treated as though it were a part of the transistor by calculating a new set of S parameters from the S parameters for the transistor and the shunt resistance acting together. The transistor S parameters are $S_{11} = 0.86\angle{-63°}$, $S_{21} = 3.08\angle 127°$, $S_{12} = .07\angle 56°$, and $S_{22} = 0.6\angle{-31°}$. The shunt resistor is to have a conductance of 0.2 (this is normalized to the input line admittance, Y_0). The conversion formulas from Eq. (3–13) are $y_{11} = [(1 - S_{11})(1 + S_{22}) + S_{12}S_{21}]/\Delta$, $y_{12} = -2S_{12}/\Delta$, $y_{21} = -2S_{21}/\Delta$ and $y_{22} = [(1 + S_{11})(1 - S_{22}) + S_{12}S_{21})]/\Delta$, where $\Delta = (1 + S_{11})(1 + S_{22}) - S_{12}S_{21}$. Substituting the values for the S parameters into these formulas, we obtain the y parameters $y_{11} = 0.515\angle 82.6°$, $y_{12} = 0.0536\angle{-86.9°}$, $y_{21} = 2.36\angle{-15.8°}$, and $y_{22} = 0.267\angle 40.9°$. To find the y parameters for the circuit, including the shunt resistor, we simply need to add the conductance 0.2 to y_{11} to get $y'_{11} = 0.515\angle 82.6° + 0.2\angle 0° = 0.576\angle 62.5°$. Now, we can use this along with the other y parameters to convert to a new set of S parameters using conversion formulas from Eq. (3–13). The new set of S parameters: $S_{11} = 0.643\angle{-66.1°}$, $S_{21} = 2.70\angle 130.8°$, $S_{12} = 0.0614\angle 59.8°$, and $S_{22} = 0.615\angle{-29.9°}$ will describe the transistor and the shunt resistor acting together as a 2 port.

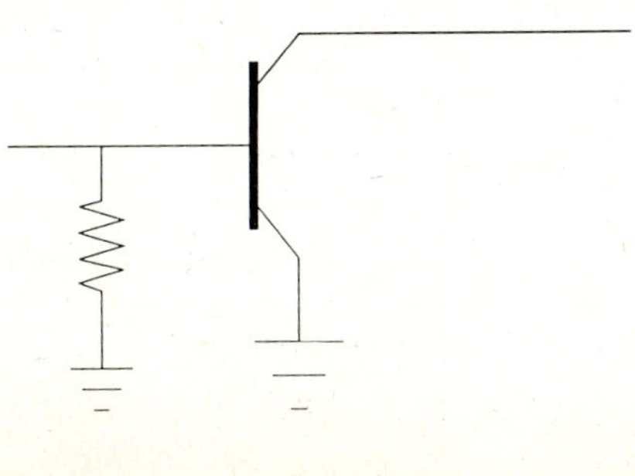

FIGURE 3–6. A Transistor Input Shunted with a Resistor.

3.6 CHAIN PARAMETERS

The *chain scattering parameters* (sometimes referred to as the *transformation parameters* or the *scattering transfer parameters*) are useful when cascading microwave circuits. These parameters are related to the wave functions as

$$\begin{pmatrix} a_1 \\ b_1 \end{pmatrix} = \begin{pmatrix} T_{11} & T_{12} \\ T_{21} & T_{22} \end{pmatrix} \begin{pmatrix} b_2 \\ a_2 \end{pmatrix}.$$

They are linearly related to the scattering parameters as

$$\begin{pmatrix} T_{11} & T_{12} \\ T_{21} & T_{22} \end{pmatrix} = \begin{pmatrix} \dfrac{1}{S_{21}} & -\dfrac{S_{22}}{S_{21}} \\ \dfrac{S_{11}}{S_{21}} & S_{12} - \dfrac{S_{11}S_{22}}{S_{21}} \end{pmatrix}$$

and

$$\begin{pmatrix} S_{11} & S_{12} \\ S_{21} & S_{22} \end{pmatrix} = \begin{pmatrix} \dfrac{T_{21}}{T_{11}} & T_{22} - \dfrac{T_{21}T_{12}}{T_{11}} \\ \dfrac{1}{T_{11}} & -\dfrac{T_{12}}{T_{11}} \end{pmatrix}.$$

For circuits that are cascaded, the overall T matrix can be calculated by multiplying the T matrices for the individual circuits (see problem 3–12).

The ABCD parameters and the h parameters are also useful on occasion. Conversion from S to these parameters and vice versa is possible. Formulas for these conversions can be derived in a manner similar to that shown above. Formulas for these parameter conversions are available.[2] We use the ABCD parameters in chapters 4 and 7.

3.7 SMITH® CHART

A common problem encountered when designing electrical equipment is that of matching circuits for maximum power transfer. This problem usually requires using reactive elements in parallel or in series with a load to result in a purely resistive load. In microwave circuit analysis, this problem is worsened by the phase changes suffered when the waves travel over the connections between circuit elements. The calculations required when compensating for these phase changes use Eq. (1–11). These calculations are best done with a programmable calculator or computer in order to obtain the accuracy needed and to determine the effects over a range of frequencies. However, when we blindly turn over the solving of such a problem to a computer, the resulting solution sometimes overlooks the obvious or is not practical. In fact, in cases where the computer is asked to optimize a solution, the computer can get lost and never arrive at the desired solution. The following discussion presents a graphical method for visualizing the effects of circuit elements, providing invaluable assistance in designing microwave circuits.

This very useful device for a preliminary analysis and design of these circuits is the *Smith® chart.* The Smith® chart is a graphical representation of Eq. (1–11). It is a plot of the circles of constant (normalized) resistance and reactance on the complex plane for the reflection coefficient Γ. If we rewrite Eq. (1–11) as Eq. (3–14), we have the transformation for mapping the normalized impedance ζ onto the *u, v* plane.

$$Z_{line}/Z_0 = \zeta = r + jx = \frac{1+W}{1-W}, \qquad \text{where } W = u + jv = \Gamma e^{j2\beta z} \qquad (3\text{–}14)$$

The fact that the plots of constant *r* and of constant *x* are circles can be proved easily (see problem 3–14). Because the reflection coefficient frequently is given in polar coordinates, the W plane is given in polar coordinates rather than rectangular. A sample Smith® chart is given in Figure 3–7. The center of the chart is the center of the polar coordinates. The reflection coefficient is measured radially outward in units as scaled at the bottom of the chart. The angle of *W* has a zero reference at the right edge of the chart and is measured positively in a counterclockwise direction from this point. The circle just inside the outermost circle of the chart corresponds to a reflection coefficient magnitude of 1. The center of the chart is for a reflection of zero. Thus, a reflection coefficient of $1\angle 0°$ corresponds to the point at the right edge where all the circles are tangent. A reflection coefficient of $0.6667\angle 180°$ is shown as a circular dot at point A.

A change in the location *z* along the transmission line corresponds to a change in the angle of *W* through the factor $e^{j2\beta z}$. Because *z* is measured negatively from the load, a negative change in angle of *W* corresponds to a position more distant from the load (towards the generator). This information is incorporated on the outer circle of the chart. A 0.1λ long transmission line terminated in a load with a reflection coefficient of $0.6667\angle 180°$ would have a reflection coefficient at its input shown at point B on Figure 3–7. This is $\Gamma = 0.6667\angle 108°$.

The family of circles that are tangential to the line $u = 1$ at the right edge of the Smith® chart where $W = 1\angle 0°$ are circles of constant *r*. The family of circles that are orthogonal to these are circles of constant *x*. The constant *x* circles in the upper half of the *W* plane correspond to inductive reactance and those in the lower half to capacitive reactance.

Example: A 50 Ω lossless transmission line, *L* meters long, is terminated in a 10 Ω resistor. We desire to find the impedance at the input of the line. The reflection coefficient calculated at the 10 Ω resistor is $-\frac{2}{3}$. This corresponds to a point on the Smith® chart of Figure 3–7 of $W = 0.6667\angle 180°$, shown as *A*. To determine the point on the Smith® chart corresponding to *L* meters from the load toward the generator, we must determine the phase constant of the line β and then calculate β*L* either in wavelengths or degrees. Assuming β*L* corresponds to λ/10, the point on the Smith® chart corresponding to the reflection coefficient at this distance from the load will lie on the intersection of a circle

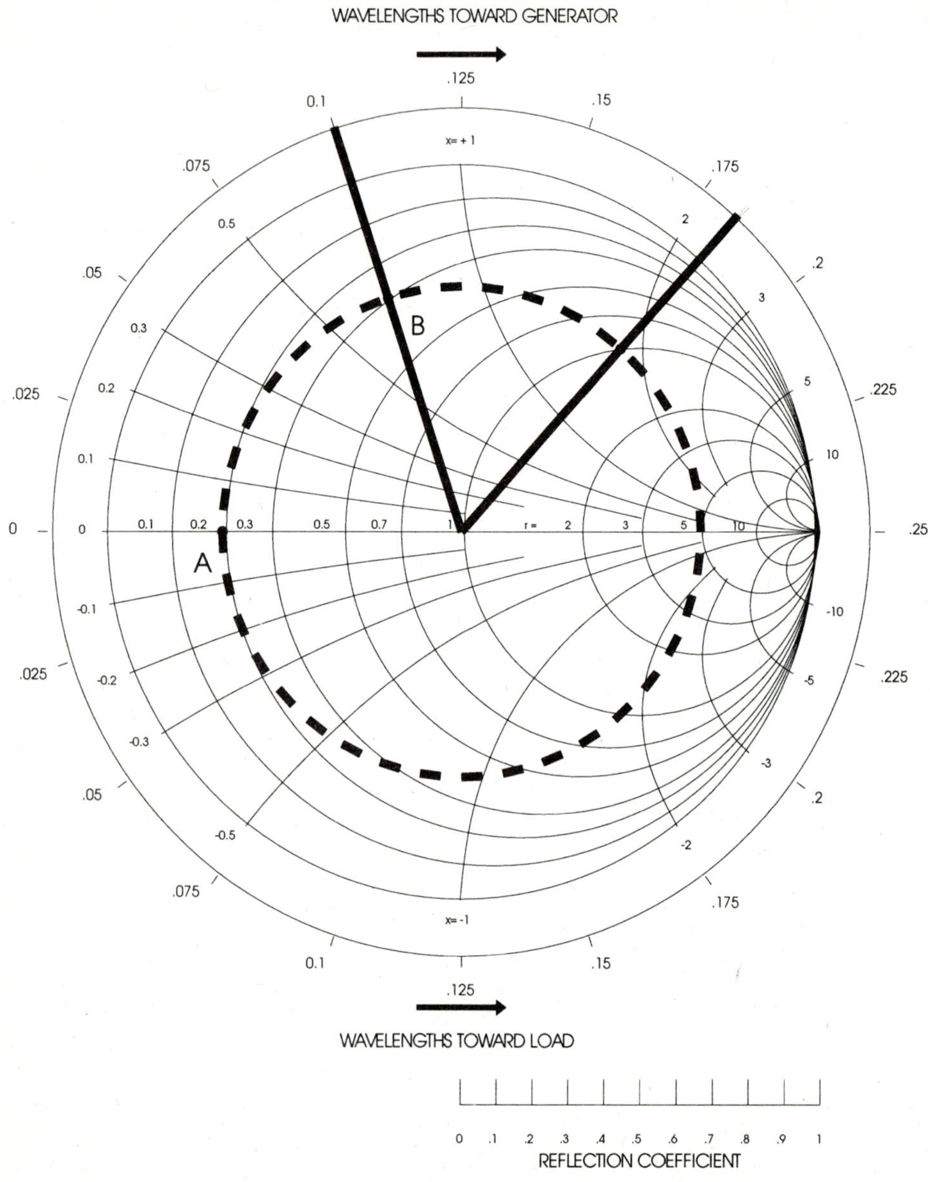

FIGURE 3–7. Smith® Chart. Permission to produce the Smith charts in this book has been granted by Analog Instruments Co., New Providence, NJ under renewal copyright issued in 1976.

and a line. The line is a radial line out to the point marked 0.1 on the outermost circle. The circle has a radius of 0.6667 and is centered at the origin. This circle is shown as a dashed line in Figure 3–7. One might think that the rotation along this circle from $W = 0.6667\angle 180°$ to the intersection would be $\frac{1}{10}$ of 360 degrees. This is not correct; the angular rotation corresponding to this is 72° rather than 36°. The reason for this is the factor of 2 in the exponent in Eq. (3–14). Consequently, moving one wavelength toward the generator corresponds to two complete clockwise revolutions on the Smith® chart.

For this example, the point $W = 0.6667\angle 180°$ at the load lies on the $r = 0.2$ circle and the $x = 0$ line. The impedance given by V_{line}/I_{line} at that point is $Z_0\,(0.2 + j0) = 10\ \Omega$. The normalized impedance given by V_{line}/I_{line} at the point 0.1λ toward the generator can be read from the circles passing through the intersection point as described above and shown in Figure 3–7 as $r = 0.3$ and $x = +0.68$. The line impedance at this point is $50(0.3 + j0.68) = 15 + j34\ \Omega$.

3.8 IMPEDANCE MATCHING WITH A SERIES ELEMENT

Once again, we consider the case of a 50 Ω transmission line loaded with a 10 Ω resistor. We wish to match this loaded transmission line to a source having a 50 Ω source impedance as shown in Figure 3–8. To match, we insert a 2 port between the source and the transmission line connecting to the load. We select the length of the transmission line from the load to the 2 port and the elements inside the 2 port to give a match. In this case, we have to match 10 Ω to 50 Ω. Looking again at Figure 3–7, we see that the normalized line impedance of 0.2 at $z = 0$ changes to different values at different points along the line as described by the circular locus of points corresponding to $|W| = 0.6667$. First, we locate the intersection of this locus with the circle for $r = 1$. At points along the circle for $r = 1$, the line impedance has a real part of 50 Ω. In particular, the normalized line impedance at the intersection between these two loci is $\zeta = 1 + j1.8$, so the actual impedance is $50 + j90\ \Omega$. If we put a capacitor in series with the line at this point and select a capacitive reactance of $-90\ \Omega$, then that reactance will subtract from the apparent inductive reactance component of the line impedance resulting in a purely resistive impedance of 50 Ω. This will match the source impedance. To find the length of transmission line necessary to produce this match, we can draw a radial line from the center of the Smith® chart through the $1 + j1.8$ impedance point to the outer circle which is calibrated in wavelengths. At this point, we read a distance of 0.183λ for the length of line from the load to the 2 port.

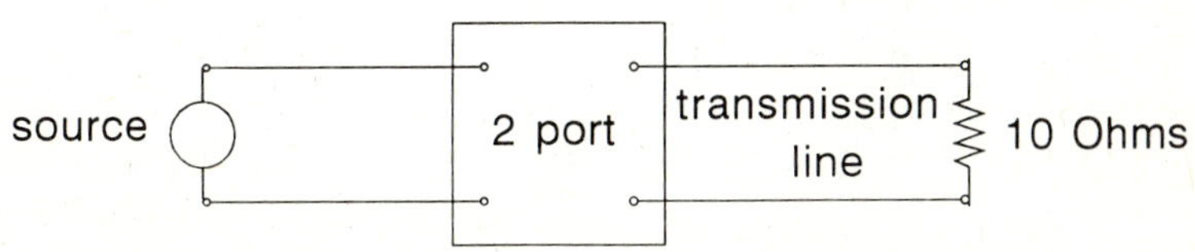

FIGURE 3–8. Matching a Source to a Load.

Referring once again to the Smith® chart of Figure 3–7, we can talk about the path taken on the Smith® chart in this matching process. A segment consisting of a circular arc centered about the origin of the W plane always corresponds to a length of lossless transmission line. The segment along the constant resistance circle always corresponds to the addition of a series reactance. Finally, the point located at the origin implies a matched impedance where the normalized impedance is $1 + j0$.

For the case discussed previously, the circle corresponding to $|W| = 0.6667$ intersects the $r = 1$ circle twice: once above the horizontal axis and once below this axis. We could achieve a match using the second intersection as well. This intersection would require a series inductance with a reactance of 90 Ω for a match, because the normalized line impedance at this intersection point has a reactance of opposite sign to the one we had before. The length of transmission line for this would be determined as before, by drawing a radial line through the intersection of the two circles out to the circle calibrated in wavelengths. This length of line would be 0.317λ.

3.9 ADMITTANCE CALCULATIONS

Frequently, we desire to use elements connected in parallel with a transmission line rather than in series. In that case, the calculations are simpler if admittances are used rather than impedances. In this discussion, we shall use admittances normalized to the characteristic line admittance Y_0. These will be denoted by a lower case y. These are directly related to the normalized impedances as $y = 1/\zeta$. The normalized line admittance is given by the Eq. (3–15).

$$y = \frac{1-W}{1+W} = \frac{1+W\,e^{j\pi}}{1-W\,e^{j\pi}} = \frac{1+W'}{1-W'} = g + jb \tag{3–15}$$

From the third step, we see that W' plays the same role as W in Eq. (3–14). So, if we know W for a given y, then we can locate that W on the Smith® chart and rotate from that point by 180° to find the corresponding y. The numerical values for g and b then correspond to the numbers assigned to the intersecting r and x circles at that point. On the Smith® chart the admittance is related to the impedance by a rotation of 180° in the W plane. Sometimes this is referred to as a "reflection through the origin." When making calculations for matching loads by the use of shunt admittances, we wish to find the length of transmission line necessary to transform the line admittance to a value of $y = 1 + jb$, rather than $\zeta = 1 + jx$. However, the locus for $\mathcal{Re}\,(y) = g = 1$ is the locus for $\mathcal{Re}\,(\zeta) = 1$ only rotated by 180°. Because of this relation, we can use the Smith® chart for admittance problems directly (or if you wish, rotate the chart by 180° and enter admittances directly using the same scales as were used for the line impedance calculations) but *we must add 180°* to the angle of the reflection coefficient when using the Smith® chart for admittances.

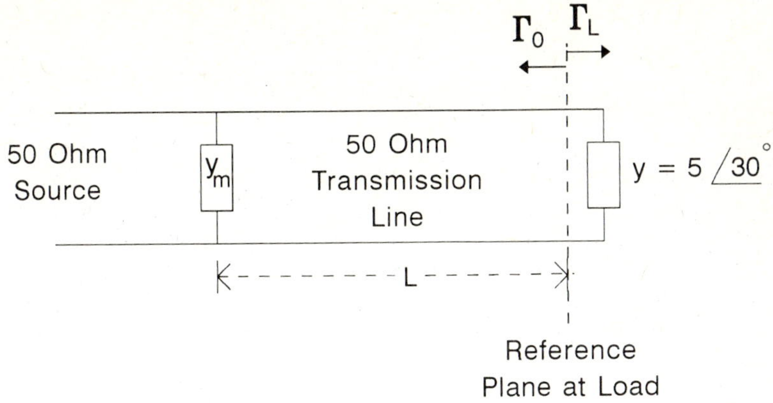

FIGURE 3–9. Shunt Matching of a Transmission Line.

Example: Matching a Line with Shunt Admittances. Referring to Figure 3–9, we desire to find the line length and the shunt admittance Y_m that will result in a matched input: $y = 1 + j0$. Locating the normalized load admittance $y = 5\angle 30 = 4.33 + j2.5$ on the Smith® chart shown as point A in Figure 3–10, we follow the arc about the center of the W plane in the direction toward the generator until the intersection with the circle $1 + jb$ is located. This intersection is shown as point B. This segment of the path is shown as a dotted line on the Smith® chart. The line admittance at this point is $1 - j2$. The line length corresponding to this segment is read on the outer calibrated circle as being $L = 0.313\lambda - 0.234\lambda = 0.079\lambda$. The admittance Y_m, which is needed to match the line at this point, corresponds to $y_m = j2$. When placed in parallel with the line admittance of $1 - j2$, the normalized admittance of the parallel combination is $1 + j0$. The segment of the path on the Smith® chart corresponding to the addition of this shunt admittance is the dotted line going from point B to the origin.

3.10 TRANSMISSION LINE STUB MATCHING

An alternative method for matching a load is to use an open circuited or short circuited transmission line in place of a lumped inductance or capacitance. Such lines are referred to as open circuited or short circuited *stubs*. The input admittance to such stubs depends upon their length. We determine the range of admittances of a stub from the Smith® chart. The admittance looking directly into a short circuit at the end of a transmission line is infinity: From Eq. (3–15), the reflection coefficient is $W = -1$. This is the point at the right edge of the 180° rotated Smith® chart. To obtain the line admittance at a point on this line away from the shorted end, we first determine the electrical distance (in wavelengths) from the end of the line to this point. Then, we rotate toward the generator from the point $y = \infty$ on a circle around the origin by this distance. We observe that this outer

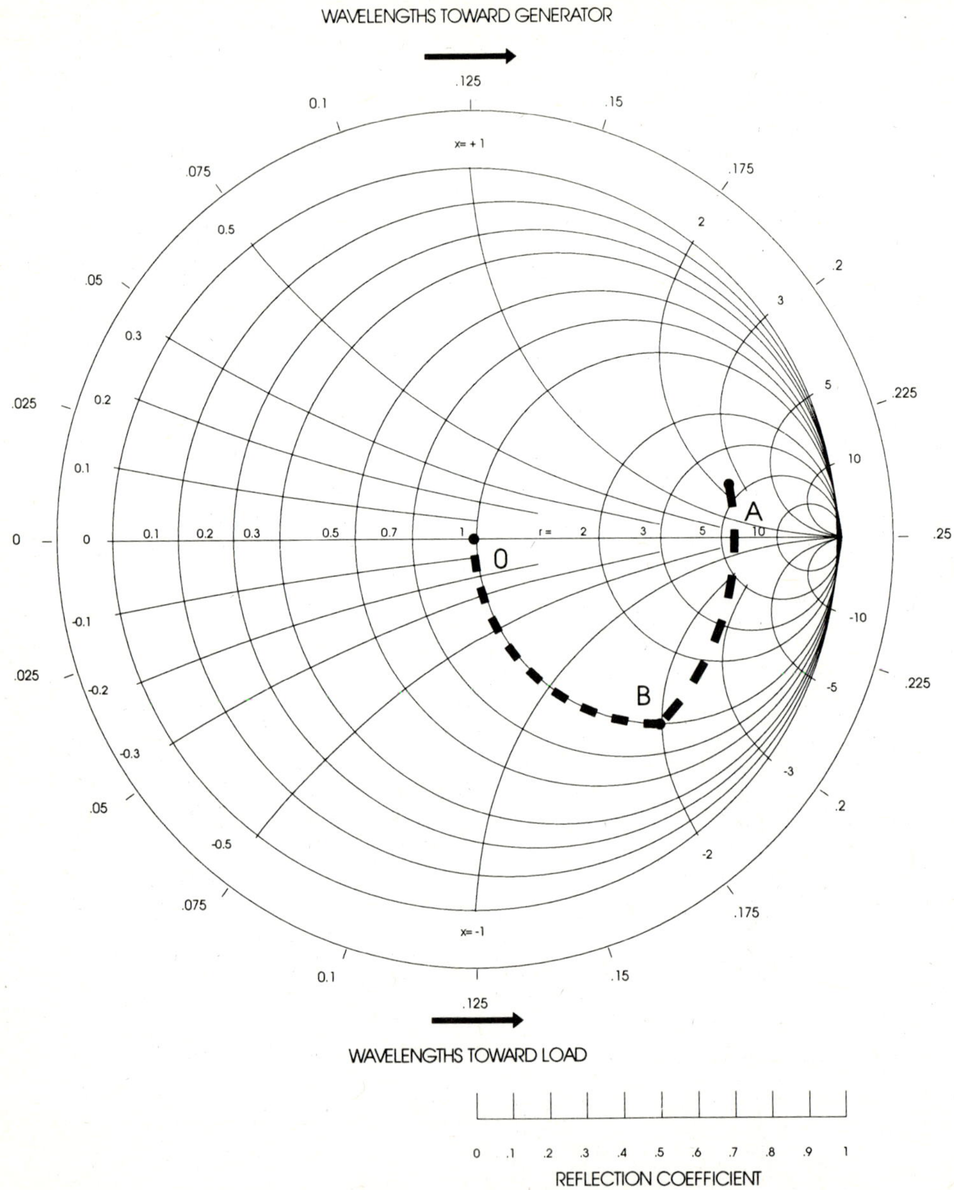

FIGURE 3–10. Smith® Chart Solution for Shunt Matching of a Line. Permission to produce the Smith charts in this book has been granted by Analog Instruments Co., New Providence, NJ under renewal copyright issued in 1976.

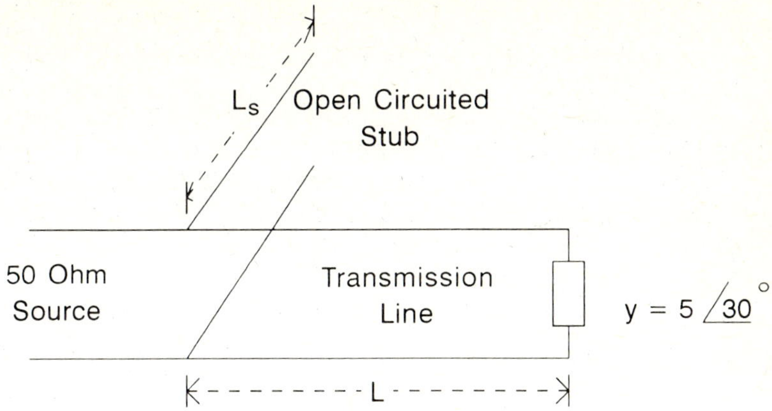

FIGURE 3–11. Matching a Load with an Open Circuited Stub.

circle is the circle for normalized conductance $g = 0$. So, the admittance for this shorted stub is purely imaginary. For the first $\frac{\lambda}{4}$ of rotation this susceptance is inductive and ranges from zero to infinity. For the next $\frac{\lambda}{4}$ rotation along the edge of the Smith® chart this susceptance is capacitive. The case for an open circuited stub is just reversed. The open circuited end of the line has $y = 0$, and rotation on the Smith® chart from that point corresponds to capacitive susceptance for the first quarter wavelength and inductive for the next.

Example: Single Stub Matching. Given the load shown in Figure 3–11, we select the lengths of the attached transmission lines to match this load to the source. The Smith® chart solution reduces to the same as in the example of section 3–9, because we have selected the same normalized load admittance. Consequently, at $L = 0.079\lambda$ from the load, the stub line shown in Figure 3–11 must present an admittance of $j2Y_0$, where Y_0 is that of the stub. We assume the characteristic impedance of the stub line to be the same as that of the main line. The admittance of an open circuited stub can be found as described above. Rotation by 0.176λ on the outer edge of a Smith® chart from the point at $y = 0$ takes us to the point at $y = j2$. So, $L_s = 0.176\lambda$ will match the system. Note that a short circuited stub also can be used, but it would have to be 0.25λ longer than the open circuited stub.

3.11 DIRECTIONS "TOWARD THE GENERATOR" AND "TOWARD THE LOAD"

Identical matching networks can have Smith® chart solutions looking very different depending upon the positive reference direction chosen for the line current. When we used the words "line impedance," we referred to the phasor determined by the ratio of the phasor line voltage to the phasor line current as in Eq. (1–11). We are assuming a positive reference direction for the line current into the impedance in question. Because this impedance is related to W as in Eq. (3–14), $\zeta = (1 + W)/(1 - W)$, and because the positive direction of rotation on the Smith® chart is that corresponding to a (positive) increase in the imaginary argument in the exponent of W, we must always take the direction "toward the load," the same as the positive reference direction of the line current. This direction may not always coincide with the direction toward the actual load in the circuit being analyzed. The convention we shall use to show the direction "toward the load" is to draw an arrow as shown in Figure 3–9. There, we have a reference plane at the load and two different arrows. For the reflection coefficient Γ_L, we show the direction arrow toward the $5\angle 30°$ impedance as the "direction toward the load." For the reflection coefficient Γ_0, the arrow is toward the left showing that as the "direction toward the load." When this circuit is matched, we have $\Gamma_L = \Gamma_0^*$. This is referred to as a *conjugate match.* Using either of these reflection coefficients with its proper "direction toward the load" results in the same value for y_m needed to match the load.

Example: Shunt Matching of a Load. In the example of Figure 3–9, we have a load on a transmission line of $y = 5\angle 30°$ (normalized to 50 Ω), which is matched to a source of 50 Ω by using a shunt admittance (again normalized) of $j2$. Implicit in that discussion is the fact that the line admittances are determined using a positive reference direction for the current from the source to the load. Suppose that we assume a positive reference direction for the line current from the load (receiver) to the source. We then solve the matching problem by starting with the fact that the line admittance as seen looking back from the receiver input toward the source has to be the conjugate of the receiver input admittance for a match. So, we locate the point $y = 5\angle -30°$ on the admittance chart shown in Figure 3–12. This is point A. Because this is an impedance determined by looking away from the load, we are assuming that the reference direction of the current for this impedance is to the left, away from the load. Consequently, the direction to the left from the load in the circuit corresponds to the direction "toward the load" on the Smith® chart. We must rotate "toward the load" on this Smith® chart until the circle intersects the $g = 1$ circle. This rotation accounts for the transmission line phase shift going from the load to the position where the matching element is to be placed (i.e., going from the load in the circuit toward the 50 Ω source). This length of transmission line is 0.079λ. The intersection is point B on Figure 3–12. At point B, the admittance $y = 1 + j2$ is seen looking from the line into the shunt element in parallel with the matched line from the source. Because the matched line has an admittance of $1 + j0$, the shunt element must have an admittance of $j2$. This describes a capacitor with a normalized susceptance of 2. Proceeding to the other side of this shunt element, the line admittance looking into the source is $1 + j0$. This corresponds to the arc on the Smith® chart from point B to point 0 in Figure 3–12.

If we now compare Figure 3–12 with Figure 3–10, we see that they are different, although they describe the same circuit. This difference is due the choice of positive reference direction for the line current which is implicit whenever the statement "the line impedance looking into . . ." is made.

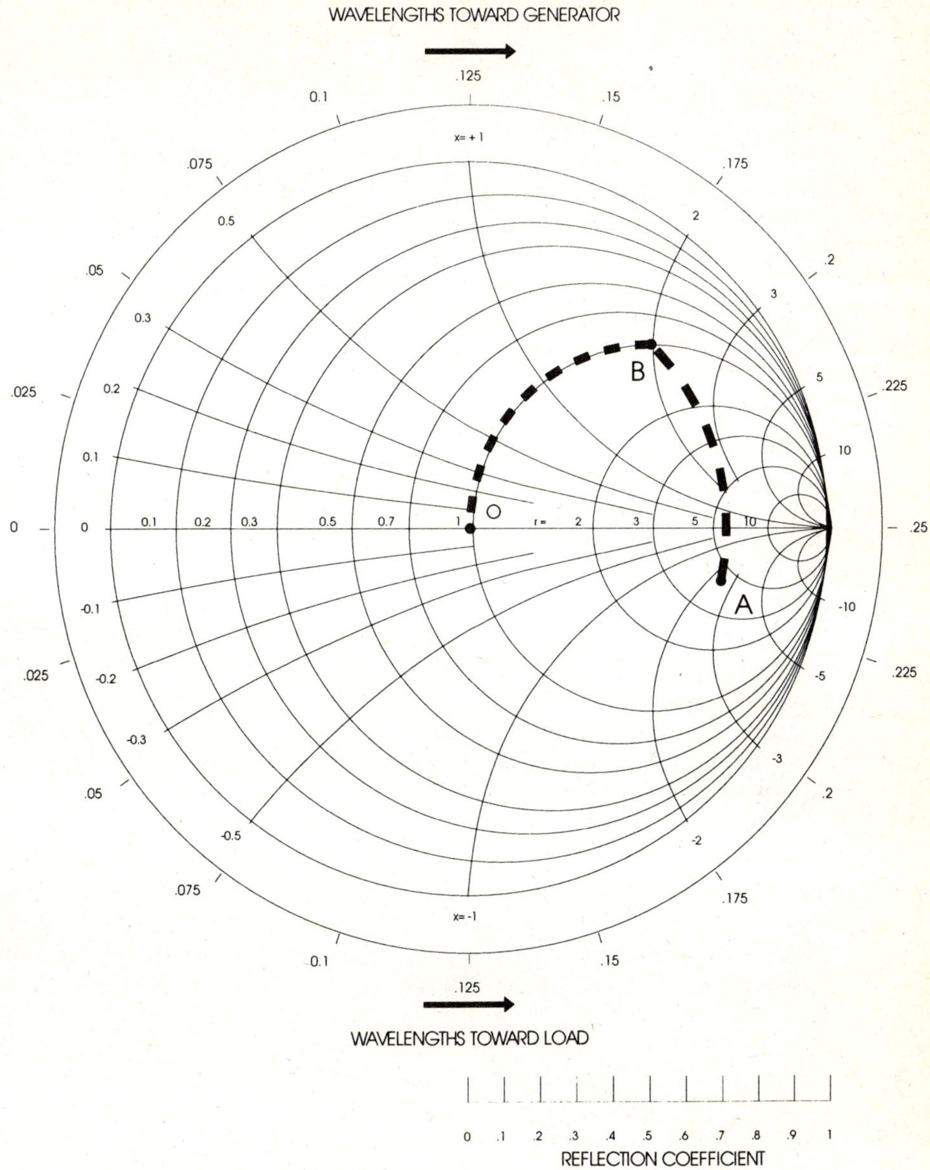

FIGURE 3–12. Alternate Smith® Chart Solution for Shunt Matching. Permission to produce the Smith charts in this book has been granted by Analog Instruments Co., New Providence, NJ under renewal copyright issued in 1976.

3.12 TWO-COLOR SMITH® CHARTS

Seldom does a real design problem allow us to work solely with impedance or solely with admittance. Usually, we must convert back and forth between these two quantities. For that reason, two-color Smith® charts are particularly useful. The two-color Smith® chart consists of the regular impedance chart in red and an admittance chart in green, which is rotated by 180° from the red chart (see Figure 3–14). With these charts, we read the impedance for a given Γ by locating it on the W plane and finding the r and x values from the intersecting red circles. We also find the admittance for that same Γ by reading the g and b values from the green circles intersecting at that same point. This admittance will be the inverse of the impedance. One advantage of the two-color chart is that the top half is a resistance plus an inductance in either a series (red circles) or parallel (green circles) combination. The bottom half represents a resistance plus a capacitance in the same manner.

Example: Pi Matching Circuit. A circuit often used at lower frequencies to match the input stage of a receiver to an antenna is called a *pi matching circuit.* It consists of parallel capacitors and a series inductor as shown in Figure 3–13. The circuit of this figure has an antenna input impedance of $30 + j30\ \Omega$, and an input impedance to the receiver of $300\angle 20°$ at a frequency of 30 MHz. Initially, we assume that the input cable from the antenna to the matching network is electrically short at 30 MHz. This leaves us with the problem of matching $300\angle 20°\ \Omega$ to $30\sqrt{2}\angle 45°$. To obtain maximum power transfer, the impedance looking into the circuit from the antenna cable must be $30\sqrt{2}\angle -45°$ (conjugate match). We normalize these impedances to 50 Ω to get $\zeta_{in} = 5.6 + j2$ and $\zeta^*_{ant} = 0.6 - j0.6$. The point A represents the receiver input impedance on the Smith® chart shown on the inside back cover, and the point D represents the conjugate of the antenna impedance. The problem of matching the circuit now corresponds to the problem of joining these two points with arcs along constant r circles, constant g circles, and constant $|\Gamma|$ circles.

Point A is located by finding the intersection of the red circles corresponding to this normalized input impedance. However, it also is located at the intersection of the green circles that correspond to the admittance $y_A = 1/\zeta_{in} = 0.157 - j0.057$. Now we can account for the capacitor C_1 by following a constant conductance path in the direction of increasing capacitive susceptance from point A until we come to point B. This follows in

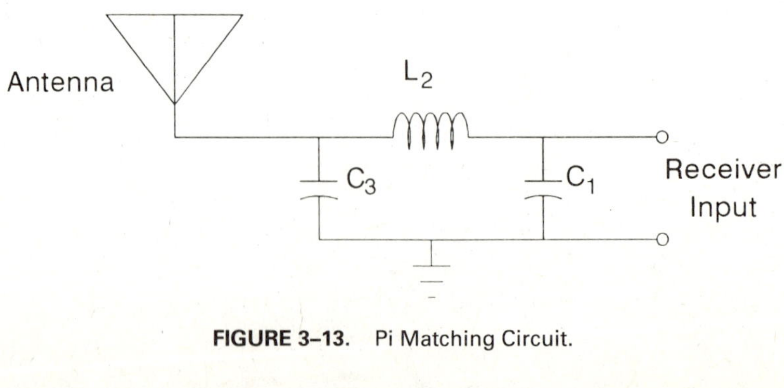

FIGURE 3–13. Pi Matching Circuit.

the direction of the green lines for constant conductance and of increasing capacitive susceptance. The admittance at point B is $y_B = 0.157 + j0.37$. This is the admittance of the parallel combination of C_1 and the receiver input admittance. The difference in susceptance between the two points A and B is $b_1 = 0.37 - (-0.057)$. This is the normalized susceptance of the capacitor C_1 so that $\omega C_1/Y_0 = 0.427$.

The arc from point B to point C accounts for the series inductor. To account for this reactance, we change from using the green lines to using the red lines. The path from B to C follows a line of constant resistance in the direction of increasing inductive reactance. The normalized impedance at point B is $\zeta_B = 0.9711 - j2.2920$ and that of point C is $\zeta_C = 0.9711 + j0.4715$. The change in impedance is $\zeta_C - \zeta_B = j2.7635 = jx_2$, so the inductance of the inductor is $L_2 = Z_0x_2/\omega = 0.733\ \mu H$.

Finally, the segment from C to D relates to the capacitor C_3, and it lies along a line of constant conductance in the direction of increasing capacitive susceptance. The admittance at point C is $y_C = 0.8333 - j0.4046$, and at point D the admittance is the conjugate of the admittance of the antenna: $y_D = 0.8333 + j0.8333$. The difference is $y_D - y_C = j1.2379 = jb_3$. We can calculate C_3 from $C_3 = Y_0b_3/\omega = 131.35$ pF.

The reactance and susceptances scaled from these path segments and their corresponding element values are $b_1 = 0.427$ and $C_1 = 45.3$ pF, $x_2 = 2.7635$ and $L_2 = 0.733$ μH, and $b_3 = 1.2379$ and $C_3 = 131$ pF.

Now, suppose that we find that the input cable from the antenna is not electrically short but is a 1 meter long 50 Ω coaxial cable. At 30 MHz, this is an electrical length of 0.1λ. We wish to recalculate the matching network to fit this revised specification. Looking at the Smith® chart shown on the inside back cover, we add in a segment between point D and E to account for this line. This segment will be a 0.1λ rotation from the point D in the direction toward the load around the center of the W plane. Now, we must connect point A to point E as before to account for the effects of the capacitors and the inductor. We do this by arbitrarily selecting the length of the segment AB. This fixes the other segments. For the segments shown on the Smith® chart on the inside back cover we have

$$b_1 = 0.28 - (-0.057) = 0.337 \text{ and } C_1 = 35.8 \text{ pF},$$

$$x_2 = 1.12 - (-2.6) = 3.72 \text{ and } L_2 = 0.987\ \mu\text{H},$$

$$\text{and } b_3 = 0.24 - (-0.29) = 0.53 \text{ and } C_3 = 56.2 \text{ pF}.$$

The Smith® chart is a visual presentation of these data in such a way that it is possible for one to estimate what a small change in element values does to the match. If one planned to use a variable capacitor for C_1, the range of variation needed in the variable capacitor can be estimated from this graph. After the initial computations are made with the use of a Smith® chart, one can resort to a computer or programmable calculator to refine these calculations.

Example: Amplifier Input and Output Matching Circuits. Consider a GaAsFET amplifier circuit as shown in Figure 3–14. The bias voltages in this circuit are inserted on the transmission line stubs marked L_B and L_E. The capacitors C_B are shunt bypass capacitors intended to short the ends of these stubs to ground at the signal frequencies. The

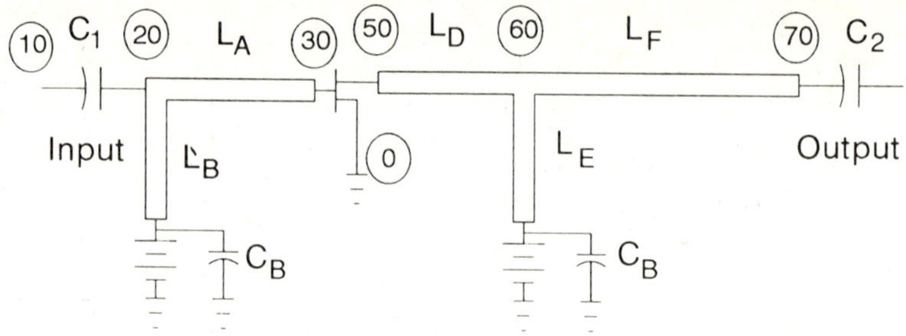

FIGURE 3–14. GaAsFET Amplifier with Input and Output Matching Circuitry. The circled numbers are the node numbers used in a CAD program.

transmission lines L_A, L_B, L_D, L_E, and L_F connecting the capacitors to the GaAsFET are to be microstrip lines. The S parameter data on the GaAsFET for this example apply at 2 GHz and are $S_{11} = 0.9744\angle{-28.48°}$, $S_{12} = 0.0$, $S_{21} = 3.3415\angle 156.43°$, and $S_{22} = 0.53553\angle{-20.734°}$. The requirement is to match the input and output of the GaAsFET by choosing the lengths of the various microstrip lines and the capacitance of the coupling capacitors C_1 and C_2.

The input circuit match is shown in Figure 3–15. For matching this input circuit, we chose to work "toward the load" on the Smith® chart. For that, we can use the reflection coefficient looking from the gate back toward the input line. This reflection coefficient must be the conjugate of S_{11} if we are to have a matched input. The steps associated with this graph on the Smith® chart are discussed below.

INPUT MATCH

Point A: $W_A = S^*_{11} = 0.9744\angle{+28.48°}$ for an input match.

Line AB: Describes the phase rotation along the transmission line L_A corresponding to going from point A on the Smith® chart "toward the load" to point B. $L_A = (-0.02338 + 0.2104)\lambda = 0.187\lambda$.

Point B: $W_B = 0.9744\angle 163.165°$, $y_B = 0.6 - j6.705$.

Line BC: Removal of shunt (shorted) stub admittance corresponds to an arc along a constant conductance circle from point B with W_B to point C with W_C:
$y_{stub} = y_B - y_C = 0.6 - j6.705 - (0.6 + j0.4899) = -j7.1949$.

Point C: $W_C = 0.3780\angle{-67.79°}$, $y_C = 0.6 + j0.4899$, $\zeta_C = 1 - j0.8165$.

Line CO: Removal of the reactance of the series capacitance C_1 corresponds to an arc along a constant resistance circle from point C to point 0.
$C_1 = 1/[50 \times 0.8165\omega] = 1.949$ pF.

Input Stub: $y_{stub} = -j7.1949$ is obtained with $L_B = 0.02198\lambda$.

The opposite viewpoint is used to match the output to show the case where we move "toward the generator" on the Smith® chart. Again, this is shown on Figure 3–15.

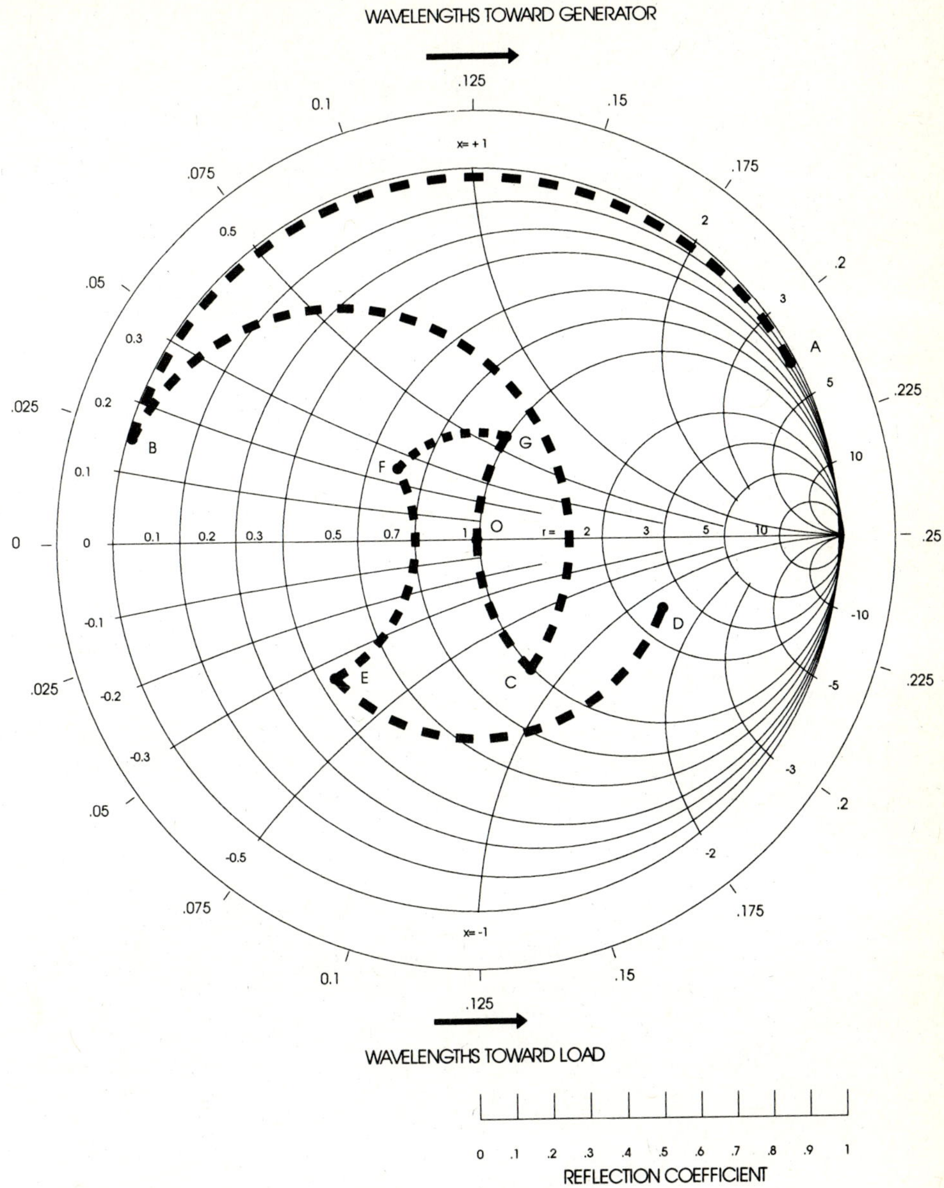

FIGURE 3–15. Smith® Chart Used for Designing Input and Output Matching Circuits for a GaAsFET Amplifier. Permission to produce the Smith charts in this book has been granted by Analog Instruments Co., New Providence, NJ under renewal copyright issued in 1976.

OUTPUT MATCH

Point D: This is $W_D = S_{22} = 0.5355\angle{-20.734°}$.

Line DE: Moving "toward the generator" on the Smith® chart accounts for the transmission line L_D: $L_D = (0.4396 - 0.2788)\lambda = 0.1608\lambda$.

Point E: $W_E = 0.5355\angle{-136.534°}$, $y_E = 1.4 + j1.4463$.

Line EF: Adds in the (shorted) stub admittance: $y_{\text{stub}} = y_F - y_E = 1.4 - j0.5865 - (1.4 + j1.4463) = j2.0328$.

Point F: $W_F = 0.2873\angle 138.027°$, $y_F = 1.4 - j0.5865$.

Line FG: Accounts for the transmission line L_F: $L_F = (0.1482 - 0.0583)\lambda = 0.0899\lambda$

Point G: $W_G = 0.2873\angle 73.301°$, $\zeta_G = 1 + jx_{C2} = 1 + j0.6$.

Line GO: Addition of the reactance of the series capacitance C_2: $C_2 = 1/[50 \times 0.6\omega] = 2.653$ pF

Output Stub: $y_{\text{stub}} = -j2.0328$ is obtained with $L_E = 0.07276\lambda$.

3.13 REGION OF NEGATIVE RESISTANCE

The Smith® chart is a mapping of the right half of the *r, x* plane into the region inside the unit circle on the *u, v* plane. Only positive values of normalized resistance are shown inside the unit circle. However, the transformation in Eq. (3–14) will transform negative values of resistance into the region outside the unit circle. The unit circle itself is that for $r = 0$. Usually, we consider that a negative value of r implies a source of signal. In the case of an active microwave circuit when the input impedance is negative usually, the circuit has positive feedback. If this feedback is sufficient (r is sufficiently negative), then the circuit will be unstable.

3.14 POWER WAVES AND GENERALIZED S PARAMETERS

The S parameters discussed to this point are based upon traveling wave functions on lossless transmission lines that have real characteristic impedances. These are most useful in the case where the devices are connected by lossless transmission lines and the signal sources have purely resistive internal impedances. In some cases, it is advantageous to use a set of S parameters, as defined by Youla, based on *power waves.*[3] To define these waves, we start with the circuit shown in Figure 3–16A and note that maximum power transfer occurs when the load impedance is equal to the conjugate of the generator impedance: $Z_L = Z_G^*$. In that case, we say that this is the *maximum available power* P_A from the source. For the matched case, we have $P_A = |E_g|^2/[4\,\mathcal{Re}\,(Z_g)]$. We define a power wave a as in Eq. (3–16) so that P_A is the square of the magnitude of a.

$$a = E_g/[2\sqrt{\mathcal{Re}\,(Z_g)}] \qquad (3\text{–}16)$$

When the circuit in Figure 3–16A has maximum power transfer, we label the voltage across the load with the symbol V^+ and calculate it by $V^+ = E_g Z_g^* / [Z_g + Z_g^*] = E_g Z_g^* / [2\,\mathcal{Re}\,(Z_g)]$. From this and Eq. (3–16), we get Eq. (3–17).

$$a = V^+ \sqrt{\mathcal{Re}\,(Z_g)} / Z_g^* \tag{3–17}$$

In the case where Z_g is real and equal to the characteristic impedance of a transmission line, we conclude that Eq. (3–17) reduces to the same expression within an arbitrary phase shift as Eq. (3–1) for the incident wave evaluated at the load.

As defined above, the power wave is closely related to the maximum power that the generator is capable of transferring to the load. When the load is not matched to the generator, we will not transfer this amount of power to the load. Then, we define a reflected power wave b so that the actual power delivered to the load is given by $|a|^2 - |b|^2$. This is the same as Eq. (3–2) for the power transfer down a lossless transmission line that is not terminated with its characteristic impedance. To facilitate the definition of b, we first define V^- to be the difference between V^+ and the load voltage: $V^- = V - V^+$. As in

$$V^- = \frac{E_g Z_L}{Z_g + Z_L} - \frac{E_g Z_g^*}{Z_g + Z_g^*} = \frac{E_g Y_g}{Y_g + Y_L} - \frac{E_g Y_g}{Y_g^* + Y_g} = -\left(\frac{E_g\, Y_g}{Y_g + Y_g^*}\right)\frac{Y_L - Y_g^*}{Y_L + Y_g} = V^+\,\Gamma_V.$$

This leads to Eq. (3–18), which is similar to the relation between the incident voltage V_i and the reflected voltage V_r. In this expression, we have defined a voltage reflection coefficient Γ_V for these power waves.

$$\Gamma_V = -\frac{Y_L - Y_g^*}{Y_L + Y_g} \tag{3–18}$$

Notice that this is the same as for traveling waves on a transmission line when the generator impedance in the circuit of Figure 3–16A is real. Using this reflection coefficient, we define the reflected power wave b as $b = (\Gamma_V V^+ \sqrt{\mathcal{Re}\,(Z_g)}) / Z_g$. This becomes Eq. (3–19).

$$b = \frac{E_g}{2\sqrt{\mathcal{Re}\,(Z_g)}}\,\frac{Z_L - Z_g^*}{Z_L - Z_g} \tag{3–19}$$

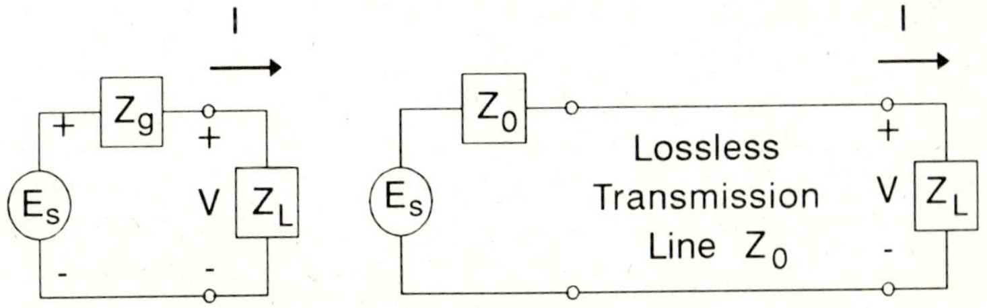

FIGURE 3–16. Circuit for Defining Power Waves.

Now, we can show that the actual power delivered to the load is given by $aa^* - bb^*$ (see problem 3–19).

As with a, this definition becomes the same as Eq. (3–1), when Z_g is real and equal to the characteristic impedance of a transmission line as shown in Figure 3–16B.

We express the terminal voltage V and current I in Figure 3–16A in terms of these power waves as shown in Eq. (3–20) and (3–21). These can be inverted to express a and b in terms of these parameters. This is the way in which Youla defined power waves, and it is shown as Eqs. (3–22) and (3–23).

$$V = \left(\frac{Z_g^*}{\sqrt{\mathcal{Re}(Z_g)}} \right) a + \left(\frac{Z_g}{\sqrt{\mathcal{Re}(Z_g)}} \right) b \tag{3–20}$$

$$I = \left(\frac{1}{\sqrt{\mathcal{Re}(Z_g)}} \right) a - \left(\frac{1}{\sqrt{\mathcal{Re}(Z_g)}} \right) b \tag{3–21}$$

$$a = \frac{V + Z_g I}{2\sqrt{\mathcal{Re}(Z_g)}} \tag{3–22}$$

$$b = \frac{V - Z_g^* I}{2\sqrt{\mathcal{Re}(Z_g)}} \tag{3–23}$$

These equations agree with those in section 3–1, when the characteristic line impedance is real (lossless line). Then, we apply these relations to find power waves at each port of an N port. An S parameter matrix follows from these column vectors of (a) and (b) as $(b) = (S)(a)$. If all the generator impedances are real and positive, then the S parameters given by these power waves agree with the S parameters given in Eq. (3–4) earlier. We can show that the S matrix using a from Eq. (3–22) and b from Eq. (3–23) has the Hermitian quality of S in Eq. (3–7). Also, we can derive transformation relations that are shown in Eq. (3–10) and (3–11) for these S parameters. In synthesizing networks, generalized S parameters are useful.[4]

3.15 COMPUTER-AIDED DESIGN

There are several programs for analyzing and optimizing microwave circuits. We have used HP-EEsof's TOUCHSTONE® for verifying the solutions of examples given in the text. This program is similar in organization to SPICE. The circuit is described by listing the components between nodes in a particular circuit block. Then, the connections between blocks are described to compose the entire circuit. In addition to the usual lumped elements and active devices such as transistors, we also have the possibility of using transmission lines of microstrip and stripline, coplanar waveguide, and others.

These analysis programs can use the S parameters of a device as the means of describing devices over a specified frequency range. They have data libraries that contain S parameters of various transistors, FETs, and diodes over a range of frequency. They also will drive network analyzers to automatically measure values of S parameters and store these measurements in a CAD file. The output form of these programs can be tabular or graphical on rectangular axes, on polar plots, or on Smith® charts. These programs will perform computations such as determining overall S parameters of networks over given frequency ranges, gain, and noise circles (see chapter 6); optimizing circuits by changing values of specified circuit elements; and calculating the dimensions of transmission lines for specified properties using specified materials. The main drawback with using such programs is that they will not do the initial design steps where the general layout of the circuit and its parts is accomplished. However, they are excellent at rapid and accurate numerical calculation.

PROBLEMS

3–1. Show that $\mathcal{Re}\,(VI^*) = |a|^2 - |b|^2$ by replacing $\mathcal{Re}\,(VI^*)$ with $(\frac{1}{2})(VI^* + V^*I)$ and substituting from Eq. (3–1).

3–2. Starting with Eq. (3–1), show that the wave functions a and b are related to the transmission line voltages and currents as in $a = (V_{line} + I_{line}Z_0)/2\sqrt{Z_0}$ and $b = (V_{line} - I_{line}Z_0)/2\sqrt{Z_0}$.

3–3. Find all S parameters for the 2 port shown in Figure 3–17.

3–4. Show that $S_{21} = 1 - S_{11}$ for the circuit of Figure 3–17, regardless of the resistance of the resistor. Start with the definitions of S_{11} and S_{21} and with the fact that the current into the #1 port is equal to the current out of the #2 port.

3–5. For the circuit in Figure 3–18, show that $S_{21} = 1 + S_{11}$, regardless of the resistance of the resistor.

3–6. Find the S parameters for the 2 port shown in Figure 3–19.

3–7. Find the S parameters for the circuit of Figure 3–20 where the capacitive reactance is –25 Ω. Check your answer by modeling this circuit on a CAD program.

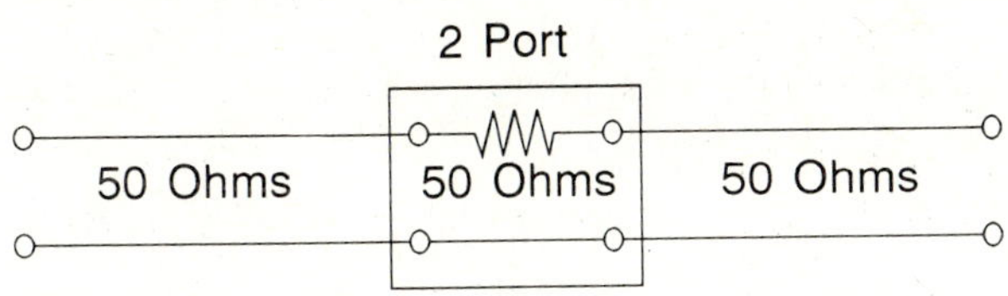

FIGURE 3–17. Circuit for Problem 3–3.

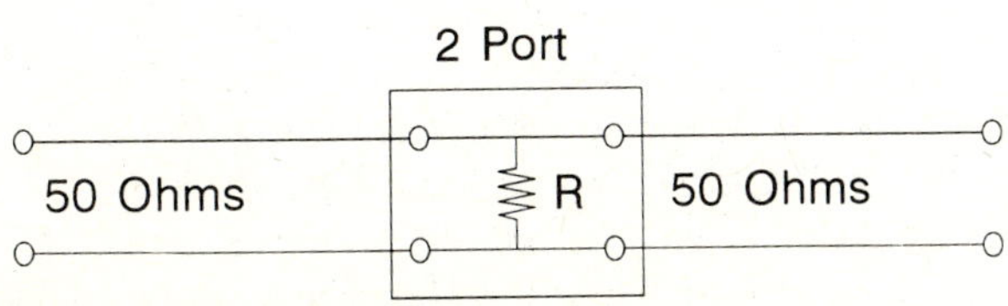

FIGURE 3–18. Circuit for Problem 3–5.

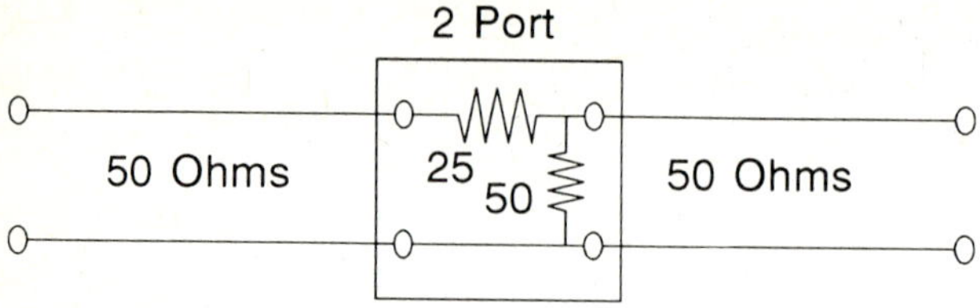

FIGURE 3–19. Circuit for Problem 3–6.

3–8. You are given a linear 2 port terminated at each end with its characteristic impedance and a voltage source V_i that supplies a current I_i. Show that the reciprocity theorem implies that $S_{21} = S_{12}$ for this 2 port by proving that $S_{21} = S_{12}$ if $I_2/V_1|_{V_2=0} = I_1/V_2|_{V_1=0}$, where the condition $V_2 = 0$ is obtained by replacing the voltage V_2 with a short circuit and the condition $V_1 = 0$ is obtained similarly, but with the source at port #2 and with V_1 shorted.

3–9. Derive the conversion formula Eq. (3–13), which converts from (Y) to (S) parameters.

3–10. Feedback has been used as a series resistor and inductor from the drain to the gate to increase the bandwidth of GaAs amplifiers.[5] To determine the optimum resistance value at low frequencies for which the capacitances and parasitic inductances were negligible, the circuit was modeled as shown in Figure 3–21. Using the conversion formulas in section 3–5, determine formulas for the S parameters in terms of G_m, G_{ds}, and G_{fb}. Assume that G_{ds} is much less than either G_m or $1/Z_0$ and use these formulas to determine the value for R_{fb} ($= 1/G_{fb}$) in terms of G_m and Z_0 that gives the least $|S_{11}|$ and $S_{22}|$.

3–11. From Eq. (3–13) show that $S_{21}/S_{12} = y_{21}/y_{12}$.

3–12. **(a)** Determine the T parameters for the circuit of a shunt resistor with a normalized admittance of 0.2 as shown in Figure 3–22.

(b) Determine the T parameters for the transistor circuit in Figure 3–22 in terms of the S parameters of the transistor.

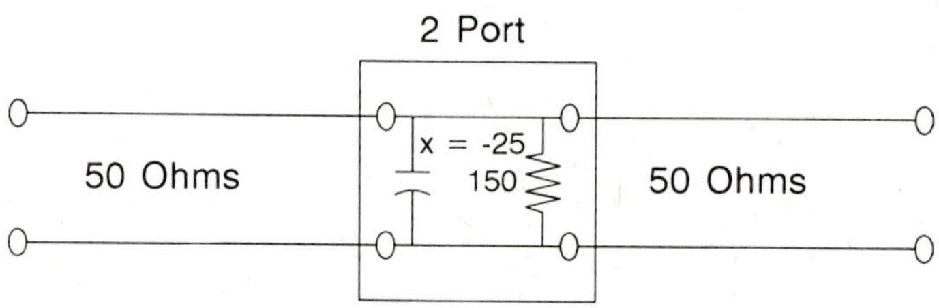

FIGURE 3–20. Circuit for Problem 3–7.

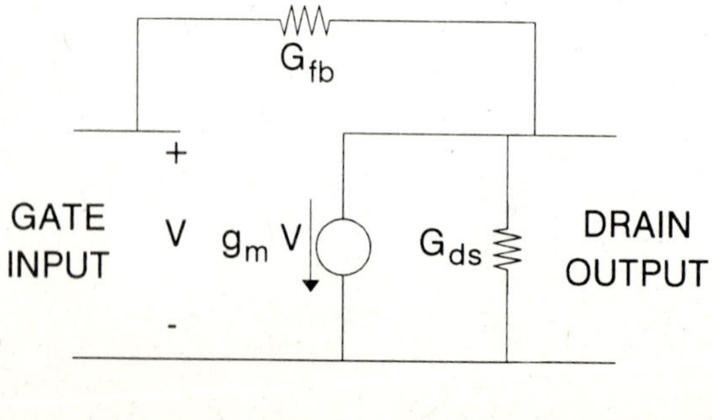

FIGURE 3–21. Circuit for Problem 3–10.

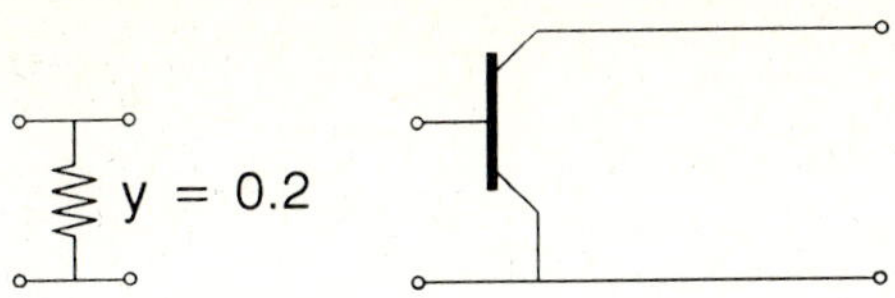

FIGURE 3–22. Circuit for Problem 3–12.

(c) Find the overall T parameters for a transistor with the 0.2 normalized shunt admittance connected to its input by premultiplying the matrix found in part B by that found in part A. You can check the answer by converting the overall T parameters to S parameters. These can be compared to the example in section 3–5.

3–13. A GaAsFET operated at 1 GHz has the approximate equivalent circuit shown in Figure 3–23. At 1 GHz, the S parameters for this device are: $S_{11} = 0.986\angle-22.5°$, $S_{21} = 3.707\angle161.9°$, $S_{12} = 0.036\angle76.9°$, and $S_{22} = .759\angle-11.4°$. From these parameters, determine R_i, C_{gs}, g_m, and g_d.

3–14. Prove that the loci of constant r are circles in the u, v plane by setting $r = (\zeta + \zeta^*)/2$, then substituting $\zeta = (1 + u + jv)/(1 - u - jv)$ into this. Then, expand this expression in u and v and complete the square in u. Show that this leads to the expression for a circle given below.

$$(u - \frac{r}{1+r})^2 + v^2 = \frac{1}{(1+r)^2}$$

Prove that the loci of constant x in the u, v plane are circles.

3–15. A 50 Ω lossless transmission line is terminated in an admittance of $Y = 0.02(3 + j2)$ S. The line is to be matched with the use of a short circuited stub. Determine the location (distance in wavelengths from the load) and length (in wavelengths) of the stub. Use the Smith® chart for making this determination.

3–16. **(a)** Match the input of the circuit in Figure 3–20 when a 75 Ω load terminates the end of a quarter wave output line. Use a shorted stub line for matching the input. Determine the stub length and position. Check your answer on a CAD program.

(b) Determine the power fed to the circuit from a 1 volt 50 Ω source when it is matched.

(c) Determine the power fed to the 75 Ω load from a 1 volt 50 Ω source connected to the input of the matched circuit.

3–17. In the example of a Pi matching circuit, the capacitors were in parallel and the inductor in series. Reverse this and put inductors in parallel and a capacitor in series and calculate their

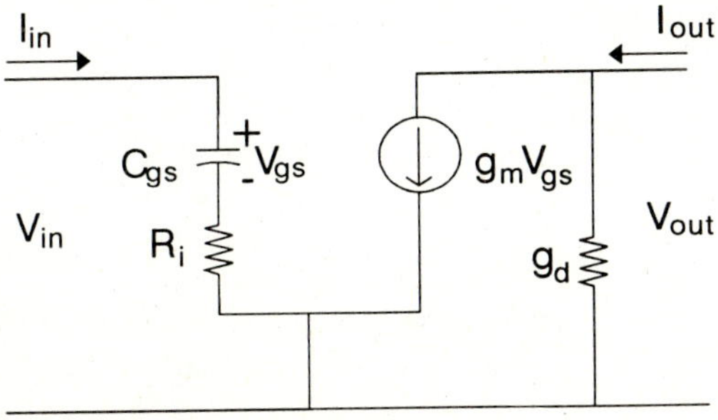

FIGURE 3–23. Circuit for Problem 3–13.

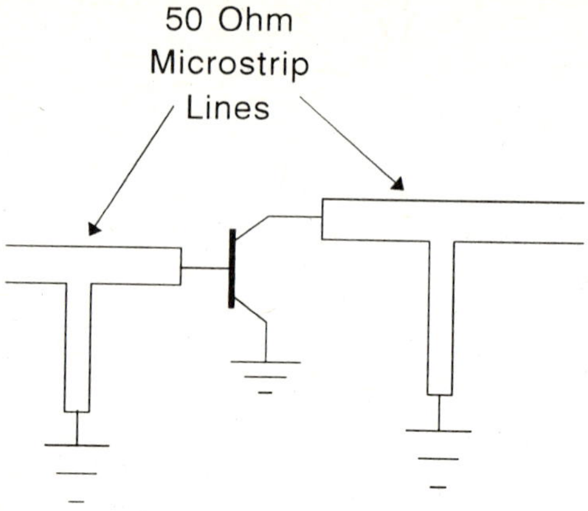

FIGURE 3–24. Circuit for Problem 3–18.

values to match the antenna to the receiver shown in the example. Check your answer on a CAD program.

3–18. The NEC transistor NE68035C has S parameters $S_{11} = 0.33\angle -143°$, $S_{21} = 8.07\angle 90°$, $S_{12} = 0.01\angle 65°$, and $S_{22} = 0.56\angle -27°$.

(a) Assuming that S_{11} is the reflection coefficient at the input to the transistor, match the input at 1 GHz using a shorted stub line as shown in the Figure 3–24. Model this circuit with a CAD program when the output is terminated with a 50 Ω load and check your answer.

(b) Match the output circuit to S_{22} of this transistor as in the figure below and check your answer by modeling the circuit with a CAD program. Recheck the S_{11} into the entire circuit (that with the input and output matching circuits) with the CAD program (when matching a transistor in which S_{12} is not zero, this feedback causes the match to change when the output is not terminated with 50 Ω).

3–19. Use Eqs. (3–16) and (3–19) in the expression $aa^* - bb^*$ and show that the result is equal to $|I|^2 \mathcal{Re}(Z_L)$. This is the power fed to the load Z_L.

ENDNOTES

1. Roger F. Harrington, *Time-Harmonic Electromagnetic Fields,* 116–119, (New York: McGraw-Hill Book Company, 1961).
2. George D. Vendelin, Anthony M. Pavio, and Ulrich L. Rohde, *Microwave Circuit Design Using Linear and Nonlinear Techniques,* 16–17, (New York: John Wiley & Sons, 1990).
3. D. C. Youla, "A New Theory of Broad-band Matching," *IEEE Trans. Circuit Theory, 11, 3* (March 1964), 30–50.
4. This subject is treated fully by Wai-Kai Chen in *Broadband Matching,* Second Edition, Chapter 2, (Singapore: World Scientific Publishing Co. Pte. Ltd., 1988).
5. Karl B. Niclas and others, "The Matched Feedback Amplifier: Ultrawide-Band Microwave Amplification with GaAs MESFET's," *IEEE Trans. Microwave Theory Tech., MTT-28, 4* (April 1980), 285–294.

APPENDIX

The following BASIC computer program is for computing the normalized y parameters of a 2 port when given the S parameters. The S parameters must be in a computer file named SPAR.BAS in the following order: Frequency, $|S_{11}|$, $\angle S_{11}$, $|S_{21}|$, $\angle S_{21}$, $|S_{12}|$, $\angle S_{12}$, $|S_{22}|$, $\angle S_{22}$. The angle must be in degrees. The y parameters will be put into a file named YPAR.BAS in the following order: Frequency, $|y_{11}|$, $\angle y_{11}$, $|y_{12}|$, $\angle y_{12}$, $|y_{21}|$, $\angle y_{21}$, $|y_{22}|$, $\angle y_{22}$. The angles will be given in degrees. The magnitudes of the *y* paramters will be normalized to the Z_0 reference impedance for the S parameters. Carefully note that the order of the 1-2 and 2-1 parameters changes between these two files.

```
10 REM ----- THIS PGM CONVERTS S-PAR TO y PARAMETERS -----
20 REM ----- FIRST READ S PARAMETERS INTO A (SPAR.BAS) FILE-----
30 REM ---The index in instr. 60 must be changed to fit dim. of Spar file--
40 REM ---The S par file name in instr. 50 must be changed ---
50 OPEN "SPAR.BAS" FOR INPUT AS #1:OPEN "YPAR.OUT"FOR OUTPUT AS #2
60 FOR I=1 TO 17
70 INPUT #1,F,MS11,AS11,MS21,AS21,MS12,AS12,MS22,AS22
80 DR = 3.141593/180
90 AS11=AS11*DR:AS21=AS21*DR:AS12=AS12*DR:AS22=AS22*DR
100 REM INSTR 120 THRU 280 CALCULATES (1-S11)(1+S22) + S12*S21 AS MAG. & ANGLE
110 RTRM1 = 1 - MS11*COS(AS11)
120 MTRM1 = SQR(RTRM1^2+(MS11*SIN(AS11))^2):ATRM1=ATN(-MS11*SIN(AS11)/RTRM1)
130 IF RTRM1>0 GOTO 150
140 ATRM1 = ATRM1 +3.141593
150 RTRM2 = 1 + MS22*COS(AS22):MTRM2=SQR(RTRM2^2+(MS22*SIN(AS22))^2)
160 ATRM2=ATN(MS22*(SIN(AS22))/RTRM2)
170 IF RTRM2>0 GOTO 190
180 ATRM2=ATRM2+ 3.141593
190 RTRMD= 1+MS11*COS(AS11):MTRMD=SQR(RTRMD^2+(MS11*SIN(AS11))^2)
200 ATRMD=ATN(MS11*SIN(AS11)/RTRMD)
210 IF RTRMD>0 GOTO 230
220 ATRMD=ATRMD+3.141593
230 NUMR=MTRM1*MTRM2*COS(ATRM1+ATRM2)+MS12*MS21*COS(AS12+AS21)
240 NUMX=MTRM1*MTRM2*SIN(ATRM1+ATRM2)+MS12*MS21*SIN(AS12+AS21)
250 MNUM=SQR(NUMR^2+NUMX^2): ANUM=ATN(NUMX/NUMR)
260 IF NUMR > 0 GOTO 290
270 ANUM=ANUM+3.141593
280 REM INSTR 300 THRU 340 CALCULATES (1+S11)(1+S22)-S12*S21
290 DENR=MTRMD*MTRM2*COS(ATRMD+ATRM2)-MS12*MS21*COS(AS12+AS21)
300 DENX=MTRMD*MTRM2*SIN(ATRMD+ATRM2)-MS12*MS21*SIN(AS12+AS21)
310 MDEN=SQR(DENR^2+DENX^2):ADEN=ATN(DENX/DENR)
320 IF DENR >0 GOTO 340
330 ADEN = ADEN + 3.141593
340 MYPAR11 = MNUM/MDEN:AYPAR11=(ANUM-ADEN)/DR
350 MYPAR12 =2*MS12/MDEN:AYPAR12=-180 +(AS12-ADEN)/DR
```

```
360 MYPAR21=2*MS21/MDEN:AYPAR21=-180+(AS21-ADEN)/DR
370 RTRM3=1-MS22*COS(AS22):XTRM3=-MS22*SIN(AS22)
380 MTRM3 =SQR(RTRM3^2+XTRM3^2):ATRM3=ATN(XTRM3/RTRM3)
390 IF RTRM3 >0 GOTO 410
400 ATRM3 =ATRM3+3.141593
410 NUMR=MTRM3*MTRMD*COS(ATRM3+ATRMD)+MS12*MS21*COS(AS12+AS21)
420 NUMX=MTRM3*MTRMD*SIN(ATRM3+ATRMD)+MS12*MS21*SIN(AS12+AS21)
430 MNUM=SQR(NUMR^2+NUMX^2):ANUM=ATN(NUMX/NUMR)
440 IF NUMR > 0 GOTO 460
450 ANUM=ANUM +3.141593
460 MYPAR22=MNUM/MDEN:AYPAR22=(ANUM-ADEN)/DR
470 PRINT #2,F,MYPAR11,AYPAR11,MYPAR12,AYPAR12,MYPAR21,AYPAR21,MYPAR22,AYPAR22
480 PRINT,F,MYPAR11,AYPAR11,MYPAR12,AYPAR12,MYPAR21,AYPAR21,MYPAR22,AYPAR22
490 NEXT
500 PRINT "y PARAMETERS IN FILE YPAR.OUT AS y11,y12,y21,y22"
510 END
```

Chapter

4

UHF and Microwave Filters

4.0 INTRODUCTION

Early interest in filters came when the problem of carrying telephone transmission and telegraphic code simultaneously along a telephone line was encountered. Van Rysselberghe was issued a series of patents from 1884 to 1887 for circuits consisting of simple two-element low pass and high pass filters for use in separating these signals.[1] This must have been the first use of diplexing electrical signals. A report of an electrical filter built by George Campbell at AT&T as early as 1906 is given by Brillouin.[2] An AT&T internal memo discussed band pass filters in 1909 and design formulas were first published in 1922.[3,4] Filter design has been of importance to practicing electrical engineers since then.

The function of the filter is to pass signals having a certain frequency range and to stop signals having a different frequency range. Filters which pass all signal components between zero Hz and some cut-off frequency f_C are called *low pass* filters. Conversely, filters designed to block signals with a frequency range from zero Hz to f_C and pass all signal components above f_C are called *high pass* filters. Filters passing signal components only in a finite range of frequencies (usually not extending to zero Hz) are called *band pass* filters. Those which are designed to block signal frequencies in a finite range (usually not extending to zero Hz) are called *band stop* filters. Finally, a filter designed to block a single frequency is called a *notch* filter.

The signal loss, expressed as the ratio of the power fed by a source to a load with and without a filter inserted between the source and the load, is called the *insertion loss* and is usually given in decibels. The ratio of the power reflected at the input to a filter to the power incident on the filter is called the *return loss.* The ratio of the power fed to a matched load at the output of a filter to the power into the filter from a source matched to the input of the filter is called the *attenuation* of the filter.

4.1 CLASSICAL FILTER DESIGN FORMULAS

Filter design formulas are usually expressed in terms of networks having lumped components. Then, to adapt the design to circuits having distributed components, some approximations are required. The design of filters is frequently predicated upon the filter network being inserted between a source having an output impedance of Z_0 and a load having an impedance of Z_0, where Z_0 is often the characteristic impedance of a transmission line. The individual circuit blocks used in filter design are the L section, the T section, and the Pi section. The circuit configuration for these blocks is shown in Figure 4–1. One can cascade these sections to produce a *ladder network* as shown in Figure 4–1. To analyze the T section, first we assume that the section is terminated in the characteristic impedance Z_0, then we derive a formula for the input impedance to the terminated T section. At that point, we require that the filter impedance elements be chosen so that the input impedance to the T section will also be Z_0. The resulting equation can be solved for a relation between Z_0 and the impedance elements in the arms of the T section.

Using series and parallel impedance combinations, one can show that the input impedance of the terminated T section is given by the formula

$$Z_{in} = \frac{Z_1}{2} + \frac{\left(\frac{Z_1}{2} + Z_0\right) Z_2}{\frac{Z_1}{2} + Z_0 + Z_2}.$$

We arbitrarily require Z_{in} to be equal to Z_0 and refer to this either as the characteristic impedance or the *image impedance*. Requiring Z_{in} to be Z_0, we can derive the formula

$$Z_0 = \sqrt{Z_1 Z_2 \left(1 + \frac{Z_1}{4Z_2}\right)}.$$

Following a similar procedure for the Pi section, one can derive the formula:

$$Z_0 = \sqrt{\frac{Z_1 Z_2}{1 + Z_1/4Z_2}}.$$

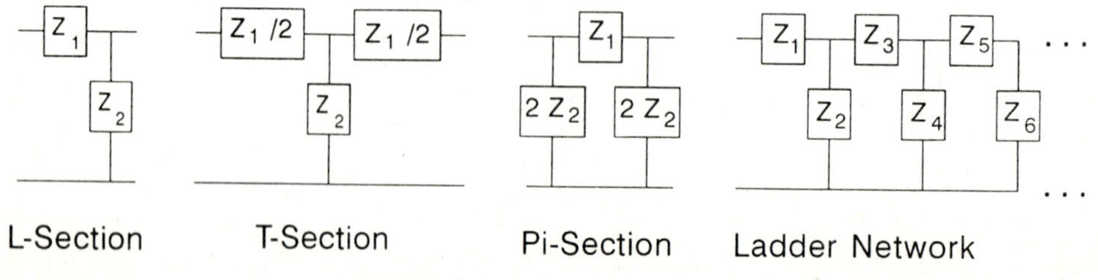

FIGURE 4–1. Filter Sections.

We note that for the filter to transmit power, the characteristic impedance Z_0 must not be a purely imaginary number. Further, we note that for lossless transmission, the impedance elements in the filter network (T section or Pi section) must be purely reactive. We see from the above formulas that the expressions for either the T section characteristic impedance or the Pi section characteristic impedance result in imaginary values for the characteristic impedance if both Z_1 and Z_2 are inductive. The same is true if both Z_1 and Z_2 are purely capacitive. From this, one can conclude that the filter should not have its branches all with the same kind of reactive element.

A low pass T-section filter can be constructed out of a pair of inductors and a capacitor as shown in Figure 4–2.

$$Z_0 = \sqrt{\frac{L}{C}\left(1 - \frac{\omega^2 LC}{4}\right)}$$

The formula for Z_0 is and the cut-off frequency occurs at the frequency where the bracketed term changes sign. Above the cut-off frequency, the impedance Z_0 is imaginary and the circuit will not accept power.

A high pass filter can be constructed out of a pair of capacitors and an inductor as shown in Figure 4–3. In this case, the Z_0 is given by

$$Z_0 = \sqrt{\frac{L}{C}\left(1 - \frac{1}{4\omega^2 LC}\right)},$$

where the cut-off frequency is again determined as the frequency when the bracketed term becomes negative.

A band pass filter can be made from combinations of inductors and capacitors as shown in Figure 4–4.

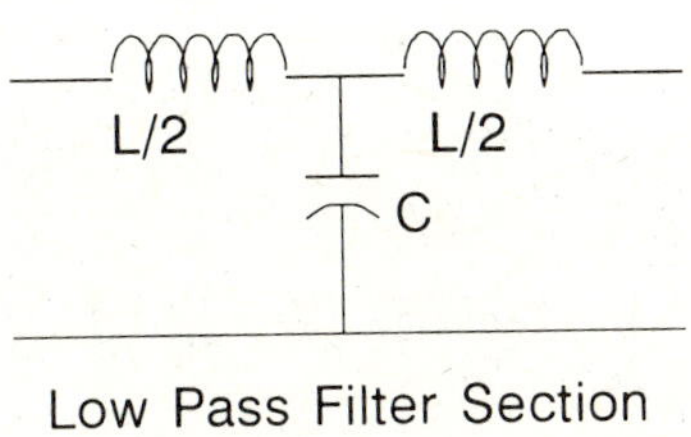

FIGURE 4–2. Low Pass Filter Section.

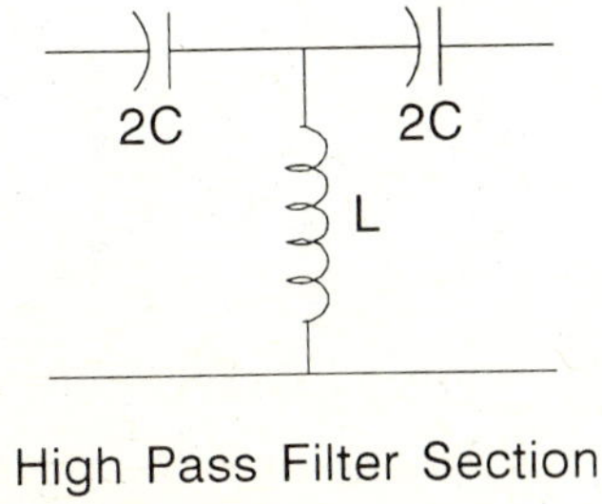

FIGURE 4–3. High Pass Filter Section.

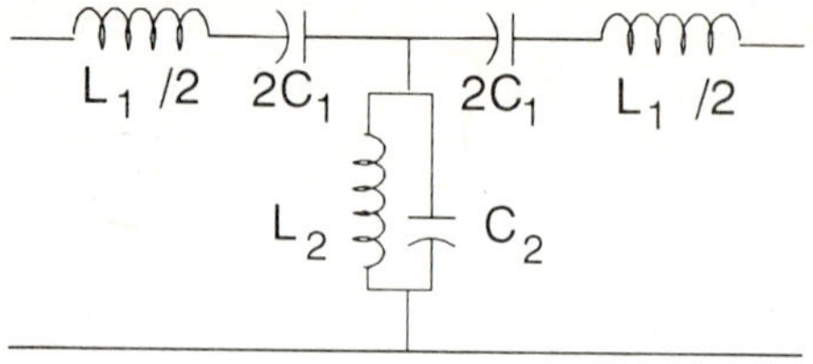

FIGURE 4–4. Band Pass Filter Section.

4.2 LADDER NETWORKS

Further investigation into the properties of networks in the 1930s through the 1950s led to network synthesis methods based largely upon properly locating the poles and zeros of the network functions to realize the desired network. We are primarily interested in ladder networks and the network functions for these ladder networks. These network functions may be any ratio of the 2-port voltages or currents, but we are interested usually in the S parameters for the ladder network. The parameter S_{21} can be expressed as a ratio of polynomials in complex frequency S. Those S_{21} that have no zeros on the finite real frequency axis ($S = j\omega$ for finite nonzero ω) and that meet certain other criteria (see section 4–11) can be realized as ladder networks such as is shown in Figure 4–5. Frequently, one uses a "handbook" approach to designing such filters

As an example of this process of determining elements of a filter using a ladder network, we choose a five-element ladder network in order to obtain a low pass filter. We choose a cut-off frequency of 0.5 GHz and we specify that the filter have a flat response in the pass band. This leads to a "Butterworth" filter. From the previous comments, we know that the topology for a low pass network is one in which the elements in the series arms of the ladder must be inductors and those in the shunt arms must be capacitors. This leaves two possible topologies for a five-element ladder as shown in Figure 4–5. Choosing the inductor input ladder for our filter, one can show that the filter transfer function is of the form of that in Eq. (4–1), where the constants a_i and K can be expressed as functions of the network elements L_1, L_3, L_5, C_2, C_4, and R_L.

$$H(S) = \frac{K}{S^5 + a_4 S^4 + a_3 S^3 + a_2 S^2 + a_1 S + a_0} \tag{4–1}$$

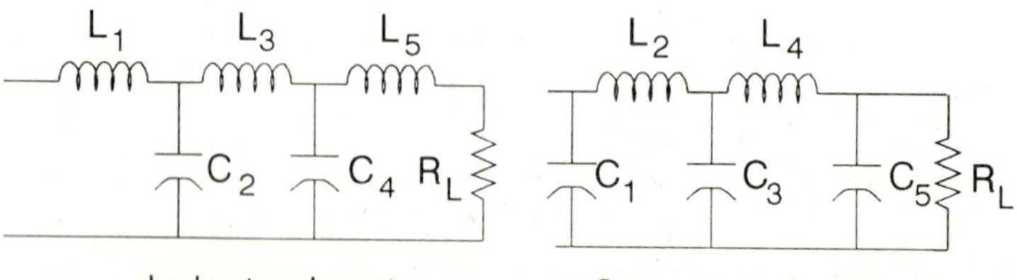

FIGURE 4–5. 5-Element Ladder Network for a Low Pass Filter.

To design the filter, we must select the values of these inductors, capacitors, and the load resistor to achieve the desired S_{21}. Usually, the load resistor is also given as part of the specifications. In many cases, the value of R_L is specified as 50 Ω. We shall use that as the load for our filter in this case. The basic problem is to specify numbers for the "constants" a_i that will produce poles of $S_{21}(S)$ at the desired locations in the S plane and then solve for inductor and capacitor values. This problem has been solved before and the results were given in a normalized form. Using these solutions, we can find that a five-element inductor input ladder network with a cut-off frequency of 1 rad/sec, a load resistance of 1 Ω, and flat (Butterworth) response has inductor and capacitor values

$$L_1 = 0.618,\ L_3 = 2.000,\ L_5 = 0.618,\ C_2 = 1.618, \text{ and } C_4 = 1.618.$$

4.3 FREQUENCY AND IMPEDANCE SCALING

The filter having these element values is a *prototype* filter. A prototype filter is one for which the cut-off frequency is 1 rad/sec and the characteristic impedance is 1 Ω. This prototype filter does not have the desired cut-off frequency of $2\pi 5 \times 10^8$ rad/sec. To adjust the cut-off frequency, we require that the reactances of the desired filter at a radian frequency of $2\pi 5 \times 10^8$ rad/sec must be the same as the reactances of the prototype filter at a radian frequency of 1 rad/sec. This is called *frequency scaling.* We see that the inductors to be used in the desired filter must be scaled down by the ratio that the frequency is scaled up and the capacitors must also be decreased by that factor. With these values and with $R_L = 1$ Ω, this filter will be matched to 1 Ω at the input.

For this filter to be matched to a load resistance of 50 Ω and to an input impedance of 50 Ω rather than 1 Ω, the impedance of all other elements in the ladder must be scaled up by a factor of 50. This is called *impedance scaling.* Thus, the inductor values in the desired network must be increased by a factor of 50 and the capacitor values must be decreased by a factor of 50.

So, the scaling of the individual element values is accomplished as in $L_1 = 0.618(50)/(2\pi 5 \times 10^8) = 9.84$ nH. The inductor at the input of the desired filter must have that inductance. Using these scale factors, the final design will have the element values $L_1 = 9.84$ nH, $L_3 = 31.8$ nH, $L_5 = 9.84$ nH, $C_2 = 10.3$ pF, and $C_4 = 10.3$ pF. It must be loaded with a 50 Ω resistance and it is matched to a 50 Ω source for frequencies in the pass band.

4.4 RICHARDS' FREQUENCY TRANSFORMATION

A practicing engineer will recognize that the inductor values in the foregoing example are quite small. Such inductors can be fabricated on integrated circuit chips,[5] or one can simulate them with a short circuited transmission line operated below the frequency where the transmission line is one-quarter wavelength long. To show this simulation, we consider the input impedance to a line D meters long as shown in the Figure 4–6. Let V_i and

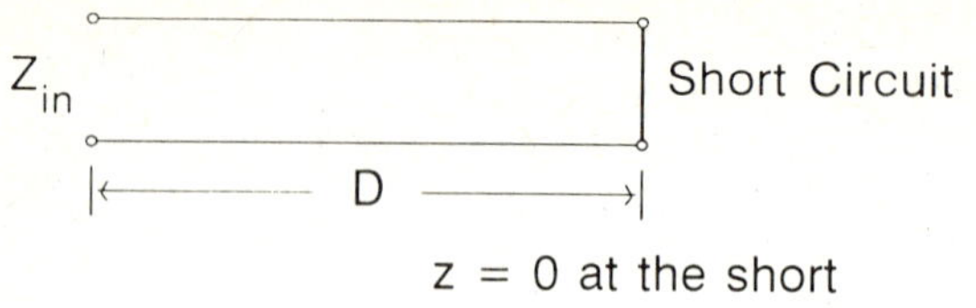

FIGURE 4–6. Shorted Transmission Line.

I_i represent the voltage and current associated with the forward traveling wave and V_r and I_r be those for the backward traveling wave. We have the voltages and currents

$$V_i = V_0 \exp(-j\beta z), \qquad V_r = -V_0 \exp(+j\beta z),$$
$$I_i = (V_0/Z_0)\exp(-j\beta z), \quad \text{and} \quad I_r = (V_0/Z_0)\exp(+j\beta z).$$

The input impedance to the line (at $z = -D$) is given by Eq. (4–2).

$$Z_{in} = \frac{V_0 \exp(+j\beta D) - V_0 \exp(-j\beta D)}{(V_0/Z_0)\exp(+j\beta D) + (V_0/Z_0)\exp(-j\beta D)} = jZ_0 \tan(\beta D) \tag{4–2}$$

The argument of the tangent function is $\beta D = \omega D/C$, where D has been chosen to be one quarter wavelength at the frequency f_0 so that the argument is $\beta D = \pi f/(2f_0)$. In this expression, we frequently use a transformation called *Richards' frequency transformation.*[6] We use the symbol Ω for radian frequency and let $\Omega = \tan(\pi f/2f_0)$. This is Richards' transformation. Now, we will refer to two different frequency domains. The notation *Ω domain* and *f domain* will be used. The frequency domain, specified by Ω, is the radian frequency ranging from 0 rad/sec to infinity as the frequency f ranges from 0 Hz to the frequency f_0. The input impedance of the shorted transmission line is now as shown in Eq. (4–3).

$$Z_{in} = jZ_0 \tan(\pi f/2f_0) = j\Omega Z_0 \tag{4–3}$$

The input impedance is a positive reactance (inductive in nature) for all frequencies below f_0 at which the line is one-quarter wavelength long. It is clear that the characteristic impedance of the shorted line Z_0 plays the role of inductance in the Ω domain. Thus, we can specify a cut-off frequency for a filter as f_C, then design the prototype filter for a radian cut-off frequency Ω_C and an inductive reactance at Ω_C of $X = Z_0 \tan(\pi f_C/2f_0)$. The shorted line used in the filter replaces the inductor required in the prototype filter.

In the same way as described above, we can simulate the action of a capacitor in the Ω domain with an open circuited one-quarter wave transmission line in the f domain. The input admittance to an open circuited line is given by Eq. (4–4).

$$Y_{in} = j(1/Z_0)\tan(\beta D) = j\Omega(1/Z_0) \tag{4–4}$$

The effective capacitance in the Ω domain is numerically equal to $1/Z_0$, when one uses the Richards' transformation between the frequency domains. Filters using one-quarter wave transmission lines of equal length as circuit elements are called *commensurate line filters.*

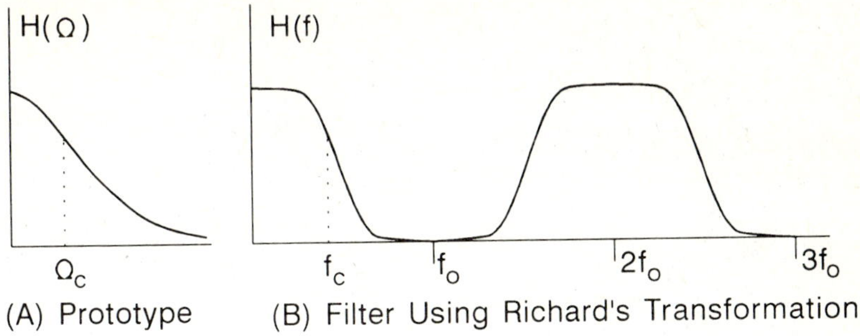

FIGURE 4–7. The Effect of Richards' Transformation on a Transfer Function.

Richards' transformation changes the range of frequencies of zero to infinity in the Ω domain into the range of frequencies from zero to f_0 in the f domain. This transform is not one-to-one, as the entire range in the Ω domain also maps into the region $2f_0$ to $3f_0$, and also into all multiples of these segments in the f domain. As an example, Figure 4–7A shows the low pass character of a prototype filter. Figure 4–7B shows the resulting filter characteristic after using Richards' transformation. It has the practical effect of producing filters with transfer functions that are periodic in the f domain.

4.5 VARIABLE LINE WIDTH FILTERS

As discussed previously, the impedance of a shorted transmission line less than $\lambda/4$ long is reactive and has an inductive nature. We also note that a narrow section of microstrip has a higher characteristic impedance than does a wide section. If the disparity in widths of transmission line is sufficient, then a narrow section of line terminated in a wide section of line will look approximately like a narrow section terminated in a short circuit. We propose to approximate a series inductance by such an arrangement. In a similar way, a short but wide section of microstrip followed by a narrow microstrip line will act like a shunt capacitance. In each of these cases, the actual normalized terminations will fall somewhere on the Smith chart inside the unit circle and so the inductive and capacitive nature at the input of these short sections will not be pure, there will be some resistive effects.[7]

To simulate the L-C ladder network of Figure 4–5, when designing a low pass filter, we start with alternating narrow and wide sections of microstrip lines connected in sequence as shown in Figure 4–8. This configuration is to simulate the L-C ladder of that figure. The short, narrow sections simulate series inductances and the short, wide sections simulate shunt capacitances. From the component values given in section 4–2, we have the values for a Butterworth filter of $g_1 = g_5 = 0.618$, $g_2 = g_4 = 1.618$, and $g_3 = 2$. These are the values of inductive reactances and capacitive susceptances of these components at the radian frequency $\Omega = 1$. Then, these were scaled to the characteristic impedance of the filter. Assuming the impedance to be 50 Ω, we have the reactance and susceptance

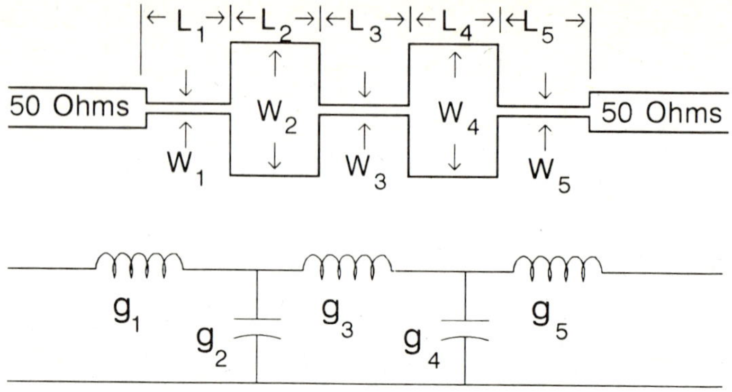

FIGURE 4–8. Configuration for a Low Pass Filter.

values of $X_{L1} = X_{L5} = 30.9$, $Y_{C2} = Y_{C4} = 0.03236$, and $X_{L3} = 100$. A low pass filter with a cut-off of 2 GHz must have these same reactances and susceptances at the cut-off frequency of 2 GHz.

The first section must have the inductive reactance approximately given by $30.9 = X_L \approx Z_o \tan(\beta L_1)$ (see Eq. (4–3)), and it must have a length L_1, which is electrically short, that is, βL_1 must be considerably less than 90°. This choice is somewhat arbitrary. However, too small an angle will usually produce a filter with lines having line impedances impossible to fabricate. Too large an angle will lead to line sections not differing greatly in impedance so that they do not present the nearly open and nearly short circuit conditions needed. Choosing $\theta_1 = \beta L_1 = 16°$, we calculated the impedance of the input line to be $Z_{01} = 30.9/(\tan(16°)) = 105.3\ \Omega$.

The second section must have an impedance much lower than 50 Ω if it is to approximate a short circuit at the end of the first section. This translates to an admittance much larger than 0.02. Using the susceptance $B_2 = 0.03236$ previously calculated and assuming the $\beta L_2 = 33°$, we determined a Z_{02} from the relation $Y \approx (j/Z_0) \tan(\theta)$ or $j0.03236 = (j/Z_0) \tan(33°)$. Solving this, we had $Z_{02} = 20.4\ \Omega$. Again, the choice of θ_2 ($= \beta L_2$) is somewhat arbitrary, but it must be considerably less than 90°.

The middle section could be calculated in the same way as the first section. Instead, we assumed that the microstrip line width was the same as the first section, the reason being that we could not reliably fabricate a line with a characteristic impedance higher than about 100 Ω. So, the phase constant and the impedance Z_{03} were also the same. From Eq. (4–3), we obtained $\theta_3 = 43.5°$. The filter is symmetrical about the middle sections so the last two sections duplicated the first in inverse order.

The filter was built on a RT/duroid® microwave laminate with a dielectric constant ε_r of 2.23 and a thickness h of 1.58 mm. For these parameters and Z_{01} of 105.3 Ω, we calculated a microstrip line width W_1 of 1.25 mm for the first section. Then, we calculated ε_{eff} of about 1.77. From this, we got a phase constant of $\beta = 32$ deg/cm. From $\beta L_1 = 16°$, we get L_1 of 5 mm. For the second section, which required $Z_{02} = 20.4\ \Omega$, we calculated a width of 15.67 mm. Then, we calculated ε_{eff} of 2.03 from that. From this, we had a phase

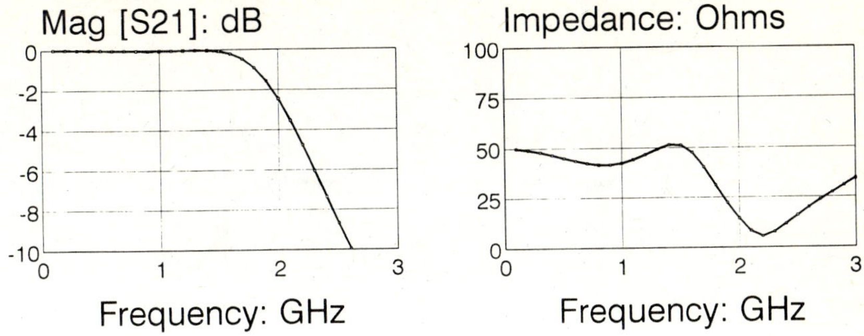

FIGURE 4–9. Transmission and Impedance Characteristics of a Low Pass Filter.

constant β equal to 34°/cm. From $\beta L_2 = \theta_2 = 33°$, we calculated the length of this section L_2 to be 9.7 mm.

For the middle section, we had $\theta = 43.5°$ and $\beta = 32$ deg/cm as determined for the first section. So, $L_3 = \theta_3/\beta = 13.6$ mm.

This filter was simulated and built on TOUCHSTONE® software. The results for the filter are shown in Figure 4–9. The transmission is down by about 2 dB at 2 GHz and drops off to about 15 dB at 3 GHz. The input impedance is similarly well behaved up to about 1.8 GHz.

This would seem to be an ideal way to design a filter. The disadvantage with this design becomes apparent when one extends the simulation to 8 GHz. The simulated parameter S_{21} recovers at higher frequencies and reaches a peak of 0 dB between 6 and 7 GHz. The impedance also varies over a wide range throughout this frequency span. Clearly, such a filter would have to rely on other parts of the circuit to supply the desired rejection at the higher frequencies. Frequently, a transistor will accomplish this result if it is connected in series with the filter.

4.6 KURODAS' IDENTITIES

We note that a microstrip line uses the ground plane for one of the conductors in the 'transmission line'. In that case, the inductor simulated by a shorted transmission line can only be realized between one conductor and the ground plane in our filter design. However, the prototype filter often requires that the inductor not be grounded. To accommodate this requirement, we can use an *impedance inverter* in the filter design. A 2 port that inverts impedance has the property that its input impedance is the inverse of the impedance connected to the output port. The notation used is $Z_{in} = K^2/Z_{load}$. A simple impedance inversion device is the quarter wave transmission line. To show how this may be used at a single frequency, consider the two circuits in Figure 4–10. The impedance of the load resistor in Figure 4–10A is inverted by the action of the output quarter wave line (see problem 1–10). Then, the capacitor is effectively in parallel with a resistor having a

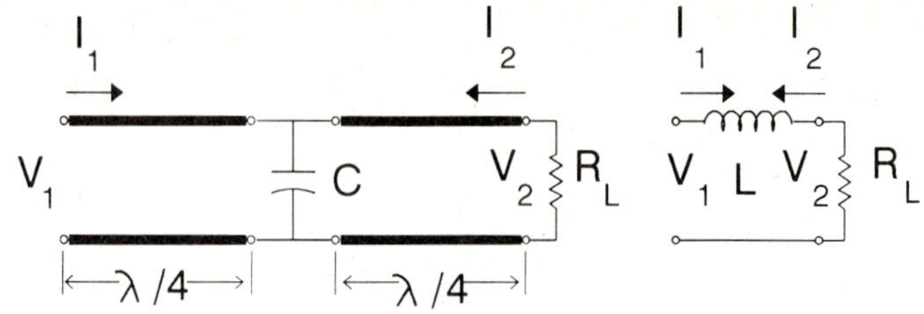

FIGURE 4–10. A: Impedance Inverting Network and B: The Equivalent Circuit.

resistance of Z_0^2/R_L. The impedance of this parallel combination $1/(j\omega C+R_L/Z_0^2)$ is then inverted and multiplied by Z_0^2 by the input quarter wave line so that the input impedance to the circuit in Figure 4–10A is given by $R_L + j\omega CZ_0^2$. If we require that the numerical value of capacitance times Z_0^2 be equal to the inductance of the inductor of the circuit in Figure 4–10B, then the input impedance of these two circuits is the same at this frequency. The circuits are equivalent in their action upon the input voltage and current as well as the relative phase and magnitude of the output voltage and output current. They are not the same in that there is a phase delay between the input voltage and the output voltage as well as between the input and output currents.

Using the concept of impedance inversion, we now can reduce the L-C ladder network to a series of shunt inductors spaced by impedance inverters as shown in Figure 4–11. The two circuits shown are equivalent at one frequency if the impedance inverter constants K_i are chosen correctly.

We must examine this impedance inversion more carefully in order to use it for filter design. This idea of impedance inversion is incorporated into a set of equivalences called *Kuroda's identities,* proposed by K. Kuroda.[8] To discuss these, we describe a transmission line in terms of the ABCD parameters in the Ω domain. The line shown in the Figure 4–12 is one-quarter wavelength long at the frequency f_0. The load is located at $z = 0$ and the input at $z = -L$. The reflection coefficient at the load is Γ. We can write the

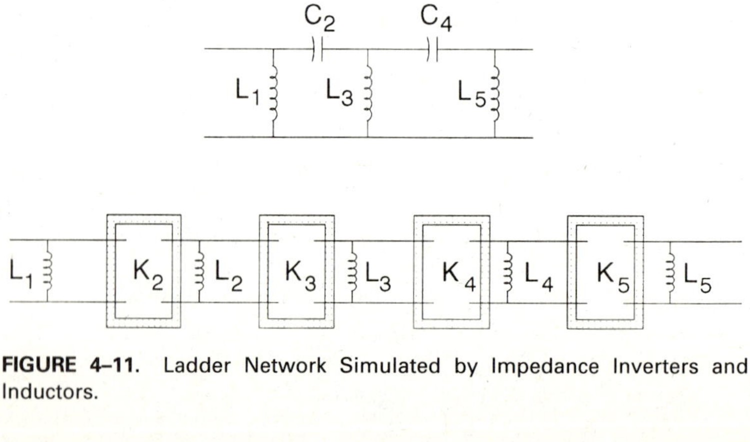

FIGURE 4–11. Ladder Network Simulated by Impedance Inverters and Inductors.

FIGURE 4–12. Transmission Line One-quarter Wavelength Long at Frequency *f*.

expression for the voltage on the line at the input as $V_1 = V(-L) = V_0 \exp(j\beta L) + \Gamma V_0 \exp(-j\beta L)$. The voltage at the output is $V_2 = V_0(1 + \Gamma)$. Letting $\theta = \beta L$, substituting $(Z_L - Z_0)/(Z_L + Z_0)$ for Γ, and substituting $V_2/(-I_2)$ for Z_L, we can rearrange the above expressions to Eq. (4–5).

$$V_1 = V_2 \cos(\theta) - jI_2 Z_0 \sin(\theta) \tag{4–5}$$

Repeating this process for the input current expression $I_1 = (V_0/Z_0)\exp(j\beta L) - \Gamma(V_0/Z_0)\exp(-j\beta L)$, we arrive at Eq. (4–6).

$$I_1 = j(V_2/Z_0)\sin(\theta) - I_2\cos(\theta) \tag{4–6}$$

Equations (4–5) and (4–6) can be expressed in matrix form and the elements of the coefficient matrix are the ABCD parameters. However, because L is $\lambda_0/4$, the parameter θ is equal to $(\pi/2)(f/f_0)$. From Richards' transformation, we see that $\tan(\theta) = S/j = \Omega$. Then, one can identify a triangle with the angle θ for which $\cos(\theta) = 1/\sqrt{1 - S^2}$ and $\sin(\theta) = S/j\sqrt{1 - S^2}$. Substituting these into Eqs. (4–5) and (4–6) and writing the result in matrix form, we arrive at the ABCD parameter description of the transmission line in the Ω domain as shown in Eq. (4–7).

$$\begin{pmatrix} V_1 \\ I_1 \end{pmatrix} = \frac{1}{\sqrt{1-S^2}} \begin{pmatrix} 1, & SZ_0 \\ S/Z_0, & 1 \end{pmatrix} \begin{pmatrix} V_2 \\ -I_2 \end{pmatrix} \tag{4–7}$$

This relation defines a *unit element* as used by Kuroda. A Unit Element (U.E.) is the Ω domain equivalent circuit for a transmission line that is one-quarter wavelength long at the frequency f_0. The inversion property of a unit element is shown in Figure 4–13. In Figure 4–13A, we show two circuits that are equivalent in terms of their ABCD matrix representations. We see that a series inductance following a unit element is equivalent to a parallel capacitance preceding a unit element.

The Kuroda identities are shown in Figure 4–13. They show circuits that are equivalent with respect to their ABCD matrices. These circuits are used to help design the prototype filter in the Ω domain. In order to show the equivalence of these circuits, we can show that they have the same ABCD matrices. As an example of this, take the circuit shown on the left in Figure 4–13A. The overall ABCD matrix of this circuit is given by the ABCD matrix of a shunt capacitor post multiplied by the ABCD matrix of the unit element. This is

$$\begin{pmatrix} 1, & 0 \\ SC, & 1 \end{pmatrix} \frac{1}{\sqrt{1-S^2}} \begin{pmatrix} 1, & SZ_0 \\ S/Z_0, & 1 \end{pmatrix} = \frac{1}{\sqrt{1-S^2}} \begin{pmatrix} 1, & SZ_0 \\ SC + S/Z_0, & S^2 Z_0 C + 1 \end{pmatrix}.$$

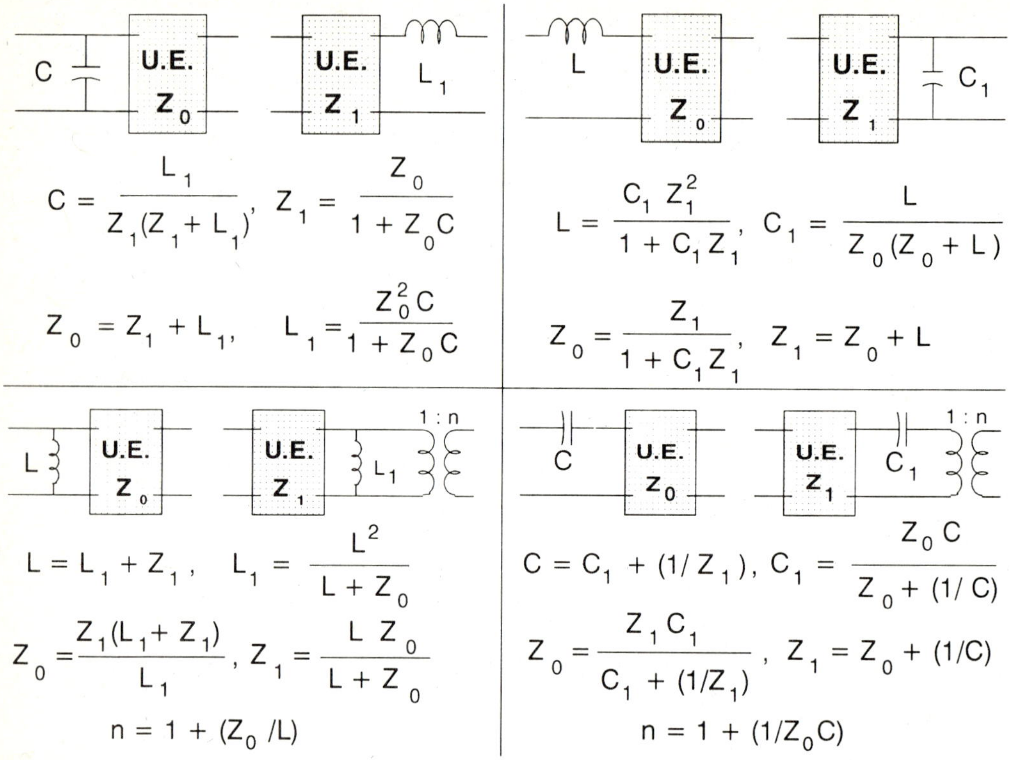

FIGURE 4–13. Kuroda's Identities.

The overall ABCD matrix for the equivalent circuit on the right in Figure 4–13A is the ABCD matrix of a unit element post multiplied by the ABCD matrix of a series inductance. This is

$$\frac{1}{\sqrt{1-S^2}}\begin{pmatrix} 1, & SZ_1 \\ S/Z_1, & 1 \end{pmatrix}\begin{pmatrix} 1, & SL_1 \\ 0, & 1 \end{pmatrix} = \frac{1}{\sqrt{1-S^2}}\begin{pmatrix} 1, & SL_1 + SZ_1 \\ S/Z_1, & S^2 L_1/Z_1 + 1 \end{pmatrix}.$$

Comparison of these two results shows that the two circuits have the same overall ABCD matrices if $L_1 = Z_0^2C/(1 + Z_0C)$ and $Z_1 = Z_0/(Z_0C + 1)$. So, they are equivalent in that regard. A more extensive set of such identities is given by Malherbe.[9]

The process of designing a low pass filter starts with a lumped element prototype filter. First, we determine the characteristics such as the cut-off frequency, the flatness in the pass band, and the roll-off rate outside of the band. Then, we choose the number, type, and impedance of the elements in the prototype filter based on these characteristics. If the transformation of the cut-off frequency in the f domain through Richards' transformation does not yield an Ω_C of unity, then these elements must be frequency scaled by the factor $1/\Omega_C$ at this point in the design process. A sufficient number of unit elements are added to this prototype at one end or the other so that each of the series inductances can be replaced with a shunt capacitance. The addition of the unit elements to the end of the prototype in the Ω domain is represented in the f domain by the addition of extra sections of transmission lines. As long as they are terminated in their characteristic impedance these sections of line change only the phase of the filter transmission coefficient and do not change the magnitude of it. Using Kuroda's identities, each of these unit elements is shifted into the filter into the place alongside of one the inductors. Again using Kuroda's identities, the combination of a unit element and series inductor can be replaced by a shunt capacitor and unit element. The result is that the ladder prototype has been transformed into a sequence of unit elements and shunt capacitors all scaled to values in the Ω domain. The resulting shunt capacitors and unit elements are transformed into open circuited stubs and transmission lines using Richards' frequency transformation. Finally, the elements for the final filter are determined by impedance scaling. This step-by-step process is shown in the following example.

Example: Using open circuited transmission lines to simulate shunt capacitors and Kuroda's identities to eliminate the need for series inductors, we discuss the design of a low pass filter with a cut-off frequency of 2 GHz. We shall require the filter to be "maximally flat" in the passband. This "maximally flat" requirement leads to a Butterworth filter. We arbitrarily chose the frequency f_0 at which the lines are quarter wave to be 3 GHz. This will lead to a rapid roll-off out to 3 GHz. With these considerations, we start with the L-C ladder network discussed before, where $L_1 = 0.618$, $L_3 = 2.0$, $L_5 = 0.618$, $C_2 = 1.618$, and $C_4 = 1.618$. Because this ladder network has a cut-off radian frequency of $\Omega = 1$, rather than $\Omega_c = \tan(\pi f_c/2f_0) = 1.732$, we must frequency scale all the components by dividing by Ω_c. The resulting ladder network is shown in Figure 4–14.

This filter has $|S_{21}|$, the same as the filter of Figure 4–15, which has unit elements added onto the input and the output of the original filter. When these unit elements are terminated in their characteristic impedance, they do not change the overall amplitude characteristic; they only change the phase delay. Using the Kuroda's identity shown in Figure 4–13A and applying it to the input U.E. and L_1, we may replace the combination of those two elements with a capacitor followed by another unit element. Next, we may

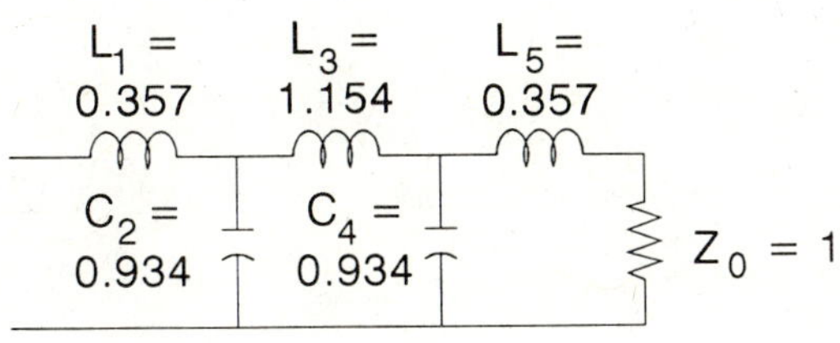

FIGURE 4–14. Prototype Butterworth Filter.

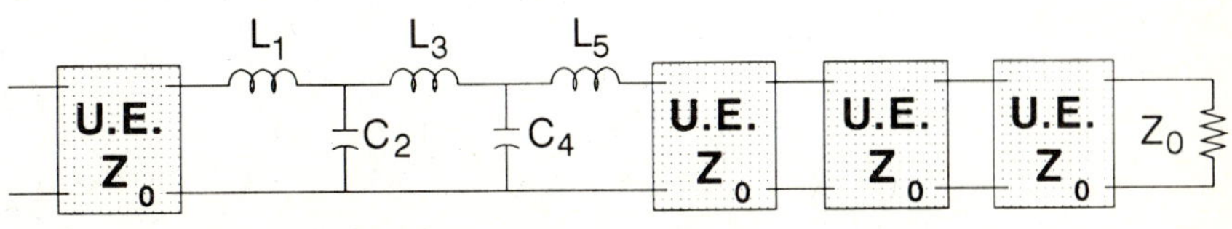

FIGURE 4–15. Modified Prototype Butterworth Filter.

replace the last inductor and the U.E. following it with a U.E. followed by a capacitor. At this intermediate stage, we have the circuit configuration in Figure 4–16. The elements C_A, Z_A, C_B, and Z_B must be determined using the formulas given in Figure 4–13. This process can be continued as follows: Replace the combination of C_4 and the U.E. with Z_B using a Kuroda identity, and similarly replace C_B and the U.E. following it using a Kuroda identity. This will result in a circuit as shown in Figure 4–17.

Now, each inductor that is followed by a U.E. can be replaced by a U.E. followed by a shunt capacitor using a Kuroda identity. The resulting filter circuit is still in the Ω domain and has a cut-off frequency of 1.732 radian per second. It has the form of five capacitors in shunt to ground separated by four unit elements, and is shown in Figure 4–18. The unit element and shunt capacitor values can be determined by following the transformation formulas in sequence. For instance, the capacitor C_F and the impedance Z_F are given in terms of the prototype filter elements as

$$C_F = \frac{L_3}{\dfrac{Z_0 + L_5}{1 + C_4(Z_0 + L_5)}\left(1 + L_3 \dfrac{Z_0 + L_5}{1 + C_4(Z_0 + L_5)}\right)} \quad \text{and} \quad Z_F = \frac{Z_0 + L_5}{1 + C_4(Z_0 + L_5)} + L_3$$

The element values for Figure 4–18 as calculated from the prototype are: $C_A = 0.2630$, $Z_A = 1.357$, $C_2 = 0.9342$, $Z_H = 1.208$, $C_F = 1.101$, $Z_g = 1.550$, $C_g = 0.6179$, $Z_H = 1.208$, and $C_H = 0.1723$.

Using Richards' transformation, the equivalent low pass filter in the f domain can be constructed by placing five open circuited stubs interspersed by four quarter wave

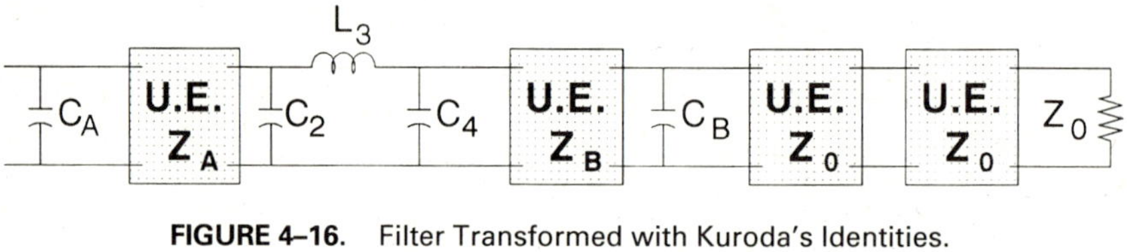

FIGURE 4–16. Filter Transformed with Kuroda's Identities.

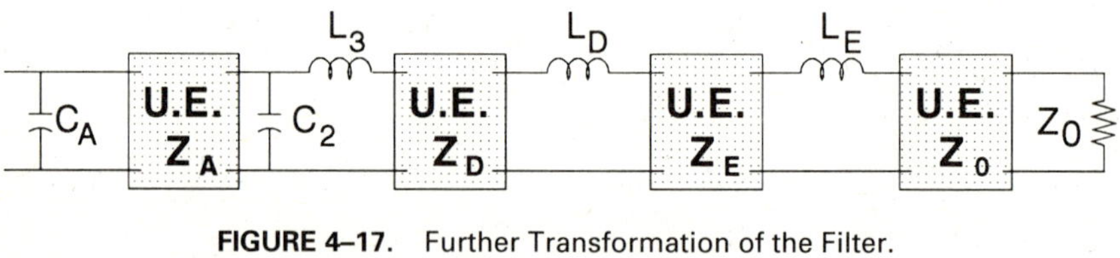

FIGURE 4–17. Further Transformation of the Filter.

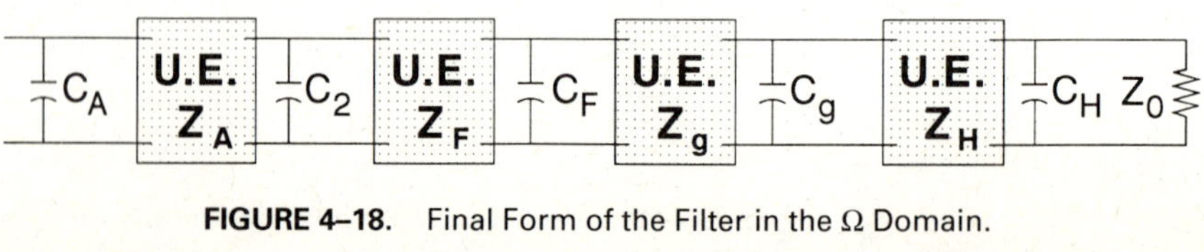

FIGURE 4–18. Final Form of the Filter in the Ω Domain.

lines. Because of the transformations that we have used, the lines will not have the same characteristic impedances. The final configuration will be as shown in Figure 4–19. To find the lengths and characteristic impedances of the lines used in this filter, we must scale the elements of the filter in Figure 4–18 in impedance. The scaling in impedance requires all of the characteristic impedances to be increased by a factor of 50 and all of the capacitors to be decreased by a factor of 50. We arbitrarily selected f_0 equal to 3 GHz as the frequency where the lines are to be quarter wave. If the filter design results in unsuitable numbers for the components, then we can choose a different value for f_0. One consideration in the choice of f_0 is that the stop band in the f domain is larger if the ratio of f_C to f_0 is smaller.

From Eq. (4–4), the admittance of the first open circuited, quarter wave stub is given by $Y_{in} = j(1/Z_1) \tan (\pi f/2f_0) = jC_A\Omega$, so the capacitance C_A must correspond to the inverse of the characteristic impedance of the open circuited stub. Thus, the characteristic impedance Z_1 is given by $Z_1 = 50/C_A = 190.1\ \Omega$. The remaining stub line characteristic impedances are calculated with the same scaling factors. They are $Z_2 = 53.52$, $Z_3 = 45.42$, $Z_4 = 80.92$, and $Z_5 = 290.1\ \Omega$. The characteristic impedances of the quarter wave lines between the open circuited stubs are obtained by scaling with a factor of 50. From this, we have $Z_A = 67.8$, $Z_F = 87.65$, $Z_g = 77.5$, and $Z_H = 60.4\ \Omega$.

The resulting filter is a low pass filter with a cut off of 2 GHz. Simulating this filter on a computer results in characteristics shown in Figures 4–20 and 4–21. One defect in

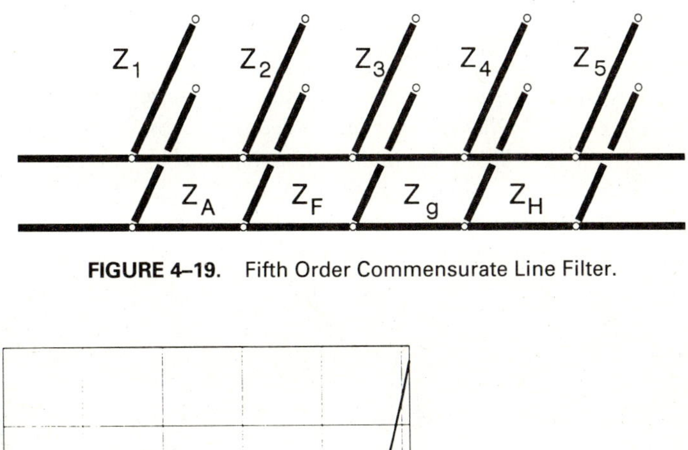

FIGURE 4–19. Fifth Order Commensurate Line Filter.

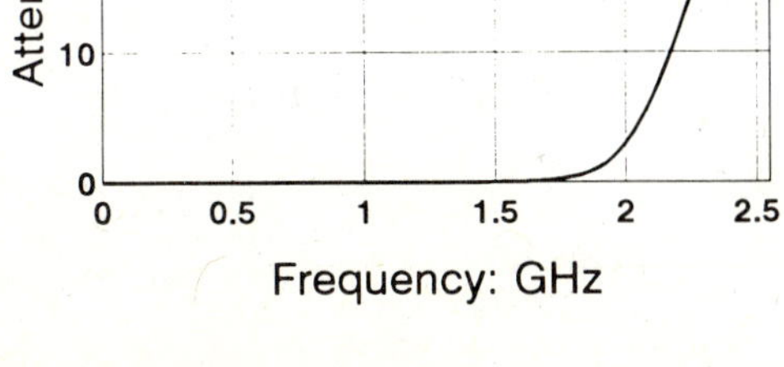

FIGURE 4–20. Attenuation versus Frequency for a Fifth Order Filter.

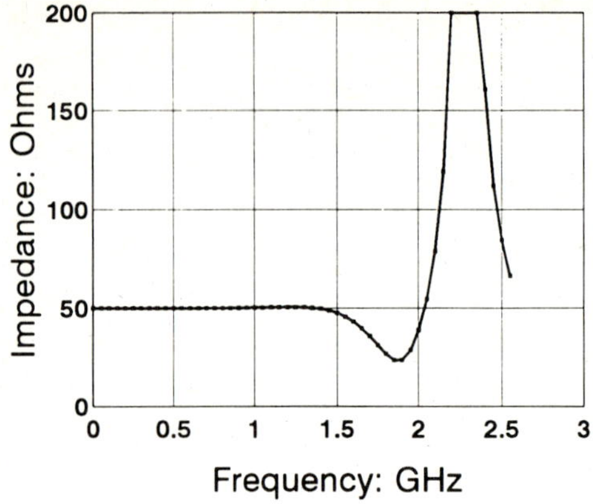

FIGURE 4–21. Characteristic Impedance vs Frequency for a Fifth Order Filter.

this design is that the line impedances are too large and cannot be fabricated readily with microstrip. These filters tend to be physically larger than necessary because of the intermediate lines required for impedance transformation. For this reason they are referred to as *redundant filters.*

4.7 THE APPROXIMATION PROBLEM

The prototype filter is usually designed as a low pass filter with a cut-off frequency of 1 rad/sec. How flat the transfer function is in the pass region and how rapidly the response decreases outside of the pass region are determined by the sizes and the number of the inductors, capacitors, and unit elements in the ladder. Usually, an engineer is given a specification on the allowable limits of the ripple in the pass band and the rejection out of the band. The engineer must then design a filter that approximates these specifications. For example, one might be given a specification such as is shown in Figure 4–22. The shaded

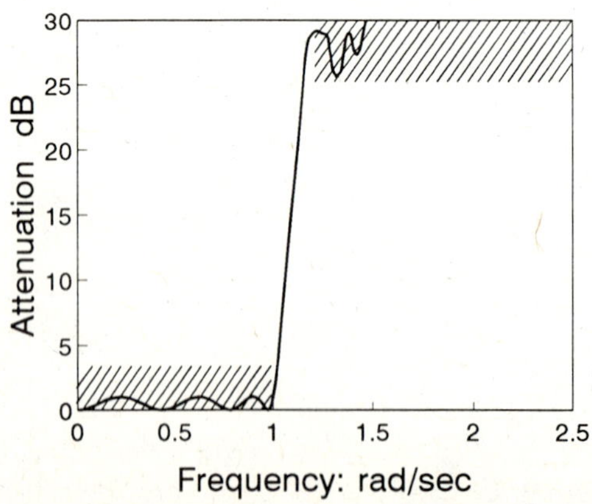

FIGURE 4–22. Sample Filter Attenuation Specification. The permissible attenuation is shown as the shaded areas. A sample attenuation curve is shown as the solid line.

areas in this figure show the allowable attenuation in the pass band and the minimum required attenuation in the rejection band as translated into the Ω frequency domain. To meet the specification, the filter components are chosen to produce an attenuation curve as shown.

The design of a lossless filter (i.e., one having only reactive elements internally) starts with a choice of S_{21} as a function of the complex frequency $S = \sigma + j\omega$. For a matched filter, the dB measure of S_{21} is the negative of the attenuation of the filter. This function can be determined by specifying its poles and zeroes. Frequently, an alternate determination of S_{21} is done by letting it be related to a polynomial such as $T_n(\Omega)$.

$\lvert S_{21}(j\Omega)\rvert^2$	Classification
$\dfrac{A}{1+\varepsilon^2 T_n^2(\Omega)}$	Chebyshev
$\dfrac{A}{1+\Omega^{2n}}$	Butterworth
$\dfrac{A}{1+\dfrac{1}{\varepsilon^2 T_n^2(\Omega)}}$	Inverse Chebyshev
$\dfrac{A}{1+\varepsilon^2 C_n^2(\Omega)}$	Generalized Chebyshev
$\dfrac{A}{1+\dfrac{1}{\varepsilon^2 F_n^2(\Omega)}}$	Elliptic Function

The properties of this polynomial are that it must be small in the pass band ($\Omega = 0$ to $\Omega = 1$ for prototype filters), and it must be large outside the pass band. The polynomials used fluctuate between -1 and $+1$ in the pass band, so the quantity ε determines the amount of ripple in the pass band. $T_n(\Omega)$ is a *Chebyshev* polynomial of degree *n*, which can be given by $T_n(\Omega) = \cos[n \arccos(\Omega)]$ for $\Omega \le 1$. The *generalized Chebyshev* is given by:

$$C_n(\Omega) = \cosh\left\{(n-3)\cosh^{-1}\left[\Omega\left(\frac{\Omega_0^2-1}{\Omega_0^2-\Omega^2}\right)^{1/2}\right] + 3\cosh^{-1}(\Omega)\right\}$$

and $F_n(\Omega)$ is related to *elliptic functions.*[10]

The difference between the responses for a Chebyshev and a Butterworth filter is shown in Figure 4–23. This figure displays the responses of two 7-element filters: one is a Butterworth filter and the other is a Chebyshev filter. The Chebyshev filter has a 1 dB ripple throughout the pass band from $\Omega = 0$ to $\Omega = 1$ and has a 1 dB loss at the cut-off frequency of $\Omega = 1$. The Butterworth filter is flat out to about 0.7 rad/sec, and then rolls off to -3 dB at the cut-off frequency of 1 rad/sec. The Chebyshev filter then rolls off

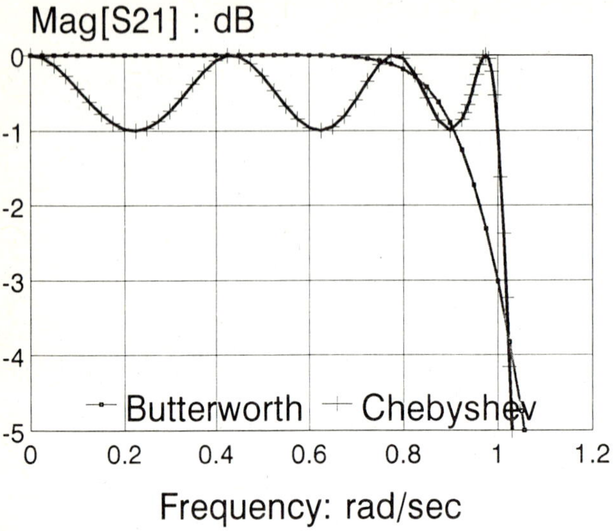

FIGURE 4–23. Butterworth and 1 dB Chebyshev Filter Passband Character.

more rapidly than the Butterworth in the stop band. The Butterworth filter has a *maximally flat* response as defined by the first n derivatives of $|S_{21}(\Omega)|$ with respect to Ω. These must be zero at $\Omega = 0$. The Chebyshev filter is sometimes referred to as an *equiripple* filter, because $|S_{21}(\Omega)|$ has ripples in the pass region that are equal in magnitude.

All the zeros of the Chebyshev and Butterworth filters occur at infinite frequency, causing a continuing roll-off to that frequency. The elliptic function, the generalized Chebyshev, and the inverse Chebyshev filters differ from the Chebyshev and Butterworth filters in that they have transmission zeros on the frequency axis. By having transmission zeros spaced on the frequency axis, these filters are capable of a more rapid roll-off at the band edge. The elliptic function filter has an equi-ripple nature in the pass band as well as in the stop band.

The Chebyshev and the Butterworth filters can be realized by a simple ladder consisting of series inductors and shunt capacitors. The inverse Chebyshev, the generalized Chebyshev, and the elliptic function filters require more complicated ladder networks consisting of series impedances having a parallel combination of an inductor and capacitor and/or shunt branches having series inductors and capacitors as shown in Figure 4–24. These lossless resonant combinations produce open and short circuits at their resonant frequencies. These resonances then produce the zeros on the real frequency axis. The theory behind their design formulas is discussed well by Rhodes.[11] Tables and graphs useful for synthesizing the elliptic function filters are given by Shepherd.[12]

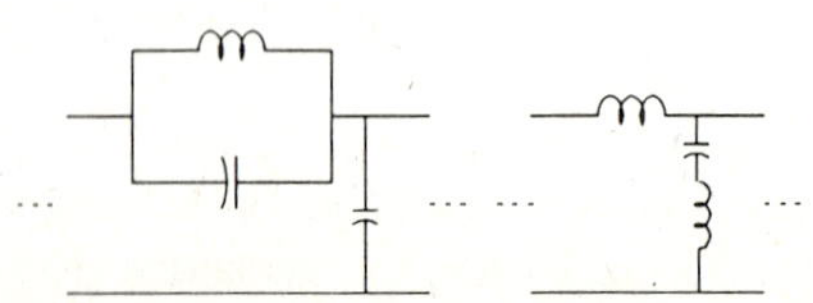

FIGURE 4–24. Elliptic Function Ladder Sections.

4.8 PROTOTYPE FILTER DESIGN

Figure 4–26 shows the attenuation rate for Butterworth filters having two through seven elements, and Figures 4–27 through 4–30 show the attenuation characteristics for Chebyshev filters of two through seven elements for different values of ripple in the passband. Corresponding to these figures are Tables 4–1 through 4–5, which list element values for these filters. The values listed in these tables are for either inductor input or capacitor input filters. The filter configuration is shown in Figure 4–25. The values labeled g_i in these tables are inductances or capacitances. The tables are for doubly terminated filters that require terminating impedances of 1 Ω at both ends of the filter in all cases, except for the case of Chebyshev filters having an even number of elements. In those cases, the terminating impedance at the output end of the filter is listed in the legend of the table.

To determine the prototype filter design, select the filter type and determine the number of elements required to obtain the desired roll-off rate outside the pass band from Figures 4–26 through 4–30. Then, the element values for the prototype filter are shown in Tables 4–1 through 4–5. More extensive tables are available in the book by Matthaei, Young, and Jones.[13]

The design of elliptic function prototype filters is discussed by Shepherd[14] and by Malherbe.[15] This design procedure is similar to that discussed above. The parameters required are the passband ripple, the frequency of the first transmission zero in the stop band, and the minimum attenuation in the stop band at frequencies above the first transmission zero. The principal difficulty with using elliptic function filters in microwave applications is that of implementing the resonant sections in the prototype filter as microwave devices. The configurations for construction of these filters and the design formulas are thoroughly discussed by Malherbe.

One method for realizing the prototype filter in the microwave frequency domain is to use coupled lines. As we discussed previously, the open circuited or short circuited stub can be used to simulate reactive elements. We can carry this further by using open and shorted ends to coupled lines to achieve low pass and band pass structures. To understand the action of these structures, chapter 5 discusses propagation in coupled lines.

(text continues on page 123)

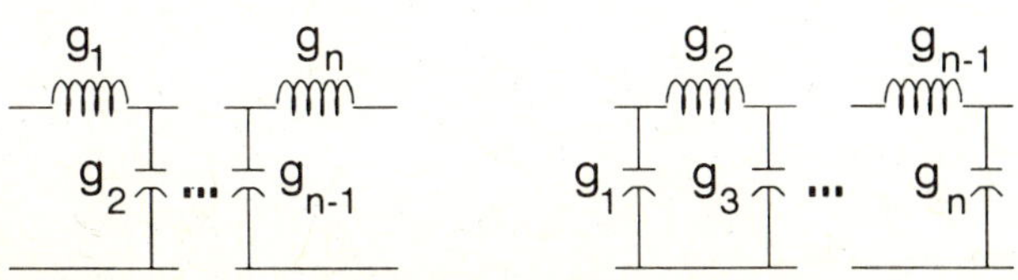

FIGURE 4–25. Inductor and Capacitor Input Filters.

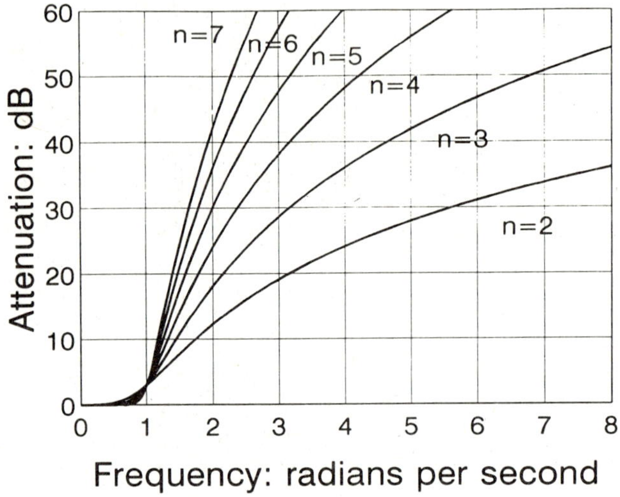

FIGURE 4–26. Butterworth Filter Attenuation Curves for $n = 2, 3, 4, 5, 6$ and 7.

TABLE 4–1 Element Values for Butterworth Filters Having n Elements.

n	g_1	g_2	g_3	g_4	g_5	g_6	g_7
2	1.414	1.414					
3	1	2	1				
4	0.7654	1.848	1.848	0.7654			
5	0.618	1.618	2	1.618	0.618		
6	0.5176	1.414	1.932	1.932	1.414	0.5176	
7	0.445	1.247	1.802	2	1.802	1.247	0.445

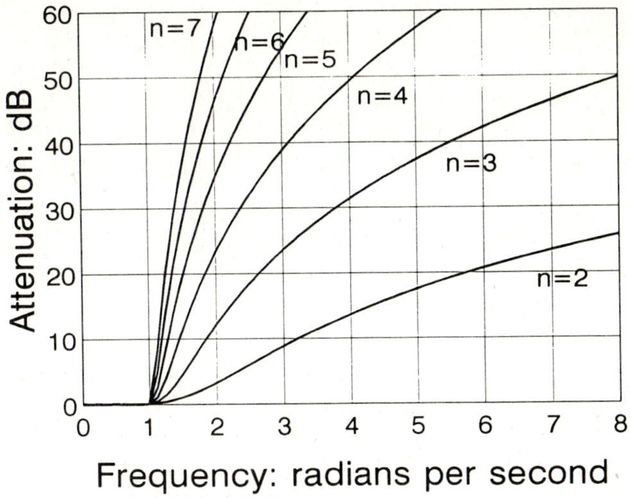

FIGURE 4–27. 0.1 dB Ripple Chebyshev Filter for $n = 2, 3, 4, 5, 6$ and 7.

TABLE 4–2 0.1 dB Ripple Chebyshev Filter Elements. For n even the terminating impedance must be 1.3554 Ω.

n	g_1	g_2	g_3	g_4	g_5	g_6	g_7
2	0.8431	0.6220					
3	1.0316	1.1474	1.0316				
4	1.1088	1.3062	1.7704	0.8181			
5	1.1468	1.3712	1.9750	1.3712	1.1468		
6	1.1681	1.4040	2.0562	1.5171	1.9029	0.8618	
7	1.1812	1.4228	2.0967	1.5734	2.0967	1.4228	1.1812

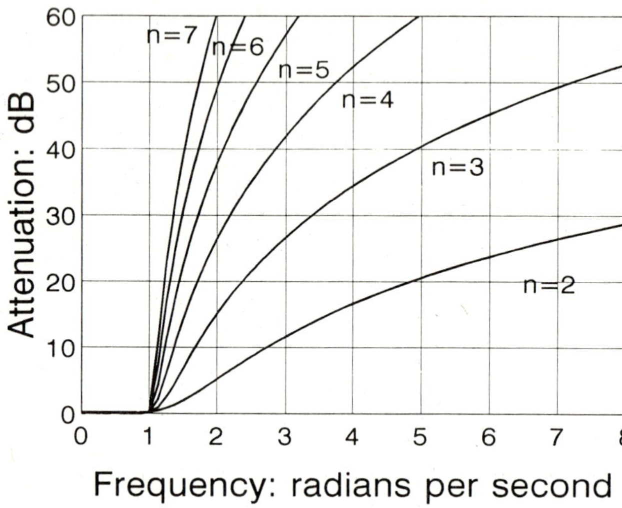

FIGURE 4–28. 0.2 dB Ripple Chebyshev Filter for $n = 2, 3, 4, 5, 6$ and 7.

TABLE 4–3 0.2 dB Ripple Chebyshev Filter Elements. For n even the terminating impedance must be 1.5386 Ω.

n	g_1	g_2	g_3	g_4	g_5	g_6	g_7
2	1.0379	0.6746					
3	1.2276	1.1525	1.2276				
4	1.3029	1.2844	1.9762	0.8468			
5	1.3395	1.3370	2.1660	1.3370	1.3395		
6	1.3598	1.3632	2.2395	1.4556	2.0974	0.8838	
7	1.3723	1.3782	2.2757	1.5001	2.2757	1.3782	1.3723

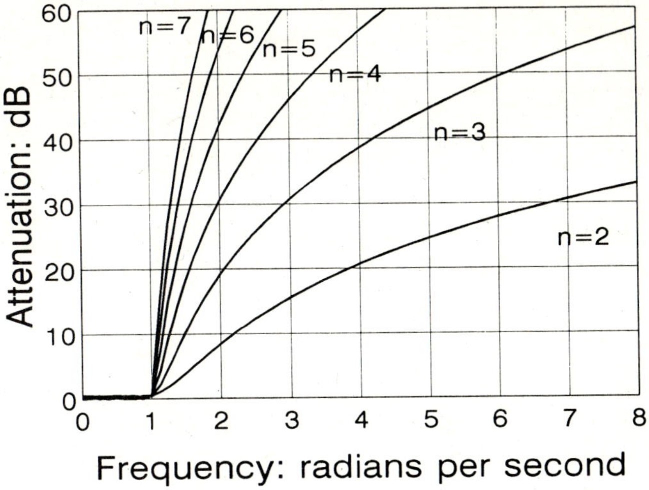

FIGURE 4–29. 0.5 dB Ripple Chebyshev Filter for $n = 2, 3, 4, 5, 6$ and 7.

TABLE 4–4 0.5 dB Ripple Chebyshev Filter Elements. For n even the terminating impedance must be 1.9841 Ω.

n	g_1	g_2	g_3	g_4	g_5	g_6	g_7
2	1.4029	0.7071					
3	1.5963	1.0967	1.5963				
4	1.6704	1.1926	2.3662	0.8419			
5	1.7058	1.2296	2.5409	1.2296	1.7058		
6	1.7254	1.2479	2.6064	1.3137	2.4759	0.8696	
7	1.7373	1.2582	2.6383	1.3443	2.6383	1.2582	1.7373

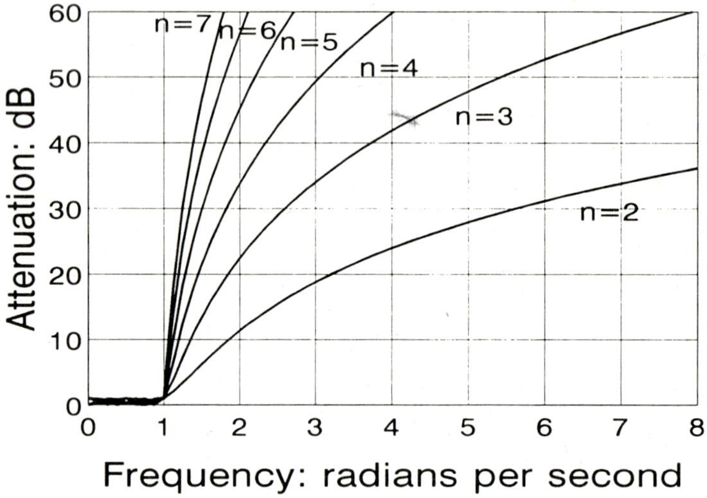

FIGURE 4–30. 1 dB Ripple Chebyshev Filter for n = 2, 3, 4, 5, 6 and 7.

TABLE 4–5 1.0 dB Ripple Chebyshev Filter Elements. For n even the terminating impedance must be 2.6599 Ω.

n	g_1	g_2	g_3	g_4	g_5	g_6	g_7
2	1.8220	0.6850					
3	2.0237	0.9941	2.0237				
4	2.0991	1.0644	2.8312	0.7892			
5	2.1350	1.0911	3.0010	1.0911	2.1350		
6	2.1547	1.1041	3.0634	1.1518	2.9368	0.8101	
7	2.1666	1.1115	3.0934	1.1735	3.0934	1.1115	2.1666

4.9 FREQUENCY TRANSFORMATIONS

Beside the Richards' transformation, there are other useful frequency transformations to change a prototype filter into one at UHF or microwave frequencies. One possible transformation changes a low pass prototype to a high pass prototype. For this, we can use the transformation $\Omega' = 1/\Omega$, and replace capacitors with inductors and inductors with capacitors in the low pass prototype. The element values g_i' must be equal to $1/g_i$. In the case of the even order Chebyshev filters, we do not invert the normalized terminating impedance. The result is a ladder with series capacitors that become short circuits at infinite frequency and shunt inductors that become open circuits at high frequencies. So, this ladder passes high frequencies and has a cut-off frequency of $\Omega_C = 1$.

Example: High Pass Filter with f_C = 1 GHz. We choose the fifth order Chebyshev filter with 0.5 dB ripple as our prototype. The element values are given in Table 4–4. Taking the inverse of these values, we have a prototype shown in Figure 4–31. When the elements are scaled to an impedance of 50 Ω and a 1 GHz cut-off frequency, the transfer characteristic is as shown in Figure 4–32A. The input impedance of the 1 GHz filter is as shown in Figure 4–32B. In practice, these characteristics would be difficult to achieve: Lumped circuit elements acting as pure inductances or pure capacitances cannot be fabricated, because of winding capacitance in the inductors and lead inductance in the capacitors.

Another useful frequency transformation is one that leads to a narrow band pass filter. Let the prototype filter be in the Ω domain and the final filter design be in the ω domain. Then, the *narrow band transformation* is $\Omega = (1/W)(\omega/\omega_0 - \omega_0/\omega)$, where W is the *fractional bandwidth* given by $W = (\omega_2 - \omega_1)/\omega_0$ where ω_1 and ω_2 are the radian frequencies at the band edges, and ω_0 is the center frequency in rad/sec of the filter given by $\omega_0 = \sqrt{\omega_1\omega_2}$. In essence, the relation transforms the frequency range −1 to 1 in the Ω domain into the range in the ω domain from ω_1 to ω_2 (see problem 4–12). A low pass prototype filter with a cut-off at Ω_C = 1 rad/sec is transformed into a filter with a pass band from ω_1 to ω_2.

The narrow band transformation transforms a series element of the prototype into a series combination of an inductor and a capacitor. The reactance of a series element in the prototype is Ωg_1. If we substitute $\omega/W\omega_0 - \omega_0/W\omega$ for Ω, the result is $(g_1/W\omega_0)\omega - (g_1\omega_0/W)/\omega$. This has the same form as the reactance of an inductor and a capacitor in series: $\omega L - 1/\omega C$. We require that this series combination have the same of reactance at $\omega = \omega_c$ as the prototype element has at $\Omega = \Omega_c$. Then, the inductance is given by $g_1/W\omega_0$

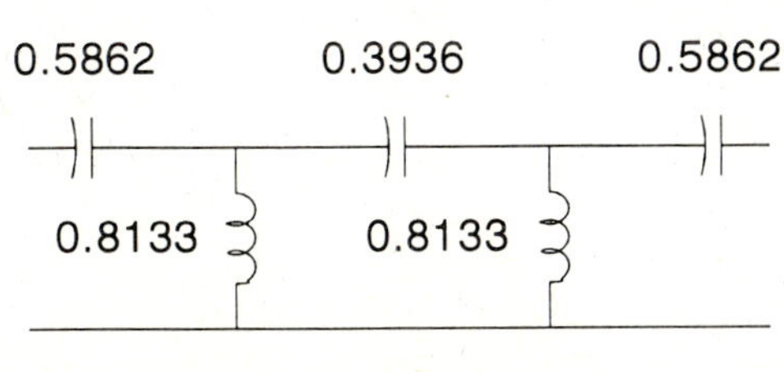

FIGURE 4–31. Prototype High Pass Filter.

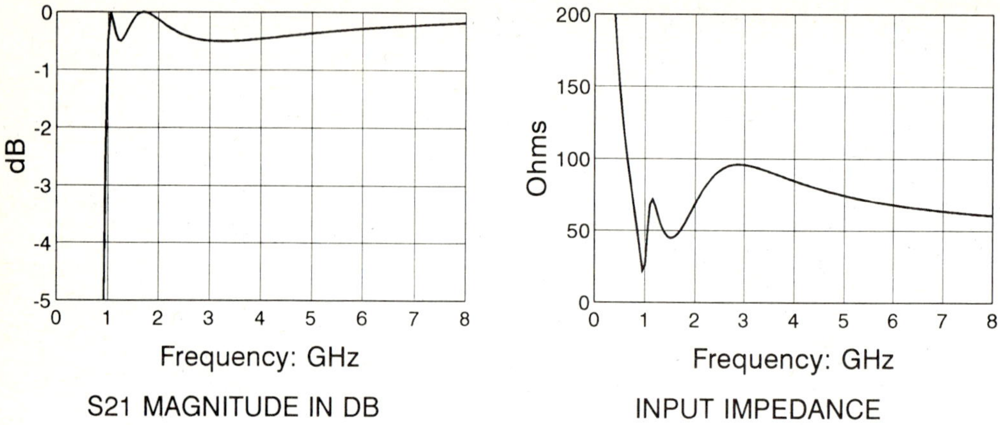

FIGURE 4–32. (A) S_{21} for a High Pass Filter with Cut-off at 1 GHz. (B) Input Impedance for the Filter.

and the capacitance is given by $W/g_1\omega_0$. These must be scaled by impedance as well, so that $L = Z_0 g_1/W\omega_0$ and $C = W/(Z_0 g_1 \omega_0)$. Similarly, the shunt elements in the prototype result in a parallel combination of L and C.

Example: Narrow Band Filter. Consider the prototype Butterworth filter with five elements discussed at the beginning of this chapter. The prototype is shown in Figure 4–33A. We desire a band pass filter with a center frequency of 1 GHz, a pass band of 20 MHz, and a termination of 50 Ω. For this, we have $\omega_1 = 2\pi(10^9 - 10^7)$, $\omega_2 = 2\pi = 2\pi(10^9 + 10^7)$, $\omega_0 = 2\pi 10^9 \sqrt{1 - 10^{-4}} \simeq 2\pi 10^9$, and $W \simeq 0.02$. The elements L_1 and C_1 are given by $L_1 = 50 \times 0.618/(2 \times 10^{-2} 2\pi 10^9) = 246 \times 10^{-9}$ and $C_1 = 0.02/(50 \times 0.618 \times 2\pi 10^9) = 1.03 \times 10^{-13}$. Similar calculations result in the circuit in Figure 4–33B. The simulated transfer characteristic and input impedance are shown in Figure 4–34. Some of these circuit elements might not be available as lumped elements and might have to be simulated by other means.

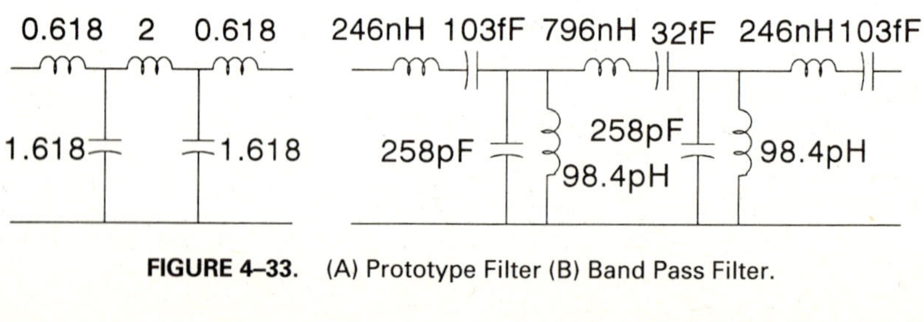

FIGURE 4–33. (A) Prototype Filter (B) Band Pass Filter.

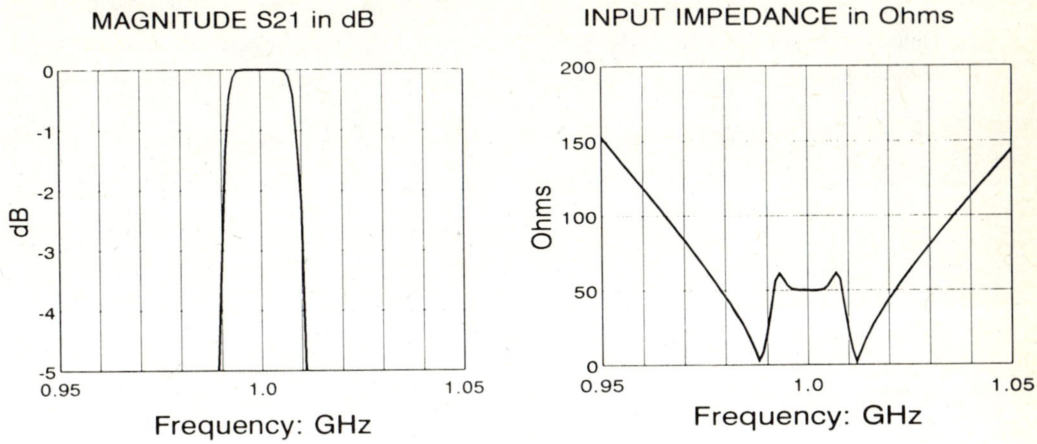

FIGURE 4–34. Narrow Band Pass Filter Transfer Characteristic and Input Impedance.

4.10 WIDE BAND FILTER

A relatively wide band filter that is sometimes acceptable can be constructed from a high pass commensurate line filter using the basic impedance inversion property of a quarter wave line rather than following through the design using Kuroda's identities. The design method is adapted from that presented by Malherbe.[16] The center frequency of the filter is f_0 of the Richards' frequency transformation. The filter will have a transfer characteristic that is periodic in the frequency domain, with a period of $2f_0$. Because of this periodicity, the frequency at which $|S_{21}|$ recovers gets closer to the upper cut-off frequency for filters with a wider bandwidth if the center frequency f_0 is fixed. This can be a distinct disadvantage.

The design of the prototype filter requires an admittance inverter shown as a two port in Figure 4–35 and has the admittance matrix given in Eq. (4–8). Matthaei, Young, and Jones give several circuits, besides quarter wave lines, that will also accomplish impedance or admittance inversion.[17] To understand Eq. (4–8), we consider the impedance inversion of a quarter wave line treated as a 2 port. The quarter wave line has the property that the load impedance and the input impedance are related as $Z_L = Z_0^2/Z_{in}$. We use

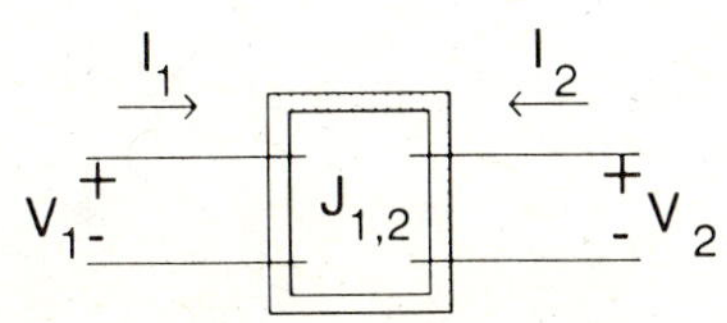

FIGURE 4–35. Ideal Admittance Inverter.

$J_{1,2} = 1/Z_0$ as the admittance inversion parameter from port 2 to port 1 so that the admittance inversion is expressed as $Y_{in} = J_{1,2}^2/Y_0$. At the frequency where a line is exactly one quarter wave length long, the input voltage V_1 and the output (port 2 input) current I_2 are related by $V_1 = -jI_2Z_0$ (see Eqs. (4–5) and (4–6)). Similarly, the port 2 (output) voltage and the port 1 (input) current are related as $V_2 = -jI_1Z_0$. These relations are shown in Eq. (4–8).

$$(Y) = \begin{pmatrix} 0 & jJ_{1,2} \\ jJ_{1,2} & 0 \end{pmatrix} \tag{4–8}$$

We shall assume at first in the following discussion that the relations between the input admittance and output admittance of ideal admittance inverters hold for all frequencies of interest.

The process of explaining the design equations is best shown in Figures 4–36 and 4–37. In these figures, we show the conversion of the high pass prototype filter to a prototype with ideal impedance inverters and internal elements all having the same inductance equal to 1/h. The next step is to equate the admittance inverter prototype to one having unit elements. This prototype can then be directly related to the final filter of Figure 4–37.

The filter to be designed has a lower cut-off frequency f_C and a center frequency f_0. The prototype filter of Figure 4–36 is to be high pass and will be simulated as a commensurate line filter. To obtain high pass filter prototype elements, we must take the inverse of the low pass prototype elements g_i. Because we are using a commensurate line filter, the cut-off frequency of the high pass prototype filter must be $\Omega_C = \tan(\pi f_C/2f_0)$ rather than 1. To accomplish that, the high pass prototype elements must be scaled in frequency by dividing them by Ω_C. So, the y_i in Figure 4–36 are $y_i = \Omega_C g_i$.

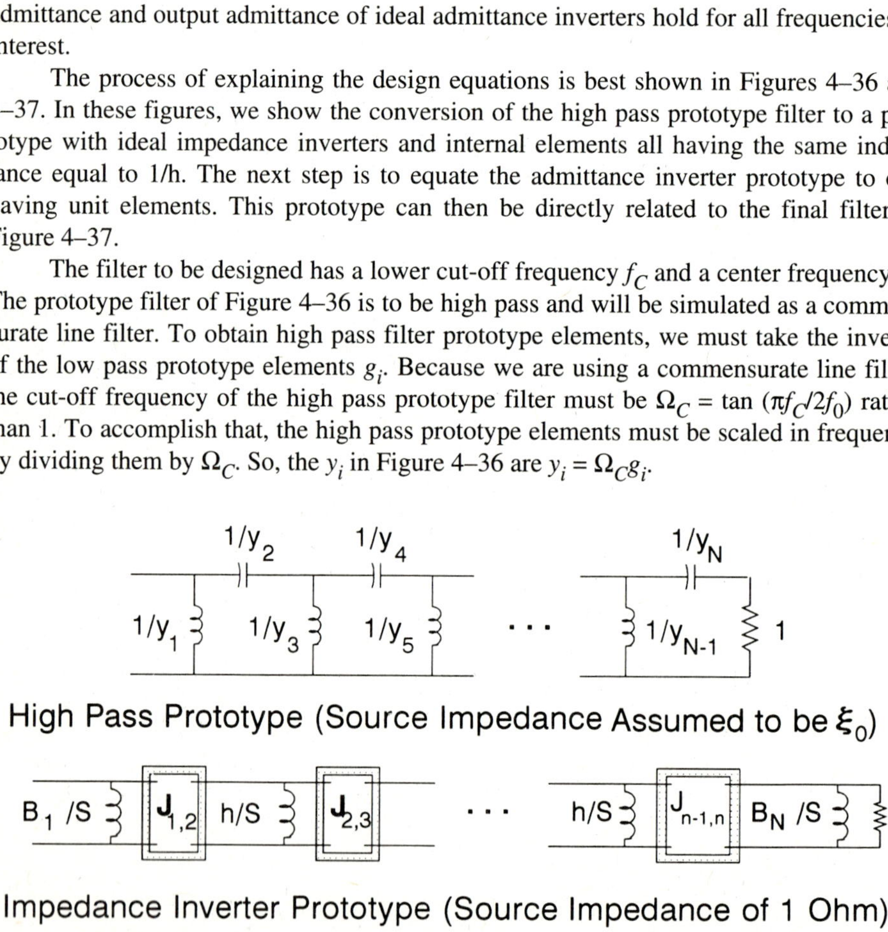

FIGURE 4–36. Equivalent Prototype Filters. Note that $1/g_i$ are values of inductance and capacitance in the high pass prototype, but that admittances are given for the elements in the impedance inverter prototype. In this figure $1/h$ is inductance.

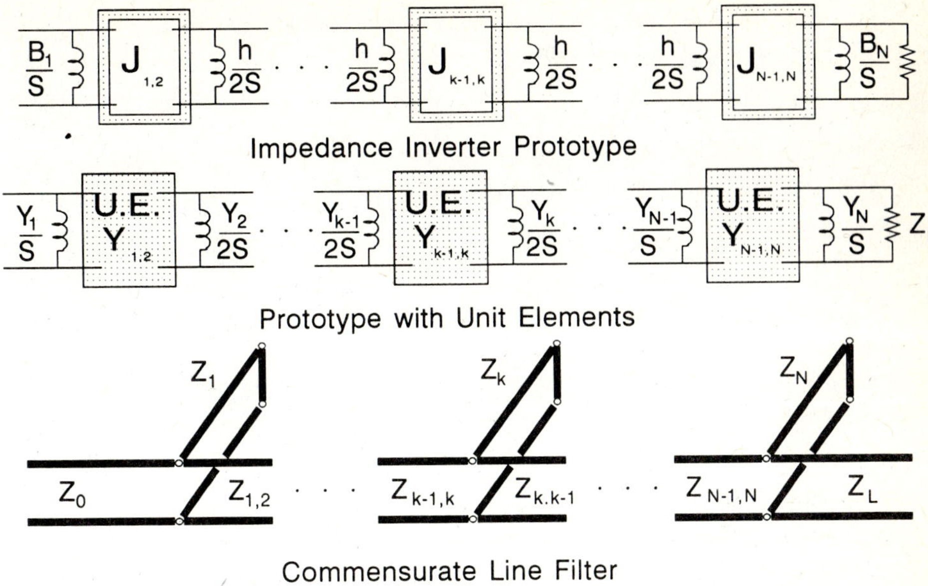

FIGURE 4–37. Filter Design Stages. Note that the Y_i and the $Y_{i-1,i}$ are the inverses of the characteristic line impedances Z_i and $Z_{i-1,i}$.

To derive the design formulas for the filter, the first step is to equate the input impedances of the terminating sections of the two prototype filters in Figure 4–36. In the impedance inverter prototype, the admittance, consisting of the inductive susceptance in parallel with a 1 S conductance, is inverted by the admittance inverter. This resulting impedance must be equivalent to the impedance of $1/y_N + 1$ of the high pass prototype. The ratio of inductive susceptance to terminating conductance in the impedance inverter prototype must be the same as the capacitive reactance to the terminating resistance in the high pass prototype if they are to have the same characteristic. From this, we have the relation $(y_N/\Omega)/1 = (B_N/\Omega)/1$, which reduces to Eq. (4–9).

$$B_N = y_N = \Omega_C g_N \tag{4–9}$$

Now, we write the input admittances to the terminating sections of the two filters in Figure 4–36. They are

$$\frac{y_{N-1}}{j\Omega} + \frac{1}{1 + y_N/j\Omega} \quad \text{and} \quad \frac{h}{j\Omega} + \frac{J_{N-1,N}^2}{1 + B_N/j\Omega}.$$

These two admittances must be equivalent in the same way as the terminating impedances are. The ratio $(\Omega/y_{N-1})/1$ of the inductive reactance term to the load resistance must be equal to the ratio of the corresponding factors in the second expression $(\Omega/h)(J^2_{N-1,N}/1)$. This results in Eq. (4–10).

$$J^2_{N-1,N} = \frac{h}{\Omega_C g_{N-1}} \tag{4–10}$$

We can equate the input admittances for an internal section of both filters as shown in Figure 4–38. The resulting expressions are

$$\frac{y_{k-1}}{j\Omega} + \frac{1}{(y_k/j\Omega) + 1/Y_{k+1}} \quad \text{and} \quad \frac{h}{j\Omega} + \frac{J^2_{k-1,k}}{(h/j\Omega) + 1/Y'_{k+1}},$$

where Y_{k+1} and Y'_{k+1} are the input admittances of the succeeding sections. In the first expression, the ratio of the impedance $y_k/j\Omega$ to the impedance $1/Y_{k+1}$ must be the same as the ratio of the corresponding factors in the second expression $hY'_{k+1}/j\Omega$. So, we have $Y_{k+1}/Y'_{k+1} = h/y_k$. Similarly, we have the expression $j\Omega Y_{k+1}/y_{k-1} = j\Omega Y'_{k+1}J^2_{k-1,k}/h$. From these two expressions, we get Eq. (4–11).

$$J^2_{k-1,k} = \frac{h^2}{g_k\, g_{k-1}\Omega^2_C} \tag{4–11}$$

Finally, we follow similar arguments to obtain Eq. (4–12) for B_1 and $J_{1,2}$ from consideration of the input section. In this relation, the source impedance driving the prototype filter is unity.

$$B_1 = \Omega_C g_1, \qquad J^2_{1,2} = \frac{h}{\Omega_C g_2} \tag{4–12}$$

If Eqs. (4–9) through (4–12) are satisfied in practice, that is, if one could actually build an ideal impedance inverter for use in the prototype, the electrical performance of the two filters shown in Figure 4–36 would be identical.

The impedance inverter filter has been redrawn in Figure 4–37A. In this figure, the internal shunt inductances have been split into two parallel inductors that combine to produce the same admittance as the internal inductors shown in Figure 4–36 for this filter. The Y_k and the $Y_{k-1,k}$ shown in Figure 4–37B are the inverse of the characteristic line impedances for the commensurate line filter in Figure 4–37C. The unit elements in

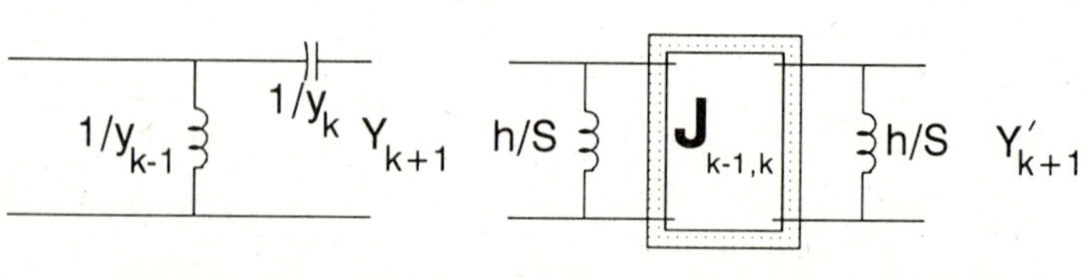

FIGURE 4–38. Internal Sections of the Prototype Filters.

Figure 4–37B correspond to quarter wave sections in the commensurate line filter (Figure 4–37C).

We require that the filter shown in Figure 4–37B be equivalent to the filter in Figure 4–37A at only one frequency Ω_c. At this frequency, the reactances in the kth section of Figure 4–37B must be equal to the kth section reactances in Figure 4–37A. Because the unit element has the same electrical nature as the ideal impedance inverter at only one frequency, these two filters will not have the same characteristic over the entire frequency range.

First, we examine the terminating sections of the two prototype filters shown in Figure 4–37. The admittance matrices for these two filter sections are

$$\begin{pmatrix} \dfrac{h}{2S}, & J_{N-1,N} \\ jJ_{N-1,N}, & \dfrac{B_N}{S}+1 \end{pmatrix} \quad \text{and} \quad \begin{pmatrix} \dfrac{Y_{N-1}}{2S}+\dfrac{Y_{N-1,N}}{S}, & -\dfrac{\sqrt{1-S^2}\,Y_{N-1,N}}{S} \\ -\dfrac{\sqrt{1-S^2}\,Y_{N-1,N}}{S}, & \dfrac{Y_N+Y_{N-1,N}}{S}+\dfrac{1}{Z_0} \end{pmatrix}.$$

We see from the 2, 2 terms that all the admittances in the left-hand matrix should be scaled to $1/Z_0$ if they are to be compared to admittances in the right-hand matrix. With this scaling, we compare the resulting 1, 2 terms directly and have the formula $jJ_{N-1,N}/Z_0 = -Y_{N-1,N}(\sqrt{1-S^2})/S$. We substitute from the Eq. (4–9) into this relation, and substitute $j\Omega_C$ for S to get a formula for $Z_{N-1,N}$ in terms of the original prototype element g_{N-1}. This is shown in Eq. (4–13).

Comparing the internal sections of the prototype filters shown in Figure 4–37, we arrive at the two admittance matrices

$$\begin{pmatrix} \dfrac{h}{2S}, & jJ_{k-1,k} \\ jJ_{k-1,k}, & \dfrac{h}{2S} \end{pmatrix} \quad \text{and} \quad \begin{pmatrix} \dfrac{Y_{k-1}}{2S}+\dfrac{Y_{k-1,k}}{S}, & -\dfrac{Y_{k-1,k}\sqrt{1-S^2}}{S} \\ \dfrac{Y_{k-1,k}\sqrt{1-S^2}}{S}, & \dfrac{Y_k}{2S}+\dfrac{Y_{k-1,k}}{S} \end{pmatrix}.$$

Normalizing the terms in the left matrix to Z_0, and equating the 2, 2 terms of the two matrices, we determine at the formula $h/2SZ_0 = Y_k / 2S + Y_{k-1,k} /S$. Now, we consider the succeeding sections of each of these filters and their corresponding admittance matrices by replacing the index k in the above matrices with the index $k + 1$. We equate the 1, 1 terms of the resulting matrices in the same way as above. We get the formula $h/2SZ_0 = Y_k/2S + Y_{k,k+1}/S$. When we add these two formulas, we have the relation $Y_k = h/Z_0 - Y_{k-1,k} - Y_{k,k+1}$. The inverse of this is the formula for Z_k given in Eq. (4–13).

Continuing this process, from the 1, 2 term of the admittance matrices, we get

$$Y_{k-1,k} = \frac{J_{k-1,k}}{Z_0}\frac{\Omega_C}{\sqrt{1+\Omega_C^2}}.$$

We substitute from Eq. (4–11) into this relation to express it in terms of the g_i parameters. The resulting formula is shown in Eq. (4–13). When we consider the input and terminating sections as well, we get the remaining design equations in Eq. (4–13).

$$\left.\begin{aligned}
Z_{1,2} &= \frac{Z_0}{h\sin\theta_C}\sqrt{hg_2\Omega_C}, \quad Z_1 = \frac{Z_0}{\Omega_C g_1 - \sqrt{\dfrac{h}{g_2\Omega_C}}\sin\theta_C} \\
Z_{k-1,k} &= \frac{Z_0\Omega_C\sqrt{g_k g_{k-1}}}{h\sin\theta_C}, \quad Z_k = \frac{1}{\dfrac{h}{Z_0} - \dfrac{1}{Z_{k-1,k}} - \dfrac{1}{Z_{k,k+1}}} \\
Z_{N-1,N} &= \frac{Z_0\sqrt{g_{N-1}\Omega_C}}{\sqrt{h}\sin\theta_C}, \quad Z_N = \frac{Z_0}{\Omega_C g_N - \dfrac{\sqrt{h}\sin\theta_C}{\sqrt{\Omega_C g_{N-1}}}}
\end{aligned}\right\} \qquad (4\text{–}13)$$

In Eq. (4–13), the impedances $Z_{i,j}$ are characteristic impedances for the lines that interconnect the shorted stubs in Figure 4–39. The impedances Z_k are the characteristic impedance of the shorted stub lines. The inductance $1/h$ is arbitrary. Choosing different values for h will result in different line impedances in the final design. Thus, one should select the value of h that results in realistic line impedances. The angle θ_C is $\pi f_C/2f_0$.

Example: Wide Band Filter. Consider the wide band filter specifications of a cen-

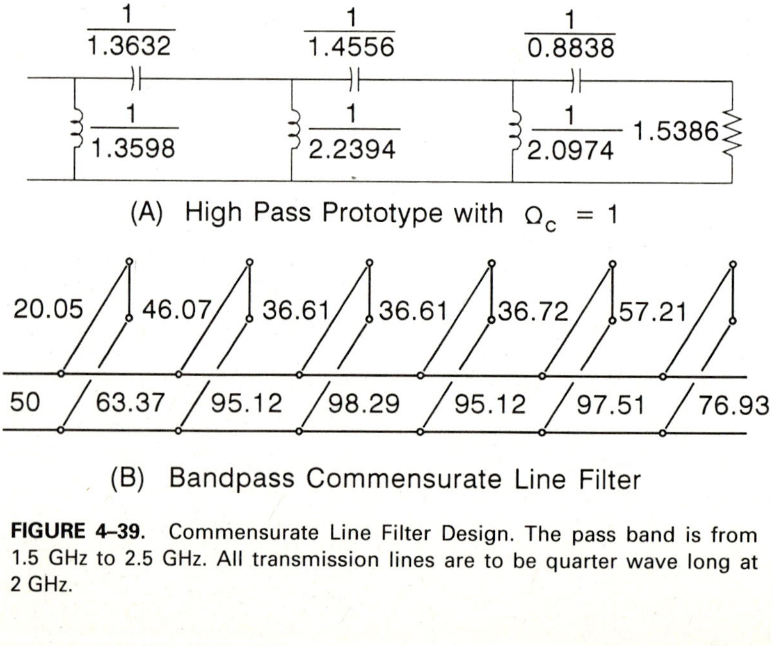

FIGURE 4–39. Commensurate Line Filter Design. The pass band is from 1.5 GHz to 2.5 GHz. All transmission lines are to be quarter wave long at 2 GHz.

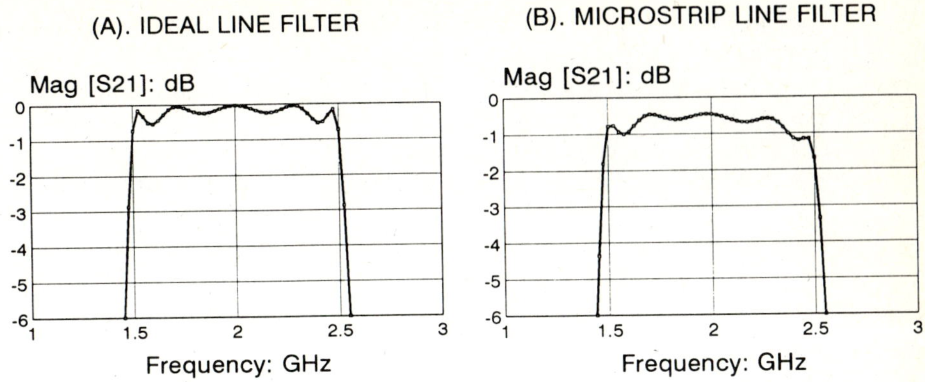

FIGURE 4–40. Transmission Parameter, S_{21}, for a Commensurate Line Filter. (A) Ideal Transmission Lines. (B) Microstrip Lines on a 25 mil Alumina substrate with $\varepsilon_r = 9.8$.

ter frequency at 2 GHz and a pass band from 1.5 GHz to 2.5 GHz. If we start the design by choosing a "folding frequency" f_0 of 2 GHz and use that in Richards' frequency transformation along with a cut-off frequency of 1.5 GHz, then the pass band will be from 1.5 GHz to 2.5 GHz. The reason for this is that the frequency transformation using $\Omega = \tan(\pi f/2f_0)$ maps the region from $f = f_0$ to $f = 2f_0$ into the region of $\Omega = -\infty$ to $\Omega = 0$. Consider a 6-element 0.2 dB ripple Chebyshev low pass prototype filter having the elements given in Table 4–3. Inverting the prototype element values and choosing an inductor input, we have the high pass prototype shown in Figure 4–39A. By using the low pass prototype filter elements in Eq. (4–13), we compute the characteristic impedance of the lines in the filter. This results in the filter shown in Figure 4–39B. The parameter h was chosen to be 2.4. The parameter Ω_c is $\tan(\pi f_c/2f_0)$. The S_{21} for the ideal transmission line case was determined from a computer simulation, and is shown in Figure 4–40. Note that the ripple does not stay constant within the pass band and it does not exceed approximately 0.5 dB. The same filter design was simulated on a computer using a microstrip design on 0.635 mm alumina having a relative dielectric constant of 9.8. The S_{21} for that is shown in Figure 4–40. This filter now shows an asymmetrical pass band staying within about 1.2 dB. This characteristic is still quite reasonable for this wide a pass band.

4.11 PROTOTYPE FILTER SYNTHESIS

The topic of network or filter synthesis has been treated extensively in numerous books. A complete account of it is not possible here in the space allotted. Consequently, we discuss L-C low pass and high pass networks and applications of Richards' theorem to realizing nonredundant commensurate line filters. Further material on this subject is presented in chapter 7 under the classification of *equalizers*. An equalizer is a matching network whose normal function is to produce a constant overall gain (and possibly a linear phase variation with frequency) over some frequency band for a network that usually contains an active device such as a transistor.

To characterize the different filter types, we use a plot of the poles and zeros of the transfer function. The Butterworth low pass filter with a cut-off frequency of 1 rad/sec. has poles spaced at equal angles in the left half plane on a unit circle.[18] It can be shown that this leads to the function $1/(1 + \Omega^n)$. The Chebyshev filter has poles placed on an ellipse whose center is at the origin in the S plane. Both of these have no zeros on the finite real frequency axis exclusive of $\omega = 0$.

A rational function is called a *positive real* (pr) function if it (1) is real for real S, (2) has no poles in the right half of the S plane, (3) has a nonnegative real part and only simple poles (with real residues) on the $j\Omega$ axis, and (4) has only complex conjugate pairs of poles off the real axis in the left half of the S plane. In general, any positive real function of S can be realized as the input impedance (or admittance) of a passive network.[19] It has been shown that the conditions listed above for positive real functions are also equivalent to the conditions $\mathcal{Re}\,[Z(S)] \geq 0$ for $\mathcal{Re}\,(S) \geq 0$ and $Z(S)$ real for real S.

In this chapter, we discuss synthesis techniques for networks whose input impedance has no poles or zeros on the finite real frequency axis, exclusive of $\omega = 0$, and whose real part is zero for at least one real frequency. Such functions are called *minimum functions.* These networks can be realized as ladder networks.

The method for synthesizing the prototype filter depends upon the termination at both ends of the filter. If the filter is driven by a zero impedance source and is terminated with the characteristic impedance, then it is referred to as a *singly terminated* filter. We use Figure 4–41 as the basis for discussing the method for synthesizing a low pass, L-C ladder network when given the location of the poles of the transfer function. The 2 port labeled L-C is to be a ladder network consisting of series inductors and shunt capacitors. This 2 port has the voltage-current relation

$$\begin{pmatrix} V_1 \\ V_2 \end{pmatrix} = \begin{pmatrix} Z_{11} & -Z_{12} \\ Z_{21} & -Z_{22} \end{pmatrix} \begin{pmatrix} I_1 \\ I_2 \end{pmatrix}$$

and the Thevenin's equivalent circuit shown. The equivalent generator voltage is given by $E_{TH} = V_1 Z_{21}/Z_{11}$. From the Thevenin's equivalent circuit, we get the output voltage of Eq. (4–14).

$$V_2 = I_2 Z_0 = I_2 1 = \frac{V_1 Z_{21}/Z_{11}}{1 + Z_{TH}}, \qquad H(s) = \frac{Z_{21}/Z_{11}}{1 + Z_{TH}} \tag{4–14}$$

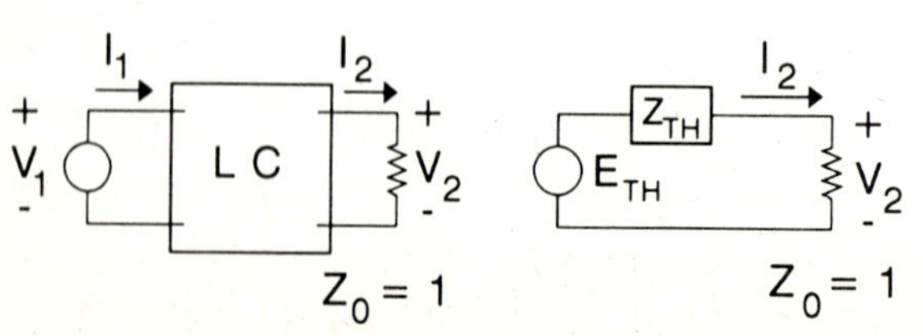

FIGURE 4–41. A 2 port and Its Thevenin Equivalent Circuit.

The impedance Z_{TH} is determined by calculating the impedance looking back into the 2 port with the voltage generator short circuited. Because the 2 port is a ladder, we can express it in the form of a *continued fraction* as in Eq. (4–15). As an example, the L-C ladder with an inductor input and five elements in Figure 4–42 has the Thevenin's impedance given in Eq. (4–15),

$$Z_{TH} = SL_5 + \cfrac{1}{SC_4 + \cfrac{1}{SL_3 + \cfrac{1}{SC_2 + \cfrac{1}{SL_1}}}} \tag{4–15}$$

which can be rewritten as

$$Z_{TH} = \frac{S^5 L_1 C_2 L_3 C_4 L_5 + S^3 (L_1 C_2 L_3 + L_1 C_2 L_5) + SL_1}{S^4 L_1 C_2 L_3 C_4 + S^2 L_1 L_2 + 1}.$$

In general, the Thevenin's impedance of an L-C ladder network must be the ratio of an even polynomial in S to an odd polynomial in S or vice versa.

If we specify the pole locations of $H(S)$, and if we specify that $H(S)$ is for a low pass filter with the zeros of $H(S)$ all at $S = j\infty$, then we can write $H(S)$ as in Eq. (4–16).

$$H(S) = \frac{K}{(S - S_1)(S - S_2)\ldots(S - S_n)} = \frac{K}{S^n + a_{n-1}S^{n-1} + \ldots + a_0} \tag{4–16}$$

The denominator of $H(S)$ is a polynomial with both even and odd parts. If we divide both the numerator and the denominator by the even part of this polynomial, then $H(S)$ has the form

$$H(S) = \frac{K/(\text{even polynomial in } S)}{1 + (\text{odd polynomial in } S)/(\text{even polynomial in } S)} \tag{4–17}$$

Comparing Eq. (4–14) and (4–17), we see that the Thevenin's impedance can be identified with the ratio of polynomials in the denominator of Eq. (4–17). By expanding this ratio in the continued fraction form of Eq. (4–15), we can determine values for the individual elements for the ladder from the coefficients in the polynomial of S shown in Eq. (4–16).

A similar argument starting with a current source driving a L-C network terminated

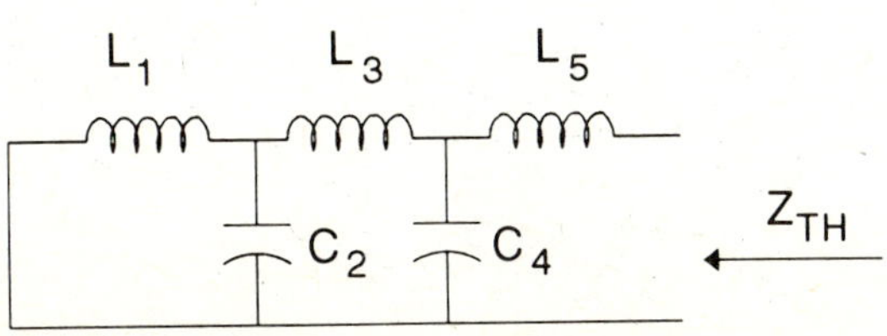

FIGURE 4–42. L-C Ladder Network with a Shorted Input.

in a 1 Ω resistor can be used to find the relations for determining the elements in a capacitor input L-C ladder network in terms of the pole locations.

In the case where the filter is terminated with an impedance Z_L and is driven by a source with an internal impedance of Z_g, the filter is said to be *doubly terminated.* The process for determining the filter network in that case starts with a specification of the forward scattering parameter, S_{21} as a ratio of polynomials in complex frequency S. If the filter is a lossless network, the reflection coefficient S_{11} is determined by Eq. (3–8): $|S_{11}|^2 = 1 - |S_{21}|^2$. The process of forming $|S_{11}|^2$ leads to a function that has poles in the right half of the S plane. This function must be factored and S_{11} must be formed using only the poles in the left half of the S plane. This usually is the most difficult step in this process. If the filter is terminated with a matched impedance, then the reflection coefficient into the filter is S_{11} and the input impedance to the filter is given by the Eq. (4–18).[20]

$$Z_{in} = Z_0 \frac{1+S_{11}}{1-S_{11}} \tag{4–18}$$

For those transfer functions with no zeros on the finite real frequency axis, exclusive of $\Omega = 0$, the ladder network of the filter can be found from Eq. (4–18) by using long division to arrive at the continued fraction form shown in Eq. (4–15). For other classes of S_{21} a partial pole extraction technique may be necessary.[21]

4.12 NONREDUNDANT FILTER SYNTHESIS

When synthesizing commensurate line filters for microwaves, unit elements are used as part of the prototype. The unit elements themselves will increase the number of poles of S_{21}, so they can be used to replace some of the inductors or capacitors of the prototype. In the extreme case, an nth order prototype can be designed with no more than a total number n of elements (including the unit elements in this count). This filter is then called a *nonredundant filter.* To effect the design of such a filter, we must use *Richards' theorem.*[22] Richards' theorem is put in the terms of pr functions. Using the ABCD parameters given in Eq. (4–7) for a unit element with the circuit shown in Figure 4–43, we get the Eq. (4–19) for the input impedance to this circuit.

$$Z_{in}(S) = Z_0 \frac{Z_L(S) + SZ_0}{Z_0 + SZ_L(S)} \tag{4–19}$$

From Eq. (4–19), we see that $Z_0 = Z_{in}(1)$. Now, we invert Eq. (4–19) and solve for Z_L as a function of Z_{in} to obtain Eq. (4–20).

$$Z_L(S) = Z_0 \frac{Z_{in}(S) - S\, Z_0}{Z_0 - SZ_{in}(S)} \tag{4–20}$$

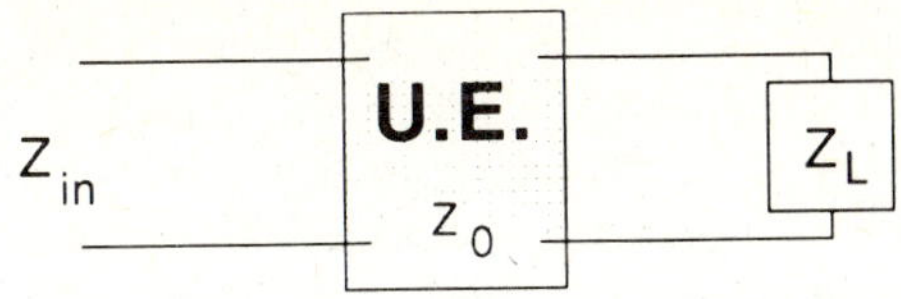

FIGURE 4–43. Richards' Theorem Network.

Richards' theorem states that, if Z_{in} (S) is a rational pr function not equivalent to S or $1/S$, then Z_L (S) is also a rational pr function. Now, consider how this might be used in the process of synthesizing a network. Let Z_{in} (S) be a network function to be realized. The network function Z_{in} (S) is for the impedance Z_{in} of Figure 4–43 and the network function Z_L (S) is for Z_L in Figure 4–43. This Z_L (S) is a new impedance function that will be a ratio of polynomials having an extra factor of $S - 1$ in both the numerator and denominator that must be divided out. Sometimes an additional factor of $S + 1$ occurs in both the numerator and denominator, which also must be divided out. After these factors are removed, Z_L (S) will be of a ratio of polynomials in S that are of lower order than Z_{in} (S). The implication of this theorem is that we can insert a U.E. at any point in synthesizing a circuit and be sure that the impedance on the other side of the U.E. is a physically realizable impedance of lower order.

A method for synthesizing a nonredundant filter was explained by Horton and Wenzel.[23]

$$|S_{21}|^2 = 1/\left(1 + |r_{21}|^2\right) \tag{4–21}$$

They showed that S_{21} satisfies Eq. (4–21), in which r_{21} can be selected as in Eq. (4–22) when working with Butterworth or with Chebyshev low pass prototype filters. In Eq. (4–22), $S_c = j\Omega_c$ is the cut-off frequency, m is the number of commensurate line stubs, and n is the number of unit elements.

$$|r_{21}|^2 = \begin{cases} \varepsilon^2\left[T_m\left(\dfrac{S}{S_c}\right)T_n\left(\dfrac{S\sqrt{1-S_c^2}}{S_c\sqrt{1-S^2}}\right) - U_m\left(\dfrac{S}{S_c}\right)U_n\left(\dfrac{S\sqrt{1-S_c^2}}{S_c\sqrt{1-S^2}}\right)\right]^2 & \text{Chebyshev} \\[2ex] \left(\dfrac{S}{S_c}\right)^{2m}\left(\dfrac{S\sqrt{1-S_c^2}}{S_c\sqrt{1-S^2}}\right)^{2n} & \text{Butterworth} \end{cases} \tag{4–22}$$

The function T_m (x) is cos (m arc cos x) and U_m (x) is sin(m arc cos x). These are mth degree Chebyshev polynomials of the first and second kind.[24] After determining $|r_{21}|^2$, one finds S_{21} by using Eq. (4–21). As in section 4–11, one finds S_{11} using Eq. (3–8): $|S_{11}|^2 = 1 - |S_{21}|^2$. We can determine S_{11} from $|S_{11}|^2$ by finding the poles of $|S_{11}|^2$ and using only those that are not in the right half of the S plane. Finally, one determines the input impedance Z_{in} using Eq. (4–18). One can determine a passive network from Z_{in} by using long division to determine series or parallel elements coupled with insertion of unit elements as explained above. This process is shown in the following example.

Example: Nonredundant Fifth Order Butterworth Filter Design. We use the same specifications for this example as in the design of the fifth order Butterworth low pass filter discussed earlier in this chapter. The cut-off frequency is to be 2 GHz and the folding frequency is to be 3 GHz. Thus, $\tan \theta_c = \Omega_c = \sqrt{3}$. In order to determine a ladder network, the prototype will require lumped capacitors or inductors interspersed with unit elements. A fifth order filter will require three lumped elements and two unit elements. So, we have $m = 3$ and $n = 2$ in Eq. (4–22). Substituting the r_{21} from this equation into Eq. (4–21), we find the prototype forward scattering parameter is given by

$$|S_{21}|^2 = \frac{1}{1+\left(\frac{S}{j\sqrt{3}}\right)^6\left(\frac{S}{\sqrt{1-S^2}}\frac{2}{j\sqrt{3}}\right)^4}.$$

From $|S_{11}|^2 = 1 - |S_{21}|^2$, we obtain the expression for

$$|S_{11}|^2 = \frac{S^{10}}{S^{10} - 15.1875\,S^4 + 30.375\,S^2 - 15.1875}.$$

This expression has twice as many poles as is required and it has half of them in the right half plane. Factoring the denominator and using only those poles in the left half plane along with half of the zeros, we determine S_{11} to be

$$S_{11} = \frac{S^5}{S^5 + 4.776\,S^4 + 11.404\,S^3 + 16.239\,S^2 + 12.528\,S + 3.897}.$$

Using this expression in Eq. (4–18), we get the expression for the input impedance to the filter.

$$Z_{in} = \frac{4.776\,S^4 + 11.404\,S^3 + 16.238\,S^2 + 12.528\,S + 3.897}{2\,S^5 + 4.776\,S^4 + 7.235\,S^3 + 6.284\,S^2 + 2.660\,S + 0.3678}$$

Because the denominator is of higher order than the numerator, we isolate a term corresponding to a shunt capacitor by inverting this expression and performing one step in long division of the two polynomials. The result is $Y_{in} = 0.4188S + 1/Z_2$, where

$$Z_2 = \frac{4.776S^4 + 11.404S^3 + 16.238S^2 + 12.528S + 3.897}{4.603S^3 + 10.992S^2 + 10.896S + 3.897}.$$

In this expression, we identify the term 0.4188S as the admittance of a shunt capacitor of 0.4188 F. This is C_1 in Figure 4–44.

The next step is to insert a unit element by using Richards' theorem. We let Z_2 in the Y_{in} expression be Z_{in} in Eq. (4–19) of Richards' theorem. The unit element has an impedance of $Z_0 = Z_{in}(1) = 1.607$. When we scale impedance by 50 Ω, this unit element impedance becomes a quarter wave line with $Z_0 = 80.36$ Ω as seen in Figure 4–44. Then, we let Z_L in Richards' theorem be called Z_3. From Eq. (4–20) we get an expression for Z_3:

$$Z_3 = \frac{-2.623S^4 - 6.264S^3 - 1.274S^2 + 6.264S + 3.897}{-2.971S^5 - 7.095S^4 - 5.500S^3 + 3.198S^2 + 8.471S + 3.897}.$$

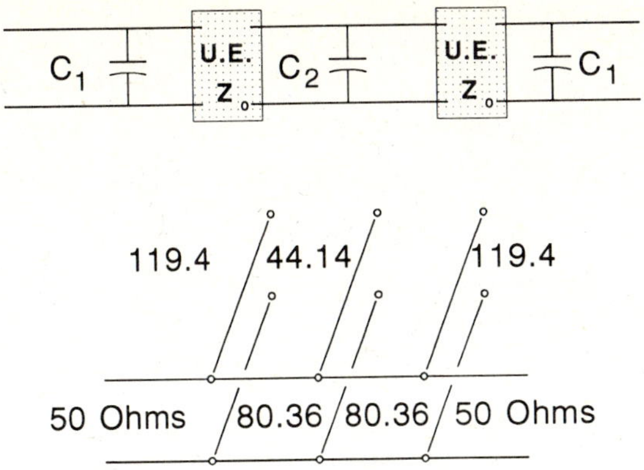

FIGURE 4–44. Nonredundant Prototype Filter and Resulting Commensurate Line Filter.

This expression has a common factor of $(S^2 - 1)$ in the numerator and denominator that must be removed. Upon dividing this factor out of the numerator and the denominator, the following expression is obtained:

$$Z_3 = \frac{-2.623S^2 - 6.263S - 3.897}{-2.971S^3 - 7.095S^2 - 8.471S - 3.897}.$$

Again, the impedance has a higher order in the denominator than in the numerator, so we can isolate a term for a shunt capacitor by inverting this expression and performing one step of long division. The result is:

$$Y_3 = 1.133S + \frac{-4.057S - 3.897}{-2.623S^2 - 6.264S - 3.897} = SC_2 + \frac{1}{Z_4}.$$

In that expression, we identify a capacitance of $C_2 = 1.133\ F$ as is shown in the prototype filter of Figure 4–44.

The remaining steps are to insert a unit element and identify the terminal shunt capacitor and resistor. However, the network is symmetrical about the central shunt capacitor C_2, so we can draw the prototype filter and the commensurate line filter (scaled to 50 Ω) as shown in Figure 4–44. The attenuation of the filter is shown in Figure 4–45.

Although only 3-place accuracy is shown in the calculations above, the calculations were made with 6-place accuracy to avoid round-off error. Also note that frequency scaling is accomplished in the first step where r_{21} was specified. The impedance scaling was done in the last step, as is normally the case. We note that the very low value chosen for the folding frequency f_0 results in a recovery in S_{21} at a frequency too close to the cut-off frequency. Usually, one desires a filter that does not recover its transmission at such a low frequency. This requires that the choice of f_0 be a much higher frequency. The obvious advantage of this method is that the space taken by the filter is considerably smaller than that required for the redundant design of the same filter shown in Figure 4–19.

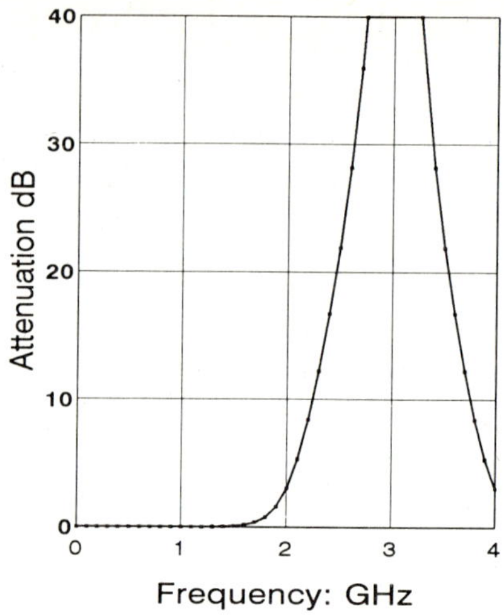

FIGURE 4–45. Attenuation for a Nonredundant Filter.

PROBLEMS

4–1. A simple low pass filter with two elements having $Z_1 = j\omega L_1$ and $Y_2 = j\omega C_2$ is to have L_1 = 9.583 nH and C_2 = 2.83 pF (*L* section filter as shown in the Figure 4–46). Although a capacitance of 2.8 pF can be purchased as a lumped component, inductors as small as 9.583 nH are not available commercially as lumped components. So, the inductor is to be simulated using a shorted line, which is a quarter wave length long at 1.4 GHz. The line is to have an input inductive reactance at 700 MHz the same as a 9.583 nH lumped inductance. Determine the required characteristic impedance for the quarter wave line.

4–2. The quarter wave shorted line in problem 4–1 is to be a coaxial cable with a ratio of diameters of the outer conductor to the center conductor of 2.3. A signal propagates in the TEM mode on the cable. What must be the relative dielectric constant if f_0 is 1.4 GHz? What value of f_0 will allow use of a 50 Ω coaxial cable?

4–3. Determine the microstrip line widths for a 5-element low pass filter to be made of different impedance lines connected in series. The filter is to have a cut-off frequency of 2 GHz. The prototype filter is to be a Butterworth filter having element values of $g_1 = g_5 = 0.618$, $g_2 = g_4 = 1.618$, and $g_3 = 2.000$. Line sections 1, 3, and 5 are to simulate shunt capacitances, and line sections 2 and 4 are to simulate series inductors. The microstrip is on 0.635 mm thick alumina ($\varepsilon_r = 9.8$) backed by a highly conducting ground plane. The line sections compos-

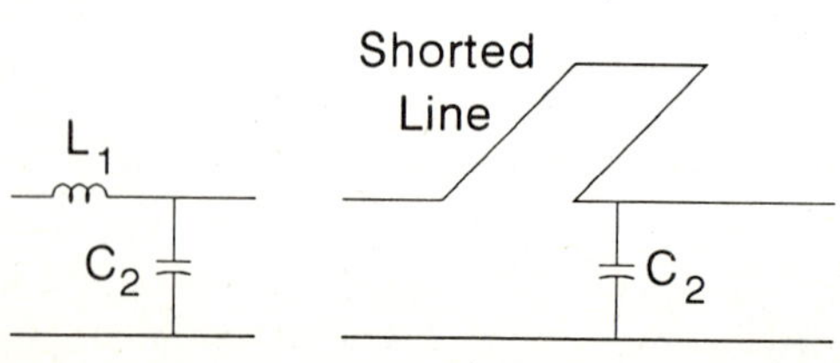

FIGURE 4–46. Circuit for Problem 4–1.

ing the filter are all to have $\beta L = 28°$ at cut-off. The characteristic impedance of the filter is to be 50 Ω.

4–4. Derive Eqs. (4–5) and (4–6).

4–5. Derive Eq. (4–7).

4–6. Prove Kuroda's second identity, shown in Figure 4–13B.

4–7. Specify numerical values for the inductors and capacitors for a low pass lumped element filter with a Chebyshev response having a 1 dB ripple and a Z_0 of 75 Ω. Assume an inductor input configuration and use a cut-off frequency of 400 MHz. This filter is to have an attenuation of at least 25 dB at 600 MHz.

4–8. Lossless inductors are not practical. Therefore, assume that the inductors in problem (4–7) have a series resistance necessary to have a Q of 50 for each of the coils. Model the lossless filter and the lossy filter (with the resistances in series with the inductors) on a suitable CAD program and compare the ripple and attenuation in the pass band.

4–9. A variable line width filter as shown in Figure 4–47 is to have a cut-off frequency of 2 GHz and must have an attenuation greater than 15 dB at 3 GHz. Using a Chebyshev design, determine the line widths and lengths for the sections in Figure 4–47, if this is to be fabricated on Duroid having $\varepsilon_r = 2.23$ and $h = 1.58$ mm.

4–10. Given the 2-element prototype filter shown in Figure 4–48, derive numerical values for the line impedances of the equivalent commensurate line low pass filter with a cut-off frequency of 0.5 GHz and a characteristic impedance of 50 Ω. Use Kuroda's identities with an $f_0 = 1$ GHz.

4–11. Calculate numerical values for inductors, capacitors, and the termination resistor for a high pass lumped element filter with a Chebyshev response having a 0.5 dB ripple. Use $n = 3$ and $Z_0 = 50$. Assume an inductor input configuration and use a cut-off frequency of $10^9/2\pi$ Hz.

4–12. Using the band pass frequency transformation and a definition for $\omega_o = \sqrt{\omega_1\omega_2}$, show that $\Omega = 0$ maps into $\omega = \omega_0$, and that $\Omega = \pm 1$ maps into $\omega = |(\frac{1}{2})[\omega_2 - \omega_1 \pm (\omega_2 + \omega_1)]|$. This shows that the band of frequencies from −1 to +1 in the Ω domain maps into the band of frequencies from ω_1 to ω_2 in the ω domain.

4–13. Calculate values for the elements of a 4-element Butterworth filter having a pass band of 20 MHz centered on 500 MHz. Assume that the input section is a shunt L-C network and that the characteristic impedance of the filter is 90 Ω.

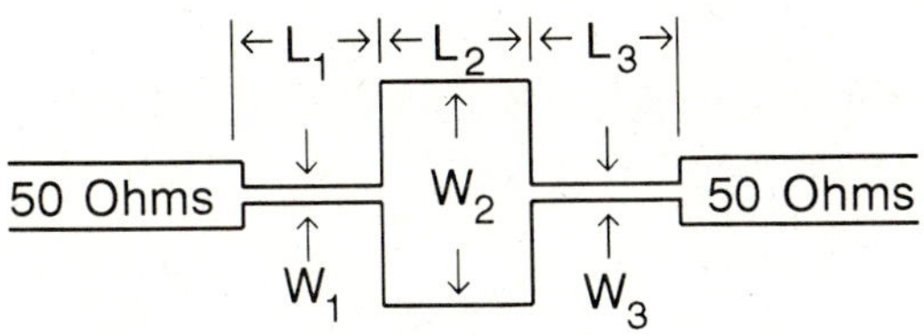

FIGURE 4–47. Filter for Problem 4–9.

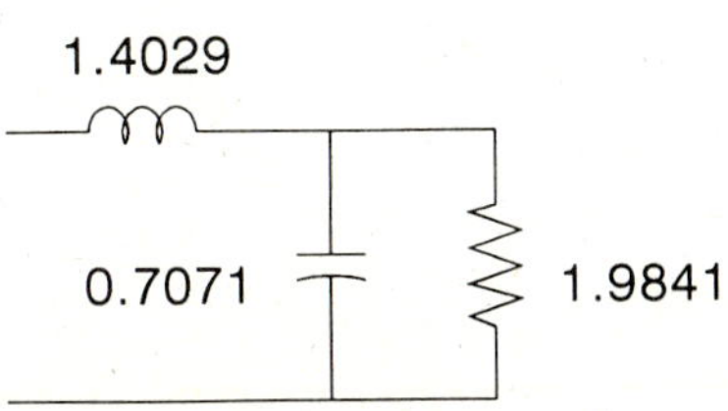

FIGURE 4–48. Circuit for Problem 4–10

4–14. Recalculate the filter component values for the filter specified in problem 4–13, assuming a Chebyshev filter with a 0.5 dB ripple. Simulate the filters in problems 4–13 and 4–14 on a CAD program and compare the 10 dB attenuation bandwidths.

4–15. The voltage on a quarter wave line is $V = V_+e^{-j\beta z} + V_+\Gamma e^{j\beta z}$ and the current is $I = (V_+/Z_0)e^{-j\beta z} - (V_+/Z_0)\Gamma e^{j\beta z}$. If the quarter wave line is treated as a 2 port, where $z = -\lambda/4$ at the input and $z = 0$ at the output, show that $V_1 = -jZ_0I_2$ and $V_2 = -jZ_0I_1$.

4–16. Design a 5-element Chebyshev commensurate line filter with a pass band from 1 to 2 GHz and a ripple of 0.5 dB. Use ideal transmission lines. The source impedance is 50 Ω.

4–17. Determine the admittance matrices for the input sections of the impedance inverter prototype and for the prototype with unit elements shown in Figure 4–37. From these, derive the relation for Z_1 given in Eq. (4–13).

4–18. Design a 4-element commensurate line band pass Chebyshev filter with 0.5 dB ripple to be driven by a 50 Ω source. The pass band is to extend from 1 to 2 GHz. The transmission line impedances are not to exceed 150 Ω.

4–19. Find the line widths and lengths for the filter designed in problem 4–18, if it is to be realized on 0.635 mm thick alumina with $\varepsilon_r = 9.8$.

ENDNOTES

1. Members of the Technical Staff, Bell Telephone Laboratories, *A History of Engineering and Science in the Bell System The Early Years,* M. D. Fagan, ed., 240, (Bell Telephone Laboratories, 1975).
2. L. Brillouin, *Wave Propagation in Periodic Structures,* 13, (New York: McGraw-Hill Book Company, Inc., 1946).
3. James E. Brittain, "Scanning the Past," *IEEE Proc., 80, 5* (May 1992), 782.
4. George A. Campbell, "Physical Theory of Electric Wave Filters", *Bell System Tech. Jour., 1, 2* (Nov. 1922) 1–32.
5. Allen Podell, Sanjay Moghe, Doug Lockie, and Fazal Ali, "GaAs Real Estate," *Microwave Jour., 30, 11* (Nov. 1987), 208–212.
6. Paul I. Richards, "Resistor-Transmission-Line Circuits," *Proc. IRE, 36, 2* (Feb. 1948), 217–220.
7. A more detailed discussion of this type of filter design is given by Terry Edwards in *Foundations for Microstrip Circuit Design,* Second Edition, 278–288, (Chichester: John Wiley & Sons, 1991).
8. The original presentation was in Japanese. A review of these is given in Kazuyuki Kuroda, "General Properties and Synthesis of Transmission-Line Networks," *Microwave Filters and Circuits,* ed. by A. Matsumoto, 22, (New York: Academic Press, 1970).
9. J. A. G. Malherbe, *Microwave Transmission Line Filters,* (Dedham, MA: Artech House, 1979).
10. Saod A. Alseyab, "A Novel Class of Generalized Chebyshev Low-Pass Prototype for Suspended Substrate Stripline Filters," *IEEE Trans. Microwave Theory Tech., MTT-30, 9* (Sept. 1982), 1341–1347.
11. J. D. Rhodes, *Theory of Electrical Filters,* 35–39, (London: John Wiley & Sons, Ltd., 1976).
12. Ford Shepherd, "Filters, Modern-Network-Theory Design," in *Reference Data for Engineers:*

Radio, Electronics, Computer, and Communications, Seventh Edition, ed. Edward C. Jordan, (Indianapolis, IN: Howard W. Sams & Co., Inc., 1985).

13. George L. Matthaei, Leo Young, and E. M. T. Jones, *Microwave Filters, Impedance-Matching Networks, and Coupling Structures,* 98–109, (Dedham, MA: Artech House, 1980).
14. Shepherd, *Reference Data. . . .*
15. Malherbe, *Microwave Filters.*
16. Malherbe, *Microwave Filters,* 118–121.
17. Matthaei, Young, and Jones, *Microwave Filters,* 144–149.
18. M. E. Van Valkenburg, *Introduction to Modern Network Synthesis,* 375, (New York: John Wiley & Sons, 1960).
19. Van Valkenburg, *Network Synthesis,* 78.
20. For the case when the filter output is not matched, the reflection coefficient looking into the filter is not S_{11}. A formula for the reflection coefficient in this case is given in section 6–2, and the filter synthesis is given under the topic of equalizer synthesis in chapter 7.
21. Van Valkenburg, *Network Synthesis,* 262–266.
22. Paul I. Richards, "A Special Class of Functions with Positive Real Part in a Half-Plane," *Duke Math. J., 14* (1947), 777–786.
23. M. C. Horton and R. J. Wenzel, "General Theory and Design of Optimum Quarter-Wave TEM Filters," *IEEE Trans. Microwave Theory Tech., MTT-13, 5* (May 1965), 316–327.
24. The equivalent polynomials are given in *Handbook of Mathematical Tables,* Second Edition, *Supplement to Handbook of Chemistry and Physics,* ed. in chief, Robert C. Weast Ph.D., ed. in charge of Math. Samuel M. Selby, Ph.D., 371, (Cleveland, OH: The Chemical Rubber Co., 1964).

Chapter

5

Coupled Line Propagation

5.0 INTRODUCTION

Coupled microstrip lines and strip lines are used extensively in the formation of directional couplers and filters in hybrid MIC and MMIC. Some of the configurations are meander line sections, comb line filters, and Lange couplers (Figure 5–1). Microstrip lines that are parallel and closely spaced experience interaction due to the common electromagnetic fields. A field traveling down a pair of parallel lines can be imagined as an electromagnetic wave guided by the lines that induces voltages and currents in each of these lines. These voltages and currents are those necessary to satisfy the boundary conditions required on the fields at the conducting surfaces of the microstrip lines. Such

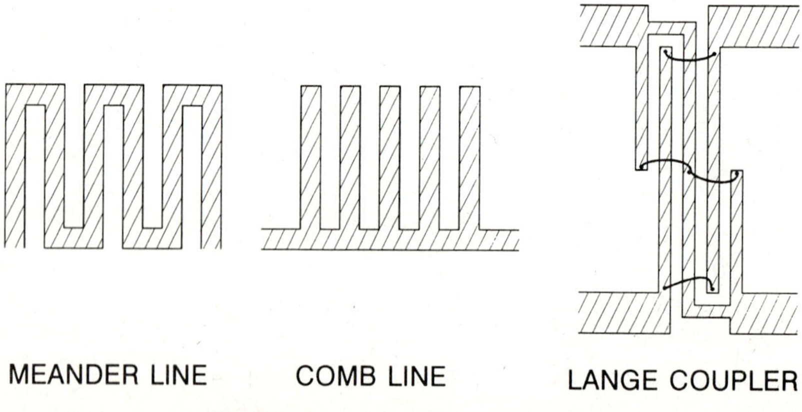

FIGURE 5–1. Coupled Line Sections.

propagation can be analyzed in a manner similar to that described in chapter 2. We shall use a simpler and more approximate description by assuming TEM propagation and analyzing the propagation in terms of the line voltages, currents, and equivalent distributed element circuits. Throughout this chapter, we shall assume that the lines are lossless.

5.1 THE WAVE EQUATION FOR COUPLED LINES

The coupled line equations are derived by considering an approximate equivalent circuit for the lines (Figure 5–2). As discussed in section 5–4, the parameters for this circuit can be derived from a TEM approximation to the actual fields. The coupled line equations are expressed in terms of the voltages and currents of the equivalent circuit. The differential equations corresponding to this circuit are Eqs. (5–1) through (5–4), where the C_i and L_i are the self-capacitance and inductance per unit length of the ith line.

$$+\frac{\partial v_1}{\partial z} + L_1 \frac{\partial i_1}{\partial t} + L_m \frac{\partial i_2}{\partial t} = 0 \tag{5–1}$$

$$+\frac{\partial i_1}{\partial z} + C_{11} \frac{\partial v_1}{\partial t} - C_m \frac{\partial v_2}{\partial t} = 0 \tag{5–2}$$

$$+\frac{\partial v_2}{\partial z} + L_2 \frac{\partial i_2}{\partial t} + L_m \frac{\partial i_1}{\partial t} = 0 \tag{5–3}$$

$$+\frac{\partial i_2}{\partial z} + C_{22} \frac{\partial v_2}{\partial t} - C_m \frac{\partial v_1}{\partial t} = 0 \tag{5–4}$$

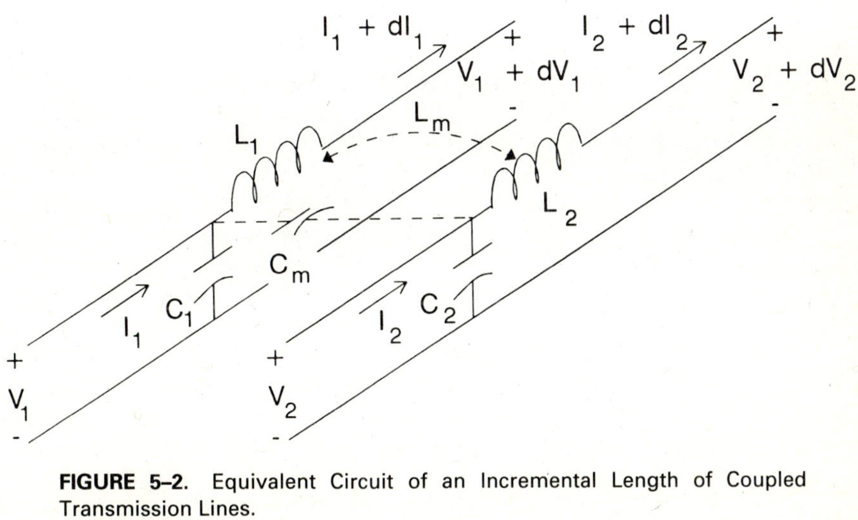

FIGURE 5–2. Equivalent Circuit of an Incremental Length of Coupled Transmission Lines.

L_m and C_m are the mutual inductance and capacitance per unit length of line. C_{ii} is the sum $C_i + C_m$. We use the notation $v = Ve^{j\omega t}$, where V is a phasor and the exp $(j\omega t)$ is the time variation. Then, the time derivatives are replaced by the $j\omega$ operator and the time variation is suppressed to result in Eqs. (5–5) through (5–8).

$$-\frac{dV_1}{dz} = j\omega L_1 I_1 + j\omega L_m I_2 \tag{5–5}$$

$$-\frac{dI_1}{dz} = j\omega C_{11} V_1 - j\omega C_m V_2 \tag{5–6}$$

$$-\frac{dV_2}{dz} = j\omega L_m I_1 + j\omega L_2 I_2 \tag{5–7}$$

$$-\frac{dI_2}{dz} = -j\omega C_m V_1 + j\omega C_{22} V_2 \tag{5–8}$$

We can differentiate with respect to the z variable and cross-substitute between these equations to arrive at the Eqs. (5–9) and (5–10), where the coefficients are $a = L_1C_{11} - L_mC_m$, $b = L_mC_{22} - L_1C_m$, $c = L_mC_{11} - L_2C_m$, and $d = L_2C_{22} - L_mC_m$.

$$-\frac{d^2V_1}{dz^2} = \omega^2 a\, V_1 + \omega^2 b\, V_2 \tag{5–9}$$

$$-\frac{d^2V_2}{dz^2} = \omega^2 c\, V_1 + \omega^2 d\, V_2 \tag{5–10}$$

Solutions of these equations are in the form of propagating waves with a propagation constant γ as in Eq. (5–11).

$$V_1 = V_{01} \exp(-\gamma z),\ V_2 = V_{02} \exp(-\gamma z) \tag{5–11}$$

If we substitute the solution given in Eq. (5–11) into Eqs. (5–9) and (5–10), we arrive at the matrix formulation

$$\begin{pmatrix} \gamma^2 + \omega^2 a, & \omega^2 b \\ \omega^2 c, & \gamma^2 + \omega^2 d \end{pmatrix} \begin{pmatrix} V_1 \\ V_2 \end{pmatrix} = \begin{pmatrix} 0 \\ 0 \end{pmatrix}.$$

A nontrivial solution of this equation exists only if the determinant of this coefficient matrix is zero. That determinant is the characteristic equation for γ: $\gamma^4 + \omega^2(a + d)\gamma^2 + \omega^4(ad - bc) = 0$. Using the quadratic formula, we have Eq. (5–12). So, the propagation constant has four possible values.

$$\gamma^2 = -\beta^2 = -\omega^2[(a + d)/2 \pm (\tfrac{1}{2})\sqrt{(a - d)^2 + 4bc}] \tag{5–12}$$

5.2 PROPAGATION MODES

Each of the values of the propagation constant in Eq. (5–12) corresponds to a distinct mode of propagation. These are the incident and reflected waves for an *even mode,* where the voltages as well as the currents on the two lines are in phase, and the incident and reflected waves for an *odd mode,* where the voltages as well as the currents on the two lines are 180° out of phase with each other. When the coupled lines are symmetric, these two modes have electric fields that are symmetric with respect to the center line between the two conductors. Approximate sketches of the electric field configurations of the two modes for that case are shown in Figure 5–3. When the two lines are identical (symmetric), then the two lines have voltages of equal magnitude and currents of equal magnitude. When the two lines are not identical, the two modes do not correspond to a case where the voltages and currents are equal in magnitude. The even mode is that for the positive sign in Eq. (5–12) and the odd mode is that for the negative sign.

To find the relation between the voltages and currents for the two modes, we consider the even mode and use the notations $V_{1e} = V^a_{ei} + V^a_{er}$ and $V_{2e} = V^b_{ei} + V^b_{er}$, where voltages on line 1 are identified by the superscript a and those on line 2 with the superscript b. The e subscript refers to the even mode, and the i and r subscripts to the incident and reflected waves. Substituting these into Eq. (5–9), we have $\beta_e^2 (V^a_{ei} + V^a_{er}) = \omega^2 a(V^a_{ei} + V^a_{er}) + \omega^2 b(V^b_{ei} + V^b_{er})$. Because the incident and reflected waves are oscillatory in z and have the opposite phase rotation with z, we can set the terms for the incident waves equal to each other without adding in the terms for the reflected waves. We can do the same for the reflected waves. Solving the resulting equations and substituting from Eq. (5–12), we arrive at $V^b_{ei} = k_e V^a_{ei}$ and $V^b_{er} = k_e V^a_{er}$, where k_e is given in Eq. (5–13). Similarly, treating the odd mode, we have $V^b_{oi} = k_o V^a_{oi}$ and $V^b_{or} = k_o V^a_{or}$, where k_o is also shown in Eq. (5–13).

$$
\begin{aligned}
k_e &= \left(\frac{d-a}{2} + \frac{1}{2}\sqrt{(d-a)^2 + 4bc} \right) / b \\
k_o &= \left(\frac{d-a}{2} - \frac{1}{2}\sqrt{(d-a)^2 + 4bc} \right) / b
\end{aligned}
\tag{5–13}
$$

We note that $d - a$ is equal to $L_2C_{22} - L_1C_{11}$. A first approximation is to assume that $L_1C_{11} = L_2C_{22}$. This causes the two values of k to equal $\pm \sqrt{|c/b|}$. The expression for c/b can be simplified. Using the approximation of $L_1C_{11} = L_2C_{22}$, we have

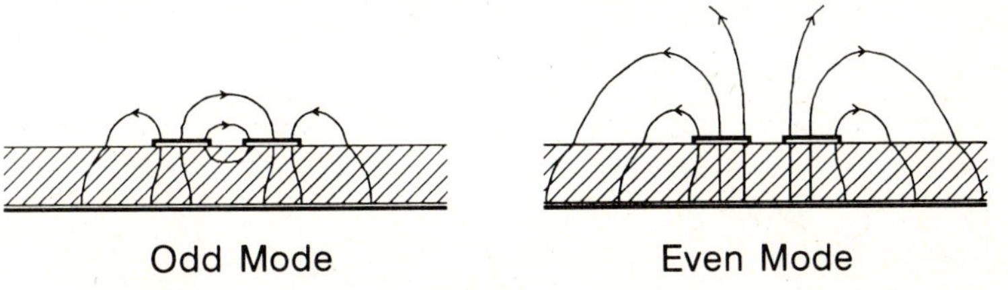

FIGURE 5–3. Even and Odd Mode Electric Field Configurations.

$c/b = (L_mC_{11} - L_2C_m)/L_mC_{22} - L_2C_m) = (L_m/L_1)(L_1C_{11} - L_1L_2C_m/L_m)/[(L_m/L_2)(L_2C_{22} - L_1L_2C_m/L_m)]$. After dividing out the two equal factors, this becomes Eq. (5–14).

$$k = \pm \sqrt{L_2/L_1} \tag{5–14}$$

A similar solution can be found for the currents. In that case one can determine that the currents are related by the inverse of k: $I_e^b = I_e^a/k$. The assumption $L_1C_{11} = L_2C_{22}$ gives the result that the product of the voltage and current for a mode on line a is equal to the product of the voltage and current in that mode on line b. Finally, one can show that this assumption also results in Eq. (5–15), relating the even and odd mode impedances and admittances of the two lines.

$$Z_{0e}^a = 1/Y_{0e}^a,\ Z_{0o}^a = 1/Y_{0o}^a,\ Z_{0e}^b = 1/Y_{0e}^b,\ Z_{0o}^b = 1/Y_{0o}^b \tag{5–15}$$

An even more restrictive condition is one that results when we have symmetrical lines for which $L_1 = L_2$ and $C_{11} = C_{22}$. In this case, Eq. (5–12) becomes Eq. (5–16) and $k = \pm 1$.

$$\gamma^2 = -\beta^2 = -\omega^2(a \pm b) \tag{5–16}$$

To show this, we let the voltage on line a be described by the phasor $V_{01}e^{-j\beta z}$, then substitute for β from Eq. (5–16) to get V_{01} exp $(-j(\omega\sqrt{a \pm b})z)$. Substituting this into Eq. (5–9), we can show that the positive sign in Eq. (5–16) corresponds to a solution where $V_1 = V_2$. The negative sign in Eq. (5–16) corresponds to $V_1 = -V_2$, which is the odd mode solution. Regardless of the approximations, the complete solution for the voltage must be the sum of the even mode and the odd mode solutions: $V = V_e + V_o$.

These are not the only modes possible on this structure. As was discussed in Chapter 2, many other modes are possible and their nature can be determined by analyzing the fields through a solution of Maxwell's equations for these fields. However, the TEM assumption made here and the resulting use of the transmission line equations masks the possibility of these other modes.

5.3 COUPLED LINE PARAMETERS

Now, we determine how the even and odd mode propagation constants and impedances of symmetrical coupled lines depend on the line inductances and capacitances. Using the even mode solutions, $V_1 = V_2$ and $I_1 = I_2$ in Eqs. (5–5) and (5–6), we obtain the relation

$$\frac{d^2V_{1e}}{dz^2} = j\omega(L_1 + L_m)(j\omega)(C_{11} - C_m)V_{1e}.$$

Then, we substitute $V_{1e} = V_{01}$ exp $(-\gamma_e z)$ into this expression to get Eq. (5–17).

$$\gamma_e^2 = -\omega^2(L_1 + L_m)(C_{11} - C_m) \tag{5–17}$$

Using a similar procedure for the odd mode excitation, we find the expression for the odd mode propagation constant given in Eq. (5–18).

$$\gamma_o^2 = -\omega^2(L_1 - L_m)(C_{11} + C_m) \tag{5–18}$$

The characteristic impedance for the even mode is obtained by substituting the even mode condition $I_1 = I_2$ into Eq. (5–5) and then substituting γ_e from Eq. (5–17) into the resulting expression. For the even mode, Eq. (5–5) becomes $\gamma_e V_1 = j\omega(L_1 + L_m)I_1$. The ratio of V_1 to I_1 is called the *even mode characteristic impedance* Z_{0e} and, substituting for γ_e from Eq. (5–17) into that results in Eq. (5–19).

$$Z_{0e} = \sqrt{\frac{L_1 + L_m}{C_{11} - C_m}} \tag{5–19}$$

From $I_1 = -I_2$, we can derive the *odd mode characteristic impedance* Z_{0o} shown in Eq. (5–20).

$$Z_{0o} = \sqrt{\frac{L_1 - L_m}{C_{11} + C_m}} \tag{5–20}$$

5.4 COUPLED LINE PARAMETER VARIATION WITH FREQUENCY

When working with lines that support TEM waves, we frequently find the line inductances and capacitances from the formulas $L = \Phi/i$ and $C = Q/v$, where Φ and Q are the magnetic flux and static charge per unit line length determined from static fields and v and i are the static line voltages and currents. Then, the propagation constants and even and odd mode impedances are calculated from L and C by the use of Eqs. (5–17) through (5–20). In the case of microstrip propagation, we do not have TEM propagation, so we usually reverse this process if we want the L and C. We first determine Z_{0e}, Z_{0o}, γ_e, and γ_o from a full wave solution for the lowest order mode (quasi-TEM), such as is discussed in chapter 2. Then, in the case of lossless lines, Eqs. (5–17) through (5–20) are solved for Eq. (5–21) where the line inductances and capacitances are found from the propagation constants and characteristic impedances.

$$\begin{aligned} L_1 &= (1/2\omega)(\beta_e Z_{0e} + \beta_o Z_{0o}), & L_m &= (1/2\omega)(\beta_e Z_{0e} - \beta_o Z_{0o}) \\ C_{11} &= (1/2\omega)(\beta_e/Z_{0e} + \beta_o/Z_{0o}), & C_m &= (1/2\omega)(\beta_o/Z_{0o} - \beta_e/Z_{0e}) \end{aligned} \tag{5–21}$$

A full wave solution using SDA for the electromagnetic fields on microstrip lines lying side by side on a planar dielectric surface (*edge coupled*) as shown in Figure 5–4 has been reported.[1,2] The solution was for unshielded coupled microstrip lines of equal width W on a dielectric with thickness h and substrate relative dielectric constant ε_r. This report presented the variation of the even and odd mode characteristic impedances and effective dielectric constants as functions of u, g, and the frequency f. The normalized line

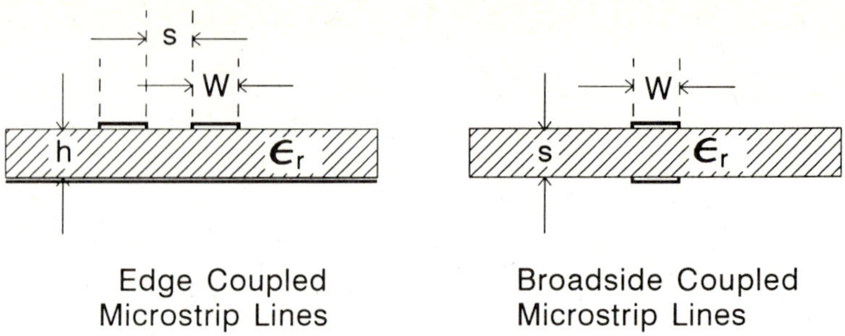

FIGURE 5–4. Coupled Microstrip Lines.

width is $u = W/h$ and the normalized line spacing is $g = s/h$, where s is the edge-to-edge microstrip line spacing. The results were a long sequence of expressions put in a form suitable for programming on a computer. Some of these expressions are listed here, and the remainder—along with a computer program written in BASIC—are included in the appendix.

The even mode effective quasi-static dielectric constant $\varepsilon_{eff,e}$ (0) is given in Eq. (5–22), where the parameters v, a_e, and b_e are given in Eq. (5–23).

$$\varepsilon_{eff,e}(0) = \frac{\varepsilon_r + 1}{2} + \frac{\varepsilon_r - 1}{2}(1 + 10/v)^{-a_e b_e} \qquad (5\text{–}22)$$

The odd mode effective quasi-static dielectric constant $\varepsilon_{eff,o}$ (0)

$$\begin{aligned} v &= u(20 + g^2)/(10 + g^2) + g/\exp(g) \\ a_e &= 1 + \frac{1}{49}\left[\ln\frac{v^4 + (v^2/50)^2}{v^4 + 0.432}\right] + \frac{1}{18.7}\ln(1 + (v/18.1)^3) \\ b_e &= 0.564\left(\frac{\varepsilon_r - 0.9}{\varepsilon_r + 3}\right) \end{aligned} \qquad (5\text{–}23)$$

is given in Eq. (5–24), where ε_{eff} (0) is the effective quasi-static

$$\varepsilon_{eff,o}(0) = ((\varepsilon_r + 1)/2 + a_o - \varepsilon_{eff}(0)) \exp(-c_o g^{d_o}) + \varepsilon_{eff}(0) \qquad (5\text{–}24)$$

dielectric constant for a single line having the same W, h, and ε_r as above (calculated using Eq. (2–9)) where the parameters a_o, c_o, and d_o are given in Eq. (5–25).

$$\begin{aligned} a_o &= 0.7287(\varepsilon_{eff}(0) - (\varepsilon_r + 1)/2)(1 - \exp(-0.179u)) \\ b_o &= 0.747/((0.15/\varepsilon_r) + 1) \\ c_o &= b_o - (b_o - 0.207)\exp(-0.414u) \\ d_o &= 0.593 + 0.694\exp(-0.562u) \end{aligned} \qquad (5\text{–}25)$$

TABLE 5–1 Computed Values for Constants of Eqs. (5–25) and 5–26).

$s(mm)$	v	a_e	$P_e(f)$	$P_o(f)$	$\varepsilon_{eff,e}(0)$	$\varepsilon_{eff,o}(0)$
0.2	2.2147	0.99974	0.24792	0.016714	7.1112	5.6927
1.2	2.0467	0.99958	0.16316	0.091398	7.0510	6.0822

The dynamic, effective, even mode, and odd mode dielectric constants are given by Eq. (5–26), where the subscript e refers to the even mode and the subscript o refers to the odd mode.

$$\varepsilon_{eff,e,o}(f) = \varepsilon_r - \frac{\varepsilon_r - \varepsilon_{eff,e,o}(0)}{1 + P_{e,o}(f)} \tag{5–26}$$

The expressions for $P_e(f)$ and $P_o(f)$ are given in the appendix. Also, expressions for computing the quasi-static and dynamic even and odd mode characteristic impedances are given in the appendix. These expressions are for calculating impedances according to the power-current definition discussed in section 2–2.

Example: Even and Odd Mode $\varepsilon_{eff}(f)$. We can compare even and odd mode effective dielectric constants calculated using these formulas with measurements by Richings and Easter.[3] They measured phase velocities at 10 GHz for two pairs of coupled lines each with $W = 0.66$ mm, $h = 0.66$ mm, and $\varepsilon_r = 9.8$. The two pairs had different line spacings of $s = 0.2$ mm and $s = 1.2$ mm respectively. The measured phase velocities for the 0.2 mm spacing were $V_{pe} = 1.09 \times 10^8$ and $V_{po} = 1.26 \times 10^8$ and for the 1.2 mm spacing were $V_{pe} = 1.10 \times 10^8$ and $V_{po} = 1.20 \times 10^8$, as best we could scale these from their figure 2. Examining Eqs. (5–23) and (5–25), one can see that only v and a_e depend on s. Thus, only those two parameters will differ for the two different cases of $s = 0.2$ and $s = 1.2$. From these equations, we have $b_e = 0.55324$, $a_o = 0.18107$, $b_o = 0.73574$, $c_o = 0.38624$, and $d_o = 0.98863$. Then, we computed these for the remaining variables with the results in Table 5–1. Finally, we used Eq. (5–26) to calculate even and odd mode effective dielectric constants. These were compared with the dielectric constants obtained from the measured phase velocities by $\varepsilon_{eff,e,o}(10) = C^2/V^2_{pe,o}$. This comparison of effective dielectric constants at 10 GHz is: for s = 0.2 mm, the even mode calculated and measured values are 7.645 and 7.52, and for the odd mode they are 5.760 and 5.62; for $s = 1.2$ mm, the even mode calculated value is 7.437, the measured value is 7.41, and for the odd mode, the calculated and measured values are 6.394 and 6.25. Considering the inaccuracies in scaling the data from figure 2 of the reference, the agreement is reasonable.

5.5 DIRECTIONAL COUPLING

Symmetrical coupled lines that are one-quarter wavelength long can be operated as a directional coupler. In this section, we shall assume that the lines have equal even and odd mode phase velocities. For directional coupling, the first port is energized from a signal generator with a generator impedance equal to $\sqrt{Z_{0e}Z_{0o}}$. The remaining ports must be terminated with an impedance given by $\sqrt{Z_{0e}Z_{0o}}$. In this case, there is zero output from one of the four ports and the other two ports have the output divided between them according to the amount of coupling between the lines. The configuration is shown in Figure 5–5.

To explain the action of these coupled lines, we first note that Eqs. (5–9) and (5–10) can be used together to get a differential equation in a single variable as in Eq. (5–27).

$$\frac{d^4V_1}{dz^4}+\omega^2(a+d)\frac{d^2V_1}{dz^2}+\omega^4(ad-bc) \tag{5–27}$$

This is a fourth order differential equation, so there must be four constants of integration A_1, B_1, A_2, and B_2 in the solution for V_1. This solution is of the form of

$$V_1(z)=A_1\exp(-j\beta_e z)+B_1\exp(j\beta_e z)+A_2\exp(-j\beta_o z)+B_2\exp(j\beta_o z),$$

where the coefficients A_i refer to the forward traveling waves in the even and odd modes, and the coefficients B_i to the reverse traveling waves in the even and odd modes. Because of the symmetry, the solution for V_2 is of the same form, except the odd mode components of the solution must be the negative of those for V_1. So, we have the relation

$$V_2(z)=A_1\exp(-j\beta_e z)+B_1\exp(j\beta_e z)-A_2\exp(-j\beta_o z)-B_2\exp(j\beta_o z).$$

The corresponding expressions for the currents are

$$I_1(z)=(A_1/Z_{0e})e^{-j\beta_e z}-(B_1/Z_{0e})e^{j\beta_e z}+(A_2/Z_{0o})e^{-j\beta_o z}-(B_2/Z_{0o})e^{-j\beta_o z}$$ and

$$I_2(z)=(A_1/Z_{0e})e^{-j\beta_e z}-(B_1/Z_{0e})e^{j\beta_e z}-(A_2/Z_{0o})e^{-j\beta_o z}+(B_2/Z_{0o})e^{j\beta_o z}.$$

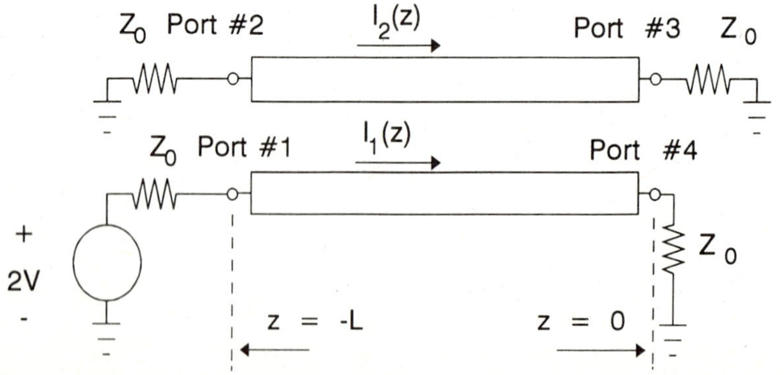

FIGURE 5–5. Coupled Microstrip Lines Forming a Directional Coupler.

The integration constants can be evaluated from the boundary conditions at the ends of the coupled lines. The load is assumed to be at $z = 0$ and the generator at $z = -L$ as shown in Figure 5–5. The boundary conditions at these ends of the lines are given in the form of Ohm's and Kirchhoff's laws for the terminals 1, 2, 3, and 4 respectively as:

At port 1, $2V = I_1(-L)Z_0 + V_1(-L)$
At port 2, $V_2(-L) = -I_2(-L)Z_0$
At port 3, $V_2(0) = I_2(0)Z_0$
At port 4, $V_1(0) = I_1(0)Z_0$

Using the expressions for the line voltages and currents in these boundary condition expressions results in four equations in the four constants of integration. The solution of these equations is given in Eq. (5–28), where we have assumed that $\beta_e = \beta_o = \beta$ and θ is equal to βL.

$$
\begin{aligned}
A_1 &= \frac{(V/2)(Z_0 + Z_{0e})}{2Z_0\cos\theta + j(Z_{0e} + Z_{0o})\sin\theta} \\
B_1 &= \frac{(V/2)(Z_0 - Z_{0e})}{2Z_0\cos\theta + j(Z_{0e} + Z_{0o})\sin\theta} \\
A_2 &= \frac{(V/2)(Z_0 + Z_{0o})}{2Z_0\cos\theta + j(Z_{0e} + Z_{0o})\sin\theta} \\
B_2 &= \frac{(V/2)(Z_0 - Z_{0o})}{2Z_0\cos\theta + j(Z_{0e} + Z_{0o})\sin\theta}
\end{aligned}
\qquad (5\text{–}28)
$$

Substituting these into the expressions for the line voltages evaluated at the ends of the line $z = 0$ and $z = -L$, we obtain $V_1 = V$, $V_2 = j(V\sin\theta)(Z_{0e} - Z_{0o})/[2Z_0\cos\theta + j(Z_{0e} + Z_{0o})\sin\theta]$, $V_3 = 0$, and $V_4 = (2VZ_0)/[2Z_0\cos\theta + j(Z_{0e} + Z_{0o})\sin\theta]$. Using the relation $Z_0 = \sqrt{Z_{0e}Z_{0o}}$ and defining the coupling coefficient C as in Eq. (5–29), we get Eq. (5–30) where θ is a function of frequency through β.

$$C = (Z_{0e} - Z_{0o})/(Z_{0e} + Z_{0o}) \qquad \text{and} \qquad \sqrt{1 - C^2} = 2\,Z_0/(Z_{0e} + Z_{0o}) \qquad (5\text{–}29)$$

Taking the case where the lines are a quarter wavelength long ($\beta_e L = \beta_o L = \pi/2$) and substituting this into Eq. (5–30), we get the port voltages of $V_1 = V$, $V_2 = CV$, $V_3 = 0$, and $V_4 = -jV\sqrt{1 - C^2}$.

$$
\begin{aligned}
V_1 &= V, \qquad V_2 = \frac{j\,V\,C\sin\theta}{\sqrt{1 - C^2}\cos\theta + j\sin\theta} \\
V_3 &= 0, \qquad V_4 = \frac{V\sqrt{1 - C^2}}{\sqrt{1 - C^2}\cos\theta + j\sin\theta}
\end{aligned}
\qquad (5\text{–}30)
$$

So, there is no output from port 3, only a small output from port 2 depending on the magnitude of the coupling coefficient, and the principal output from port 4. Similarly, energizing the lines at port 4 results in outputs at ports 1 and 3, but no output at port 2. Thus, the pair of coupled lines acts as a directional coupler. We note that the output at port 4 is 90° out of phase with the input. Couplers with that phase relation are called *quadrature* couplers.

If we substitute the values of the integration constants from Eq. (5–28) into the expression for I_1 $(-L)$ and let $-\beta L = -\pi/2$, we then have the input current at the port 1 as $I_1 = V_1\, Z_0/Z_{0e}Z_{0o}$. So, the requirement of $Z_0 = \sqrt{Z_{0e}Z_{0o}}$ results in an input impedance V_1/I_1 equal to the square root of the product of the even and odd mode characteristic impedances.

Generally, the designer is faced with the problem that the even and odd mode impedances required for a design are known, but the line dimensions, substrate thickness, and dielectric constant that will yield these impedances are not known. A method for computing these has been published.[4] Frequently, the designer wishes to obtain tighter coupling than is readily obtained with edge coupled lines. In that case, broadside coupled lines are used. Methods for calculating the impedances of broadside coupled and offset broadside coupled lines have been published.[5,6,7] Usually, the design engineer has CAD software available that incorporates coupled line computations of this type.

Example: 20 dB Directional Coupler. A 20 dB directional coupler will ideally have a power output at one port that is one percent of the input power coming from one direction on the through line and no power from a wave coming from the other direction on the through line. Considering a pair of coupled lines as a 4 port (Figure 5–5), we choose the first port as the input, the fourth port as the principal output, and port 2 as the reduced output. To obtain directionality, the output from the third port must be zero. Then, because the device is symmetrical, the output from port 2 is zero when the input is at port 4. The phrase "20 dB coupling" implies that the voltage ratio must be $C = 0.1$ to achieve a power ratio of –20 dB. Substituting this value of C into Eq. (5–29), results in an equation relating Z_{0e} and Z_{0o}. Matching the input to 50 Ω will provide another equation. These are

$$0.1 = \frac{Z_{0e} - Z_{0o}}{Z_{0e} + Z_{0o}} \qquad \text{and} \qquad 50 = \sqrt{Z_{0e}Z_{0o}}$$

and can be solved for $Z_{0e} = 55.28\ \Omega$ and $Z_{0o} = 45.23\ \Omega$. Using CAD software to calculate the line widths and lengths for broadside coupled strip lines embedded in a material with a relative dielectric constant of 2.23, we get $W = 1.66$ mm, $S = 2.63$ mm, $B = 4.04$ mm, and $L = 2.51$ cm, where s is the separation distance of the two lines, W is their widths, B is the distance between the surrounding ground planes, and L is their lengths. A simulation of this coupler on CAD software results in the graph of Figure 5–6. This graph shows that the coupled output is maintained well over a relatively wide band of frequencies.

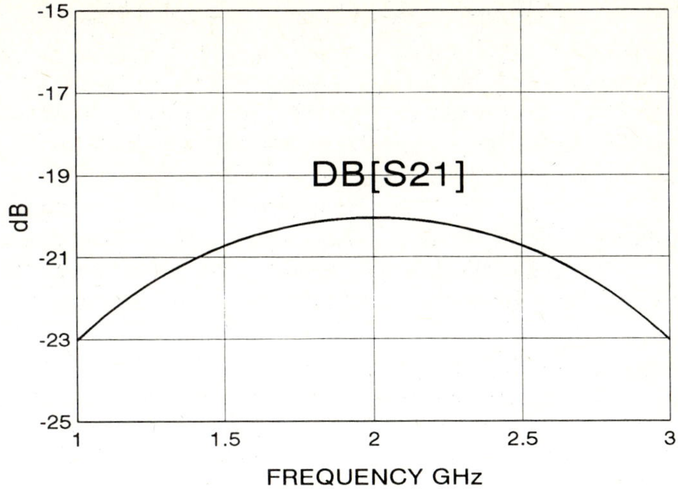

FIGURE 5–6. Directional Coupler Output. S_{21} is the nominal –20 dB output.

5.6 THE LANGE COUPLER

A quadrature coupler that is used frequently in microstrip designs is the Lange coupler, shown in Figure 5–7A. It consists of multiple edge coupled interlaced lines with the alternate lines connected electrically at the middle and at the ends.[8] The "unfolded" Lange coupler, shown in Figure 5–7B, is electrically the same as the Lange coupler, but is easier to relate to the mathematical model.[9] The Lange coupler has the advantage that it is planar, and because of the coupling of both edges of a line to its neighbors, it can have a

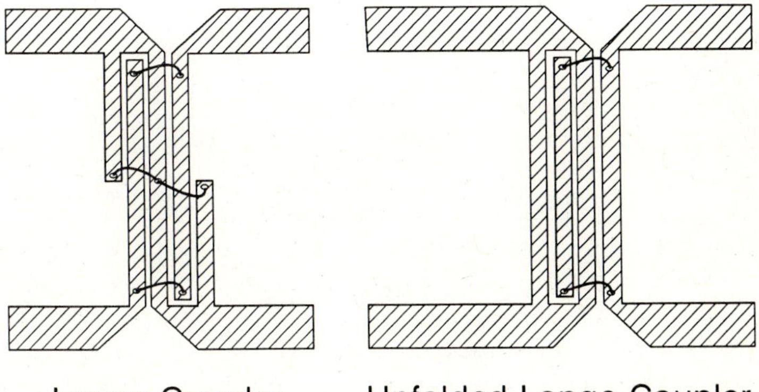

FIGURE 5–7. The Lange Coupler and Its Equivalent Unfolded Version. © 1972 IEEE.

greater amount of coupling. A disadvantage is that alternate individual lines in the coupler must be connected with air bridges.

The coupler can be analyzed approximately by assuming TEM propagation, equal dimension lines, that coupling between nonadjacent lines can be neglected, and that even and odd mode phase velocities V_p are equal. Then the even and odd mode characteristic impedances can be related to the line to ground capacitances in these modes through $C_e = 1/(V_p Z_{0e})$ and $C_o = 1/(V_p Z_{0o})$. The coupler is modeled by an equivalent, 2-conductor line as shown in Figure 5–8. The capacitances C_1 and C_{21} must be such that the 2-conductor line has characteristic impedances that are the same as the multiconductor line. In the multiconductor line, we assume C_{s1} is approximately equal to $C_s + C_{s1}C_m/(C_{s1} + C_m)$.[10] We assume that the coupler has an even number of N interdigitated lines. The characteristic impedance of any pair of lines in the coupler is in parallel with that of the other lines where it is connected. Because these impedances are related to the inverse of the even

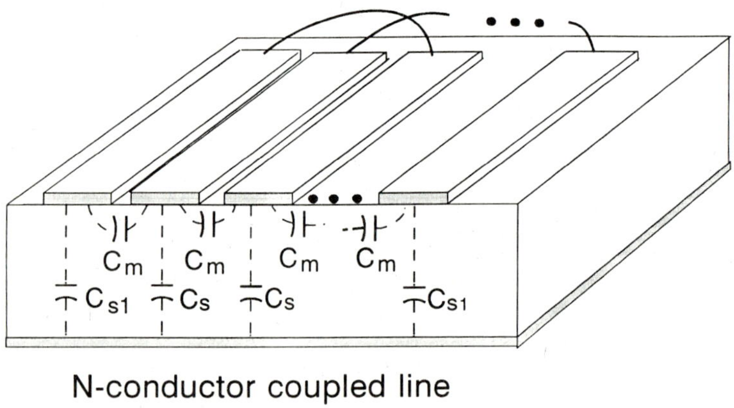

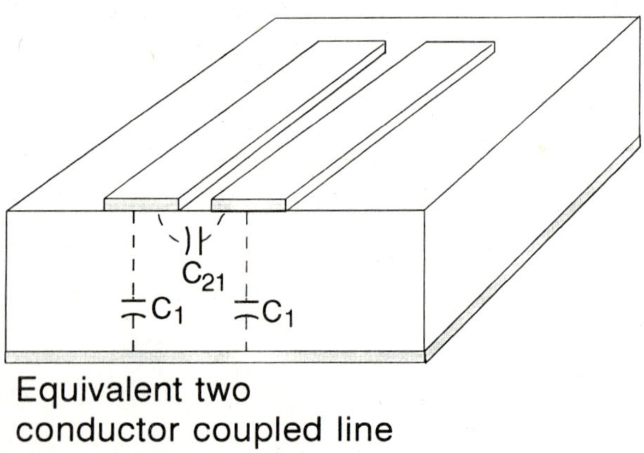

FIGURE 5–8. Lange Coupler Capacitances and an Equivalent 2-Conductor Model.

and odd mode capacitances and because impedances in parallel combine as the sum of their inverses, we can determine the coupler characteristic impedances by determining the sums of the capacitances. The equivalent line has an even mode capacitance to ground of C_{en}, which must be equal to the effective capacitance to ground of $N/2$ conductors of the composite line. In the even mode, adjacent conductors are at equal potentials, so there is no capacitive current due to C_m. Thus, we have $C_{en} = C_1 = (N/2 - 1)C_s + C_{s1}$. The coupler in the odd mode has a voltage between lines that is twice the line voltage to ground. So, the interline capacitance C_m produces a capacitive effect from line to ground of $2C_m$. Thus, the effective line to ground capacitance of the $N/2$ lines in the odd mode is $(N/2 - 1)(C_s + 4C_m) + C_{s1} + 2C_m$. This is the odd mode capacitance C_{on} of the equivalent line. The characteristic impedances of the equivalent line are $Z_{0en} = 1/(V_pC_{en})$ and $Z_{0on} = 1/(V_pC_{on})$. From these, we get Eq. (5–31).

$$Z_{0en} = \frac{2(C_m + C_{s1})}{V_pC_{s1}[NC_{s1} + 2C_m]}$$
$$Z_{0on} = \frac{2(C_m + C_{s1})}{V_p(C_{s1} + 2C_m)[N(C_{s1} + 2C_m) - 2C_m]} \qquad (5\text{–}31)$$

From section 5–5, the required matching impedance for this equivalent line is $Z_{0n} = \sqrt{Z_{0en}Z_{0on}}$. This is the impedance to which the coupler is matched. From Eq. (5–29), we have the coupling coefficient of this equivalent line (and also the coupler) as $C = (Z_{0en} - Z_{0on})/(Z_{0en} + Z_{0on})$. Because microwave engineers are better acquainted with what line impedances are practical, we put these relations in terms of the even and odd mode impedances of a coupled pair of lines having the same spacing, width, dielectric constant, and thickness as the lines in the coupler.[11] Such a coupled pair will have $Z_{0e} = 1/(V_pC_{s1})$, $Z_{0o} = 1/[V_p(C_{s1} + 2C_m)]$, and $Z_0 = \sqrt{Z_{0e}Z_{0o}}$. Equation (5–32) converts Eq. (5–31) to these parameters.

$$Z_{0en} = Z_{0e}\frac{Z_{0e} + Z_{0o}}{(N-1)Z_{0o} + Z_{0e}} \qquad Z_{0on} = Z_{0o}\frac{Z_{0e} + Z_{0o}}{(N-1)Z_{0e} + Z_{0o}} \qquad (5\text{–}32)$$

Then, the coupling coefficient of the coupler in terms of these impedances is

$$C = \frac{\dfrac{Z_{0e}}{(N-1)Z_{0o} + Z_{0e}} - \dfrac{Z_{0o}}{(N-1)Z_{0e} + Z_{0o}}}{\dfrac{Z_{0e}}{(N-1)Z_{0o} + Z_{0e}} + \dfrac{Z_{0o}}{(N-1)Z_{0e} + Z_{0o}}}.$$

Using $R = Z_{0o}/Z_{0e}$ in this relation, we can solve for the equation: $R^2 + 2R\frac{C}{(N-1)(C+1)} + \frac{C-1}{C+1} = 0$.[12] Using the quadratic formula to solve this equation, we arrive at Eq. (5–33).

$$Z_{0o}/Z_{0e} = R = \frac{C}{(C+1)(N-1)}\left(-1 + \sqrt{1 + (1/C^2 - 1)(N-1)^2}\right) \qquad (5\text{–}33)$$

Similarly, we can use Eq. (5–32) in the formula $Z_{0n} = \sqrt{Z_{0en} Z_{0on}}$ to determine Eq. (5–34).

$$\sqrt{Z_{0e} Z_{0o}} = Z_0 = Z_{0n}\sqrt{[N-1+R][(N-1)R+1]/(1+R)} \tag{5–34}$$

Because of the assumptions, these relations are not overly accurate, but they are sufficient for first order calculations. The design procedure is to choose C and N then to calculate R using Eq. (5–33). From this and the specification of Z_{0n} one calculates Z_0, Z_{0e}, and Z_{0o} using Eq. (5–34). Then, one must determine the dimensions of the coupler lines and substrate and the dielectric constant of the substrate to achieve these impedances.

Example: 3 dB Lange Coupler. To make a 50 Ω 3 dB coupler operating at 4 GHz with four interdigitated lines, we use Eq. (5–33) with $C = 1/\sqrt{2}$ and $N = 4$. From this, we get $R = 0.2985$. From Eq. (5–34), with $Z_{0n} = 50$, we get $Z_0 = 96.28\ \Omega$. Then, $Z_{0e} = \sqrt{Z_0^2/R} = 176.22\ \Omega$ and $Z_{0o} = \sqrt{Z_0^2 R} = 52.61\ \Omega$. These are the even and odd mode impedances of any pair of lines in the coupler when operated by themselves. To determine the dimensions of the lines in the coupler, we used the equations of section 5–4 (see the program in the appendix) iteratively to find that a coupler on a substrate of $h = 3.175$ mm, having a relative dielectric constant of 9.1, with microstrip lines having widths of 0.489 mm and a spacing of 0.425 mm, would have the dynamic even and odd mode impedances $Z_{0e} = 176.25$ and $Z_{0o} = 52.53\ \Omega$. To find the dimensions of the 50 Ω input and output lines, we use the section 2–10 equations for computing characteristic impedances and obtain the result that a microstrip line having the width of 3.6 mm on this same substrate has a characteristic impedance of 49.98 Ω. The interdigitated fingers of Lange couplers are usually very narrow and closely spaced relative to the width of the 50 Ω lines connecting to them.

The Lange coupler has the circuit representation shown in Figure 5–9. The S parameters for this coupler can be derived from Eq. (5–30) and are shown in Eq. (5–35), where θ is the usual $\pi f/(2f_0)$ and f_0 is the quarter wave frequency of the lines in the coupler.

$$\begin{aligned} S_{21} &= S_{12} = S_{43} = S_{34} = j\frac{C\sin\theta}{\sqrt{1-C^2}\cos\theta + j\sin\theta} \\ S_{31} &= S_{13} = S_{24} = S_{42} = 0 \\ S_{41} &= S_{14} = S_{23} = S_{32} = \frac{\sqrt{1-C^2}}{\sqrt{1-C^2}\cos\theta + j\sin\theta} \end{aligned} \tag{5–35}$$

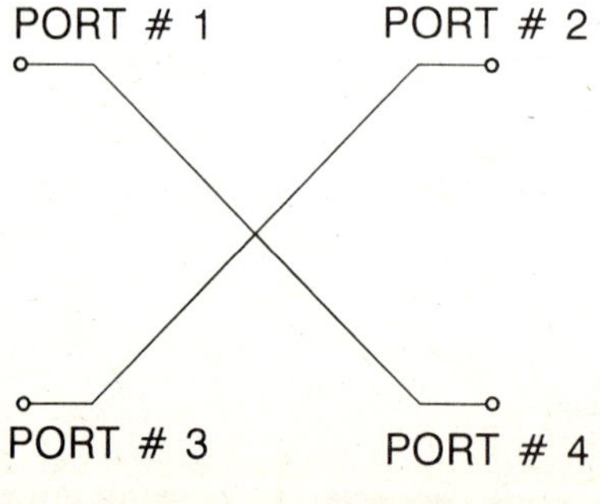

FIGURE 5–9. Circuit Representation of a Lange Coupler.

5.7 COUPLED LINE PAIR TREATED AS A 4 PORT

As discussed by Jones and Bolljahn, we can use pairs of coupled lines to create some band pass, band stop, and other filter networks.[13] The building block used to form these filter networks is a pair of coupled lines that are open or short circuited at two of the terminals, while the other two form the input and output terminals. We will not assume that the lines have the same dimensions, so they will have different even and odd mode characteristic impedances and unequal even and odd mode voltage magnitudes.

In order to analyze the performance of coupled lines when operated in this manner, we first look at an open circuited coupled pair of lines with even excitation at a pair of ports. Then, we repeat this procedure using other excitations and terminations. Using the superposition principle with these solutions, we find general solutions for the coupled lines with arbitrary excitations. Finally, examining a pair of coupled lines that are suitably open circuited or short circuited, we see that they have an inductive or capacitive property at a frequency below $\lambda/4$ wavelength. These properties can be used to construct filters in the same way that single stubs were used.

We are considering the transmission lines shown in Figure 5–10 as a 4 port. Because the lines are not identical, the even and odd modes are both excited by each of the different source configurations shown in the Figure 5–11. As in section 5–2, we will assume that $L_1C_{11} = L_2C_{22}$, so the two modes are related by $k = \pm\sqrt{L_2/L_1}$ as shown in Eq. (5–14). We have four possible excitations that we can use separately and then superimpose the resulting solutions. These are

(1) even excitation at ports 1 and 2
(2) odd excitation at ports 1 and 2
(3) even excitation at ports 3 and 4
(4) odd excitation at ports 3 and 4

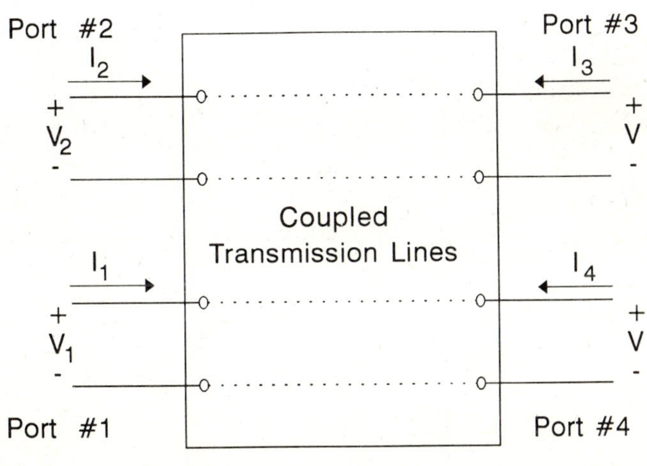

FIGURE 5–10. Diagram Representing a Pair of Unequal Coupled Lines as a 4 Port.

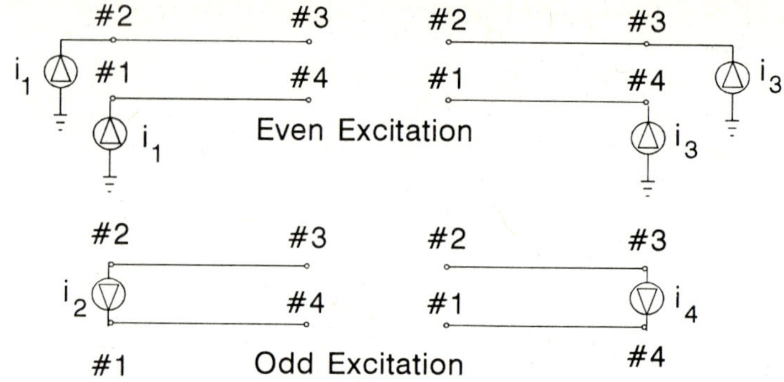

FIGURE 5–11. Even and Odd Excitation of Coupled Transmission Lines.

Consider the diagram for case (1) as shown in Figure 5–12. We choose to measure distance from the right end of the lines so the excitation is at $z = -L$. We label the lines a and b and use the superscripts a and b to identify the associated line voltage, current, or impedance. The lines are open circuited at ports 3 and 4 so, there is a reflected wave set up at these ports. The lines are not symmetrical, so we have both even and odd mode solutions, even though the excitation is even with respect to the lines. Equation (5–36) is for the voltage and current along line a.

$$
\begin{aligned}
I^a(z) &= A_1\, e^{-j\beta_e z} + B_1\, e^{j\beta_e z} + A_2\, e^{-j\beta_o z} + B_2\, e^{j\beta_o z} \\
V^a(z) &= A_1 Z_{0e}^a\, e^{-j\beta_e z} - B_1 Z_{0e}^a\, e^{j\beta_e z} + A_2 Z_{0o}^a\, e^{-j\beta_o z} - B_2 Z_{0o}^a\, e^{j\beta_o z}
\end{aligned} \qquad (5\text{–}36)
$$

Equation (5–37) is for line b, where k is that from Eq. (5–14). We evaluate the constants in this solution by using the boundary conditions.

$$
\begin{aligned}
I^b(z) &= \left(A_1\, e^{-j\beta_e z} + B_1\, e^{j\beta_e z} - A_2\, e^{-j\beta_o z} - B_2\, e^{j\beta_o z}\right)/k \\
V^b(z) &= \left(A_1 Z_{0e}^b\, e^{-j\beta_e z} - B_1 Z_{0e}^b\, e^{j\beta_e z} - A_2 Z_{0o}^b\, e^{-j\beta_o z} + B_2 Z_{0o}^b\, e^{j\beta_o z}\right)/k
\end{aligned} \qquad (5\text{–}37)
$$

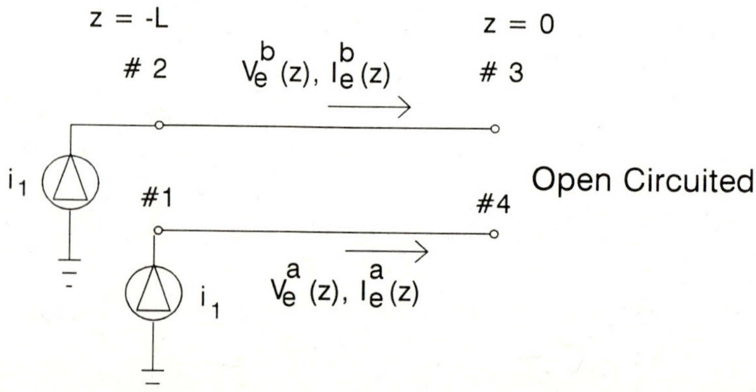

FIGURE 5–12. Even Excitation of a Pair of Coupled Lines.

At the open circuit end ($z = 0$), the currents along the lines must be zero. Using that condition in Eqs. (5–36) and (5–37), we can determine that $B_1 = -A_1$ and $B_2 = -A_2$. At port 1, the line current must be equal to the source current i_1. Using that in Eq. (5–36) and substituting $-A_1 = B_1$ and $-A_2 = B_2$, we find $A_1 = i_1\,(1 + k)/(4j \sin \beta_e L)$ and $A_2 = i_1\,(1 - k)/(4j \sin \beta_e L)$. Substituting these back into Eq. (5–36), we have $V^a(z) = -ji_1\big((1 + k)Z^a_{0e} \cos(-\beta_e z)\big)/(2 \sin(\beta_e L) - ji_1\big((1 - k)Z^a_{0o} \cos(-\beta_o z)\big)/(2 \sin(\beta_o L)$. We obtained a similar expression for $V^b(z)$.

We are interested principally in the voltages and currents at the ports. So, we introduce the subscript notation that relates the first subscript to the port where the voltage (or current) is observed, and the second subscript relates to the current configuration exciting the lines. Using that notation, we can express our solution for the voltage at port 1 as $V_{11} = V^a(-L) = -ji_1\big((1 + k)Z^a_{0e} \cot(\beta_e L) + (1 - k)Z^a_{0o} \cot(\beta_o L)\big)/2$. Similarly, the voltages at the other ports caused by this even excitation are

$$V_{21} = -ji_1\big((1 + 1/k)Z^b_{0e} \cot(\beta_e L) + (1 - 1/k)Z^b_{0o} \cot(\beta_o L)\big)/2$$

$$V_{31} = -ji_1\big((1 + 1/k)Z^b_{0e} \csc(\beta_e L) + (1 - 1/k)Z^b_{0o} \csc(\beta_o L)\big)/2$$

$$V_{41} = -ji_1\big((1 + k)Z^a_{0e} \csc(\beta_e L) + (1 - k)Z^a_{0o} \csc(\beta_o L)\big)/2$$

These are the expressions for the voltages at the four ports when the lines are excited at the first and second ports in an even mode excitation with the current i_1.

We now consider case (2), where the lines have an odd excitation with a current connected between ports 1 and 2. The diagram representing this case is shown in Figure 5–13. As before, we assume both even and odd mode currents and voltages on both lines and evaluate the constants from the boundary conditions at $z = 0$ and $z = -L$. Using the subscript notation introduced when discussing the even excitation, the voltage at port 2, due to this odd excitation, is $V_{22} = V^b(-L)$. The port voltages for the i_2 of Figure 5–13 are

$$V_{12} = -ji_2\big((1 - k)Z^a_{0e} \cot(\beta_e L) + (1 + k)Z^a_{0o} \cot(\beta_o L)\big)/2$$
$$V_{22} = -ji_2\big((-1 + 1/k)Z^b_{0e} \cot(\beta_e L) - (1 + 1/k)Z^b_{0o} \cot(\beta_o L)\big)/2$$
$$V_{32} = -ji_2\big((-1 + 1/k)Z^b_{0e} \csc(\beta_e L) - (1 + 1/k)Z^b_{0o} \csc(\beta_o L)\big)/2$$
$$V_{42} = -ji_2\big((1 - k)Z^a_{0e} \csc(\beta_e L) + (1 + k)Z^a_{0o} \csc(\beta_o L)\big)/2.$$

A similar process can be followed for the even and odd excitations at ports 3 and 4 to find the remaining eight expressions for the voltages at the four ports due to these exci-

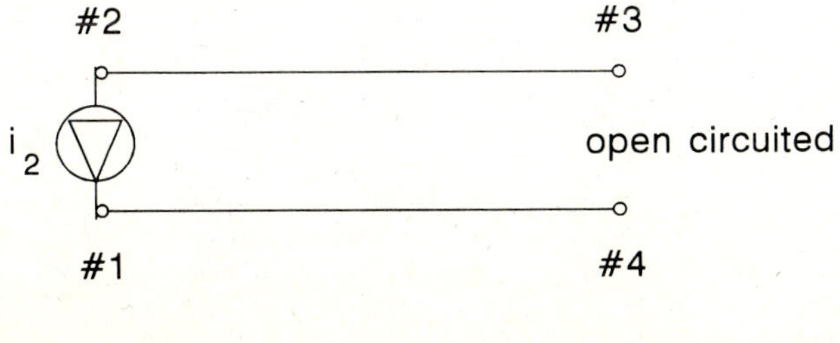

FIGURE 5–13. Odd Excitation of Coupled Lines.

tations. Now, we use superposition with these sixteen expressions for the port voltages due to all these even and odd excitations. The superposition relation for each of the four voltages and the individual ports is given by $V_i = \Sigma_{j=1}^{4} V_{ij}$, where the single subscript on V denotes the port where the voltage occurs. Applying this summation to find each port voltage, we arrive at Eq. (5–38), where (Z) is a matrix with terms obtained from the 16 expressions. Equation (5–38) is a relation between the port voltages and the excitation currents, but the excitation currents are not equal to the port currents.

$$\begin{pmatrix} V_1 \\ V_2 \\ V_3 \\ V_4 \end{pmatrix} = (Z) \begin{pmatrix} i_1 \\ i_2 \\ i_3 \\ i_4 \end{pmatrix} \tag{5–38}$$

The relation we desire is one that holds between the voltages and the currents at the four ports. The previous relation can be converted into the desired relation by considering the configuration with all the current sources present. This is shown in Figure 5–14. The relation between the port currents and the even and odd excitation currents found by applying Kirchhoff's law at the inputs is

$$\begin{pmatrix} I_1 \\ I_2 \\ I_3 \\ I_4 \end{pmatrix} = \begin{pmatrix} 1 & 1 & 0 & 0 \\ 1 & -1 & 0 & 0 \\ 0 & 0 & -1 & 1 \\ 0 & 0 & 1 & 1 \end{pmatrix} \begin{pmatrix} i_1 \\ i_2 \\ i_3 \\ i_4 \end{pmatrix}.$$

We can invert this matrix to obtain the relation

$$\begin{pmatrix} i_1 \\ i_2 \\ i_3 \\ i_4 \end{pmatrix} = (1/2) \begin{pmatrix} 1 & 1 & 0 & 0 \\ 1 & -1 & 0 & 0 \\ 0 & 0 & -1 & 1 \\ 0 & 0 & 1 & 1 \end{pmatrix} \begin{pmatrix} I_1 \\ I_2 \\ I_3 \\ I_4 \end{pmatrix}.$$

Substituting this into Eq. (5–38), we arrive at the relation between the port voltage matrix (V) and the port current matrix (I) of $(V) = (Z)(I)$ where the impedance matrix (Z) has the elements shown in Eq. (5–39).

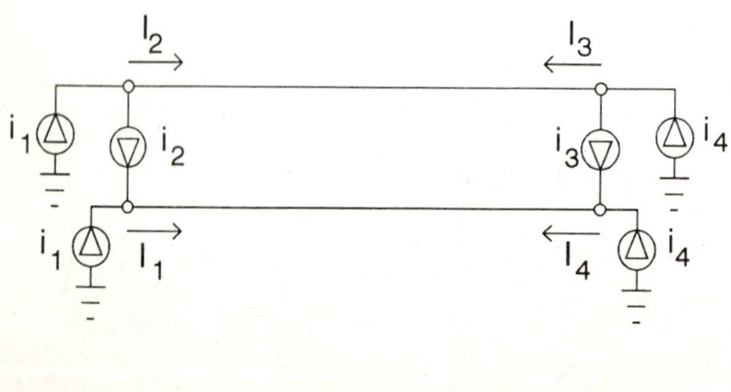

FIGURE 5–14. Coupled Transmission Lines with Even and Odd Excitation.

$$\left.\begin{aligned}
Z_{11} = Z_{44} &= -(j/2)[Z^a_{0e}\cot(\beta_e L) + Z^a_{0o}\cot(\beta_o L)]\\
Z_{21} = Z_{34} &= -(j/2k)[Z^b_{0e}\cot(\beta_e L) - Z^b_{0o}\cot(\beta_o L)]\\
Z_{31} = Z_{24} &= -(j/2k)[Z^b_{0e}\csc(\beta_e L) - Z^b_{0o}\csc(\beta_o L)]\\
Z_{41} = Z_{14} &= -(j/2)[Z^a_{0e}\csc(\beta_e L) + Z^a_{0o}\csc(\beta_o L)]\\
Z_{12} = Z_{43} &= -(jk/2)[Z^a_{0e}\cot(\beta_e L) - Z^a_{0o}\cot(\beta_o L)]\\
Z_{22} = Z_{33} &= -(j/2)[Z^b_{0e}\cot(\beta_e L) + Z^b_{0o}\cot(\beta_o L)]\\
Z_{32} = Z_{23} &= -(j/2)[Z^b_{0e}\csc(\beta_e L) + Z^b_{0o}\csc(\beta_o L)]\\
Z_{42} = Z_{13} &= -(jk/2)[Z^a_{0e}\csc(\beta_e L) - Z^a_{0o}\csc(\beta_o L)]
\end{aligned}\right\} \quad (5\text{–}39)$$

In order to derive the elements of the admittance matrix, which also relates the currents and the voltages at the ports of a coupled line pair, we can follow a procedure similar to that above. However, the lines are considered to be driven by sets of voltage generators as shown in Figure 5–15. Using each of these voltage generators separately or in combination, as was done with the current generators, and then summing the individual responses to superpose solutions, we can derive the admittance elements that describe the admittance matrix of the equation $(I) = (Y)(V)$.[14] These elements are shown in Eq. (5–40). These expressions apply generally to microstrip coupled lines.

$$\left.\begin{aligned}
Y_{11} = Y_{44} &= (-j/2)[Y^a_{0e}\cot(\beta_e L) + Y^a_{0o}\cot(\beta_o L)]\\
Y_{21} = Y_{34} &= (-jk/2)[Y^b_{0e}\cot(\beta_e L) - Y^b_{0o}\cot(\beta_o L)]\\
Y_{31} = Y_{24} &= (+jk/2)[Y^b_{0e}\csc(\beta_e L) - Y^b_{0o}\csc(\beta_o L)]\\
Y_{41} = Y_{14} &= (+j/2)[Y^a_{0e}\csc(\beta_e L) + Y^a_{0o}\csc(\beta_o L)]\\
Y_{12} = Y_{43} &= (-j/2k)[Y^a_{0e}\cot(\beta_e L) - Y^a_{0o}\cot(\beta_o L)]\\
Y_{22} = Y_{33} &= (-j/2)[Y^b_{0e}\cot(\beta_e L) + Y^b_{0o}\cot(\beta_o L)]\\
Y_{32} = Y_{23} &= (+j/2)[Y^b_{0e}\csc(\beta_e L) + Y^b_{0o}\csc(\beta_o L)]\\
Y_{42} = Y_{13} &= (+j/2k)[Y^a_{0e}\csc(\beta_e L) - Y^a_{0o}\csc(\beta_o L)]
\end{aligned}\right\} \quad (5\text{–}40)$$

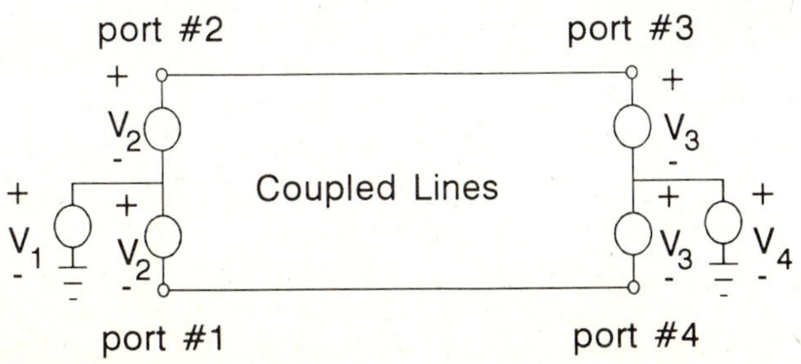

FIGURE 5–15. Even and Odd Mode Excitation of a Coupled Line Pair by Voltage Sources.

5.8 COUPLED LINE PAIR OPERATED AS A 2 PORT, ASSUMING $\Theta_e = \Theta_o$

Normally, a pair of coupled lines is used in filter design with two of the ports as the input and output ports, and the other two ports either short circuited or open circuited. For instance, we can short circuit port 2, open circuit port 3, then use port 1 as the input and port 4 as the output as shown in Figure 5–16. To design filters using coupled lines in these configurations, we use their equivalent circuits in the Ω domain. With the assumption that the lines are identical and have equal even and odd mode phase velocities so that $\beta_e L = \beta_o L$, the equivalent circuits are relatively simple. Some of the possible circuits are shown in Figure 5–16 for the case of the symmetrical lines. The relations between components required for equivalence of these circuits are:

Circuit A: $Z_{0e} = Z_0 + L + \sqrt{L(Z_0 + L)}$ and $Z_{0o} = Z_0 + L - \sqrt{L(Z_0 + L)}$,
Circuit B: $Y_{0e} = 2/Z_0 + 1/L$ and $Y_{0o} = 1/L$, and
Circuit C: $Z_{0e} = 2Z_0 + 1/C$ and $Z_{0o} = 1/C$.

Note that Y_{0e} must be smaller than Y_{0o} so the circuit B is nonphysical.

These relations can be derived by using Richards' transformation along with either Eq. (5–39) or (5–40). Circuit A in Figure 5–16 shows the case of equal coupled lines operating as a 2 port and its Ω domain equivalent circuit. To show the equivalence of these two circuits, we use Eq. (5–39) and set both V_2 and I_3 equal to zero as

$$\begin{pmatrix} V_1 \\ 0 \\ V_3 \\ V_4 \end{pmatrix} = (Z) \begin{pmatrix} I_1 \\ I_2 \\ 0 \\ I_4 \end{pmatrix}.$$

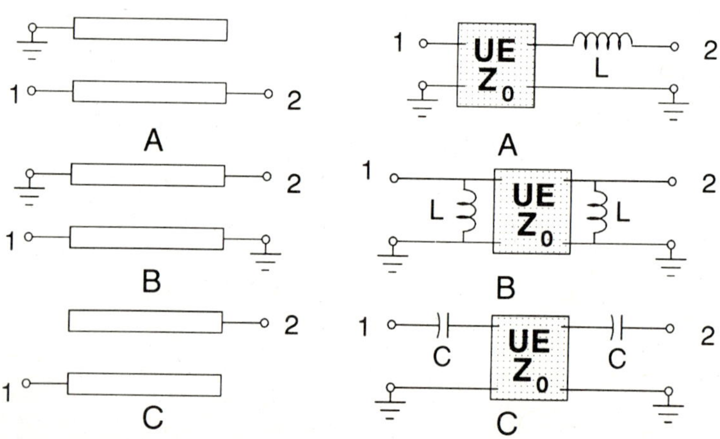

FIGURE 5–16. Coupled Lines and Their Prototype Equivalent Circuits. The lines have equal dimensions.

Now, we solve the equation represented by the second line in this matrix for the relation $I_2 = -(Z_{21}/Z_{22})I_1 - (Z_{24}/Z_{22})I_4$ and substitute this into equations for V_1 and V_4 obtained from this matrix. Assuming $k = 1$, $\beta_e L = \beta_o L = \theta$, $Z^a_{0e} = Z^b_{0e}$, and $Z^a_{0o} = Z^b_{0o}$ in Eq. (5–39), we substitute the resulting expressions for Z_{ij} into these expressions to find

$$\begin{pmatrix} V_1 \\ V_4 \end{pmatrix} = -\frac{j}{2}(Z)\begin{pmatrix} I_1 \\ I_4 \end{pmatrix}, \text{ where } (Z) =$$

$$\begin{pmatrix} \left(Z_{0e} + Z_{0o} - \dfrac{(Z_{0e} - Z_{0o})^2}{Z_{0e} + Z_{0o}}\right)\cot\theta, & \left(Z_{0e} + Z_{0o}\right)\csc\theta - \dfrac{(Z_{0e} - Z_{0o})^2 \csc\theta}{Z_{0e} + Z_{0o}} \\ \left(Z_{0e} + Z_{0o} - \dfrac{(Z_{0e} - Z_{0o})^2}{Z_{0e} + Z_{0o}}\right)\csc\theta, & \left(Z_{0e} + Z_{0o}\right)\cot\theta - \dfrac{(Z_{0e} - Z_{0o})^2 \csc^2\theta}{(Z_{0e} + Z_{0o})^2 \cot\theta} \end{pmatrix}.$$

We employ Richards' transformation to obtain $\cot\theta = 1/\Omega = j/S$ and $\csc\theta = (\sqrt{1 + \Omega^2})/\Omega = (j\sqrt{1 - S^2})/S$. Substituting these for $\cot\theta$ and $\csc\theta$ and simplifying the expression, we get Eq. (5–41).

$$\begin{pmatrix} V_1 \\ V_4 \end{pmatrix} = \frac{1}{2}\begin{pmatrix} \dfrac{4\,Z_{0e}Z_{0o}}{Z_{0e} + Z_{0o}}\dfrac{1}{S}, & \dfrac{4\,Z_{0e}Z_{0o}}{Z_{0e} + Z_{0o}}\dfrac{\sqrt{1 - S^2}}{S} \\ \dfrac{4Z_{0e}Z_{0o}\sqrt{1 - S^2}}{(Z_{0e} + Z_{0o})\,S}, & \dfrac{4\,Z_{0e}Z_{0o} + (Z_{0e} - Z_{0o})^2 S^2}{(Z_{0e} + Z_{0o})S} \end{pmatrix}\begin{pmatrix} I_1 \\ I_4 \end{pmatrix} \tag{5–41}$$

If we let Z_0 be $2\,Z_{0e}\,Z_{0o}/(Z_{0e} + Z_{0o})$ and L be $\frac{1}{2}\,(Z_{0e} - Z_{0o})^2/(Z_{0e} + Z_{0o})$ then Eq. (5–41) becomes Eq. (5–42), which is identical to that for a U.E. in series with an inductor as shown in Figure 5–16.

$$\begin{pmatrix} V_1 \\ V_4 \end{pmatrix} = \begin{pmatrix} \dfrac{Z_0}{S}, & Z_0\dfrac{\sqrt{1 - S^2}}{S} \\ Z_0\dfrac{\sqrt{1 - S^2}}{S}, & \dfrac{Z_0 + S^2 L}{S} \end{pmatrix}\begin{pmatrix} I_1 \\ I_4 \end{pmatrix} \tag{5–42}$$

Comparing Eqs. (5–41) and (5–42), we can arrive at the formulas in Eq. (5–43) for the even and odd mode impedances of the coupled lines in terms of the prototype elements.

$$Z_{0e} = Z_0 + L + \sqrt{L(Z_0 + L)}, \qquad Z_{0o} = Z_0 + L - \sqrt{L(Z_0 + L)} \tag{5–43}$$

We can use this pair of coupled lines as a building block for a low pass filter design similar to that designed in chapter 4.

5.9 LOW PASS FILTER DESIGN, ASSUMING $\theta_e = \theta_o$

As shown previously, the equations used to design coupled line filters are considerably simplified if the assumption can be made that the even mode phase delay θ_e is equal to the odd mode phase delay θ_o. This is approximately true for microstrip coupled line sections operating at lower frequencies where the quasi-static approximation holds. Generally, this is less accurate at higher frequencies. However, coupled strip lines with a uniform dielectric medium will satisfy this condition. Using this assumption greatly simplifies the equivalent circuits for the coupled lines.

Ozaki and Ishii discussed the design of low pass filters using coupled lines.[15] This design process is essentially the same as in chapter 4. One of the principal drawbacks with the design of commensurate line filters as discussed in section 4–6 was the inability to model the action of a series inductor with an open circuited stub when working with microstrip or strip line. That problem arose because one side of microstrip stubs must be the ground plane, so the series inductor, where neither side was grounded, could not be simulated with open circuited microstrip stubs. However, this difficulty can be circumvented by using coupled lines as shown in the top circuit of Figure 5–16. The following example uses this feature.

Example: Low Pass Filter Design. We specified a low pass filter having a cut-off frequency of 1.5 GHz and at least a 10 dB insertion loss at 3 GHz. Further, we required that the filter be matched to 50 Ω at the input and output. Inspecting the Chebyshev filter attenuation curves in Figure 4–28, we chose a third order Chebyshev filter with 0.2 dB ripple.

To design a third order Chebyshev low pass filter, we started with a prototype filter as shown in Figure 5–17A. For this, we have the elements $g_1 = 1.2276$, $g_2 = 1.1525$, and $g_3 = 1.2276$. We chose f_0 equal to 5 GHz and then frequency scaled by $\tan(\pi f_c/2f_0) = \tan(0.15\pi) = 0.5095$. Then, we inserted three unit elements at the output end of the filter and successively moved them into the interior using Kuroda's identities as in chapter 4. We arrived at the prototype shown in Figure 5–17B, where the elements were determined using the Kuroda identity formulas given in Figure 4–13. Scaling these element values to 50 Ω yielded

$$Z_1 = 17.8542,\ Z_2 = 13.8747,\ Z_3 = 35.3591,$$

$$L_1 = 109.91,\ L_2 = 71.459,\ \text{and}\ L_3 = 14.641.$$

Each of these unit elements paired with the following inductor can be simulated with coupled line sections where port 2 is grounded and the input is at port 1 with the output at the port 4 as described in Eq. (5–42) and pictured in Figure 5–16A. The final form of the filter is shown in Figure 5–17C. Using Eq. (5–43), we got the even and odd mode impedances: first pair $Z_{0e} = 190.41$ and $Z_{0o} = 65.12$; second pair $Z_{0e} = 163.42$ and $Z_{0o} = 7.245$; and third pair $Z_{0e} = 77.06$ and $Z_{0o} = 22.94\ \Omega$.

To test the design of this filter, we simulated it on CAD software by calculating the width, the spacing, and length for the coupled lines. To overcome the unequal θ_e and θ_o

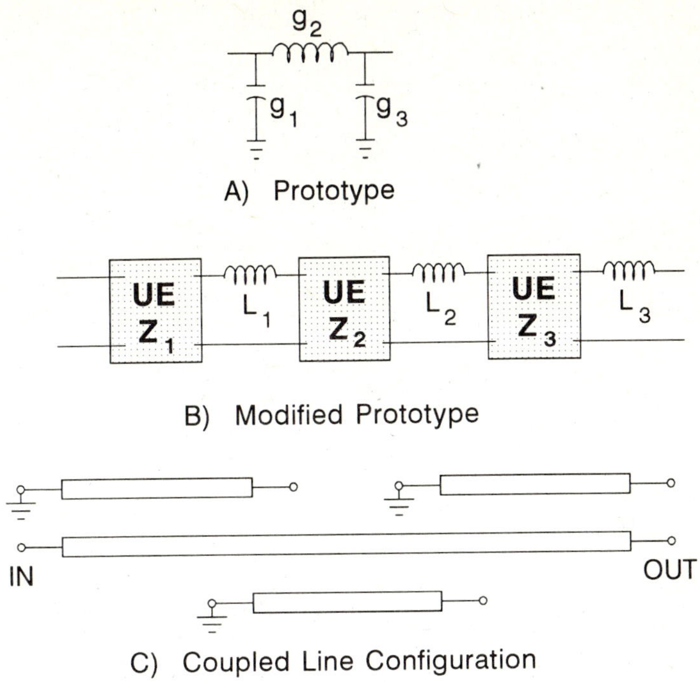

FIGURE 5–17. Design Stages in a Third Order Chebyshev Filter Using Coupled Line Sections.

of edge coupled lines, we calculated these dimensions for broadside coupled strip lines as listed in Table 5–2. Because the alternate broadside coupled lines would abut each other, this design could not be realized physically without some modification. The computer simulated response for this filter with these dimensions is shown in Figure 5–18.

TABLE 5–2 Dimensions and Impedances for Coupled Lines in a Filter.

Line	Width mm	Length mm	Spacing mm	Ground Plane Spacing mm	Z_{0e} Ω	Z_{0o} Ω
#1	0.371	10.04	0.545	5.08	190.4	65.12
#2	1.356	10.04	0.0897	5.08	163.4	7.24
#3	3.831	10.04	1.092	5.08	77.06	22.94

The dimensions are in mm and impedances in Ω. The lines are located equidistant between a pair of ground planes in a uniform dielectric with $\varepsilon_r = 2.23$.

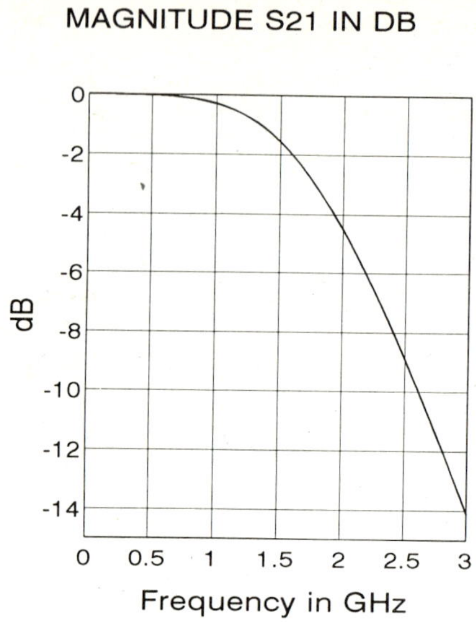

FIGURE 5–18. Transfer Function of a Low Pass Filter Using Coupled Lines.

5.10 COUPLED LINE PAIR ANALYZED AS A 2 PORT: θ_e NOT EQUAL TO θ_o

In the case where θ_e is not equal to θ_o but the lines are equal in dimension, the circuits are not so simple. Zysman and Johnson have reported equivalent circuits and ABCD parameters for the various possible 2-port connections of coupled lines.[16] We show how to derive the ABCD parameters and the resulting equivalent circuit for the case where ports 2 and 4 are open circuited and ports 1 and 3 are the input and output ports. The connections and the resulting equivalent circuit are shown in Figure 5–19.

To analyze the action of this pair of coupled lines, we shall use the 4-port equations and impedance matrix of Eq. (5–39) as in Eq. (5–44).

$$\begin{pmatrix} V_1 \\ V_2 \\ V_3 \\ V_4 \end{pmatrix} = (Z) \begin{pmatrix} I_1 \\ 0 \\ I_3 \\ 0 \end{pmatrix} \tag{5–44}$$

In this equation, we have set the currents I_2 and I_4 equal to zero because these two terminals are open circuited. The first and third equations represented by this matrix form can be written as Eq. (5–45), which represents a 2 port with the input voltage and current as V_1 and I_1, and the output voltage and current as V_3 and I_3, where θ_e is $\beta_e L$ and θ_o is $\beta_o L$.

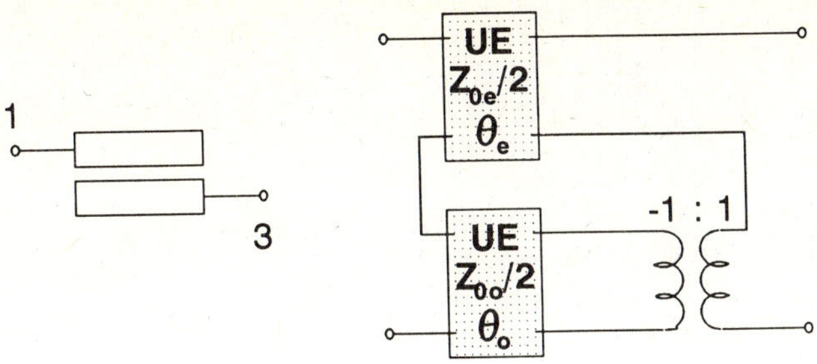

FIGURE 5–19. A: Coupled Strip Line and B: Equivalent Circuit.

$$\begin{pmatrix} V_1 \\ V_3 \end{pmatrix} = -j(Z_{0e}/2)\begin{pmatrix} \cot\theta_e & \csc\theta_e \\ \csc\theta_e & \cot\theta_e \end{pmatrix}\begin{pmatrix} I_1 \\ I_3 \end{pmatrix} - j(Z_{0o}/2)\begin{pmatrix} \cot\theta_o & -\csc\theta_o \\ -\csc\theta_o & \cot\theta_o \end{pmatrix}\begin{pmatrix} I_1 \\ I_3 \end{pmatrix} \quad (5\text{–}45)$$

If we characterize V_1 as being the sum of two voltages V_{1A} and V_{1B}, then V_{1A} is due to the first term on the right-hand side of Eq. (5–45). However, this matrix is identical to that for a section of line L meters long with a phase constant of β_e (see Eqs. (4–5) and (4–6)). The second term on the right-hand side of Eq. (5–45) is similar to this except for the signs. If we were to reverse the polarity of V_3 and the direction of I_3, then the signs for this matrix would be the same as for the first matrix. This reversal of polarity and direction can be accomplished with an ideal inverting transformer. So, Eq. (5–45) corresponds to the circuit in Figure 5–19B. The corresponding strip line representation is shown in Figure 5–19A. Now, we can solve Eq. (5–45) to obtain the ABCD parameters, which relate the input voltage V_1 and current I_1 to the output voltage V_3 and output current I_3. These parameters are given in Eq. (5–46).

$$A = D = \frac{Z_{0e}\cot\theta_e + Z_{0o}\cot\theta_o}{Z_{0e}\csc\theta_e - Z_{0o}\csc\theta_o}, \qquad C = \frac{2j}{Z_{0e}\csc\theta_e - Z_{0o}\csc\theta_o}$$
$$B = j\frac{Z_{0e}^2 + Z_{0o}^2 - 2Z_{0e}Z_{0o}(\cot\theta_e\cot\theta_o + \csc\theta_e\csc\theta_o)}{Z_{0e}\csc\theta_e - Z_{0o}\csc\theta_o} \quad (5\text{–}46)$$

The ABCD parameters for the different possible input-output terminals and terminations were derived by Zysman and Johnson.[17] Two of these are shown in Figure 5–20. Their corresponding ABCD parameters are:

Meander Line Section

$$A = \frac{Z_{0e}\cot\theta_e - Z_{0o}\tan\theta_o}{Z_{0e}\cot\theta_e + Z_{0o}\tan\theta_o}$$
$$B = \frac{2j\,Z_{0e}\,Z_{0o}\cot\theta_e\tan\theta_o}{Z_{0e}\cot\theta_e + Z_{0o}\tan\theta_o}$$
$$C = \frac{2j}{Z_{0e}\cot\theta_e + Z_{0o}\tan\theta_o}$$
$$D = A$$

Comb Line Section

$$A = \frac{Z_{0e}\cot\theta_e + Z_{0o}\cot\theta_o}{Z_{0e}\cot\theta_e - Z_{0o}\cot\theta_o}$$
$$B = \frac{-2jZ_{0e}\,Z_{0o}\cot\theta_e\cot\theta_o}{Z_{0e}\cot\theta_e - Z_{0o}\cot\theta_o}$$
$$C = \frac{2j}{Z_{0e}\cot\theta_e - Z_{0o}\cot\theta_o}$$
$$D = A$$

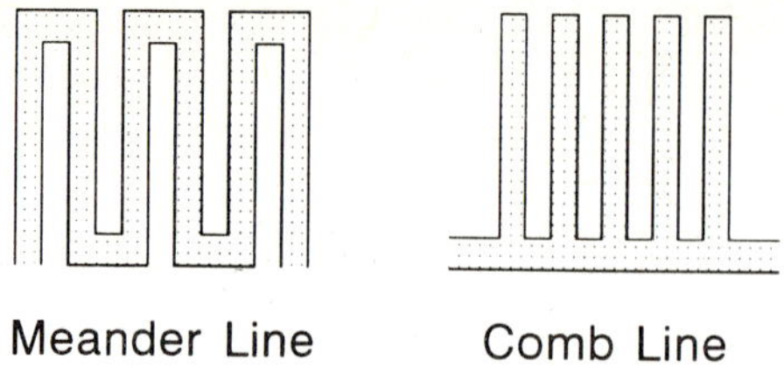

FIGURE 5–20. Coupled Line Sections.

5.11 NARROW BAND FILTERS USING COUPLED RESONATORS

Following Cohn, we can design a narrow band pass coupled line filter by using the ideas and formulas developed for the case where a sequence of resonators are coupled with impedance inverters to form a band pass filter.[18] Design equations are determined by making an approximate equivalence between the circuit representing the coupled resonators and the circuit representing a pair of coupled lines.

To develop formulas for the case of coupled resonators, we start with the prototype low pass filter having elements g_i for $i = 1$ to n. We assume that the prototype is matched at the input to unit impedance. Then, we use the band pass transformation of section 4–9: $\Omega = (1/W)(\omega/\omega_r - \omega_r/\omega)$, where ω_1 and ω_2 are the edges of the band, ω_r is the center of the band such that $\omega_r = \sqrt{\omega_1 \omega_2}$, and $W = (\omega_2 - \omega_1)/\omega_r$. We arrive at the filter shown in Figure 5–21A, where the series elements are given by $L_i' = g_i/W\omega_r$ and $C_i' = W/g_i\omega_r$ and the parallel elements are given by $L_j' = W/g_j\omega_r$ and $C_j' = g_j/W\omega_r$. We require that this be equivalent to the circuit model for coupled resonators as shown in Figure 5–21B. For this model, we simulate the resonators in the neighborhood of ω_r with shunt LC resonant circuits. To find the impedance inverter scaling in terms of the equivalent inductance L_r and capacitance C_r, we use the technique of section 4–10 and equate the impedances Z_{in} and Z_{in}' (scaled by Z_0) of these circuits to obtain $Z_{in}'Z_0 = Z_0[j(\omega L_1' - 1/\omega C_1') + Z_2'] = Z_{in} = K_1^2[j(\omega C_{r1} - 1/\omega L_{r1}) + Y_2]$. From this with $1/C_{r1} = \omega_r^2 L_{r1}$, we get the impedance inverter scaling factor of K_1 as shown in Eq. (5–47).

$$K_1 = \sqrt{\frac{Z_0\, L_{r1}\, g_1\, \omega_r}{W}} \tag{5–47}$$

In a similar way, we can find the remaining impedance inverter scaling factors as shown in Eq. (5–48).

$$K_j = \frac{\omega_r}{W}\sqrt{g_{j-1}\, g_j\, L_{rj-1}\, L_{rj}} \qquad K_{n+1} = \sqrt{\frac{Z_0\, g_n\, \omega_r\, L_{rn}}{W}} \tag{5–48}$$

To obtain the required impedance inversion, one must know the effective inductances of each of the resonators. Some method of simulating the impedance inversion must then be found.

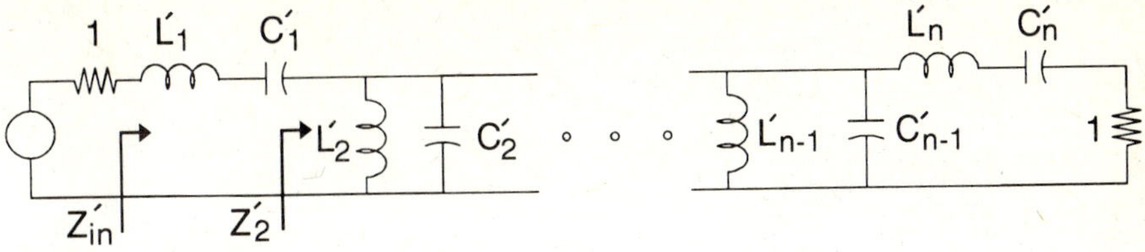

A Band Pass Prototype

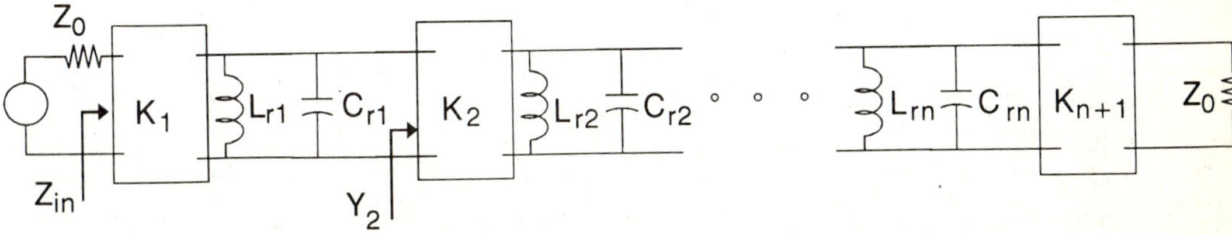

B Coupled Resonator Filter with Impedance Inverters

FIGURE 5–21. Coupled Resonator Filter and Its Prototype.

Our interest lies in using half wave lines to model the resonators shown in Figure 5–21B.[19] For this purpose, we rewrite Eqs. (4–5) and (4–6) as Eq. (5–49), where $\theta = \pi f/2f_0$, and identify this with a T network as shown in Figure 5–22.

$$\begin{pmatrix} V_1 \\ V_2 \end{pmatrix} = -j\, Z_0 \begin{pmatrix} \cot\theta & \csc\theta \\ \csc\theta & \cot\theta \end{pmatrix} \begin{pmatrix} I_1 \\ I_2 \end{pmatrix} \tag{5–49}$$

The T network has the voltage current relation: $V_1 = I_1\, Z_1/2 + I_1\, Z_2 - I_2\, Z_2$. Using the trigonometric identity $\cot\theta = \cot(\theta/2) - \csc\theta$, the voltage V_1 in Eq. (5–49) can be written as $V_1 = -j(Z_0 \cot\theta/2)I_1 + j\, Z_0 \csc\theta\, I_1 - j\, Z_0 \csc\theta\, I_2$. Letting $\theta = 180° + \delta$, and comparing these two expressions for the case when δ is small, we get $Z_1/2 = -j\, Z_0 \cot\theta/2 \cong j\, Z_0 \sin\delta/2 \cong 0$ and $Z_2 = j\, Z_0 \cot\theta \cong j\, Z_0 / \sin\delta$. From this, we see that the circuit of the half wave transmission line is approximately represented by a shunt admittance $Y_p \cong j\, (\sin\delta)/Z_0 \cong j\, \delta/Z_0$. This is to represent the shunt inductor-capacitor combination in Figure 5–21B. Because $\delta = \pi f/2f_0 - \pi$, Y_p can be written as $Y_p \cong j\, (\omega - 2\omega_0)/(4f_0 Z_0)$. For the transmission line resonator model, we can form $d(Y_p)/d\omega$. Noting that f_r of the transmission line resonators is $2\, f_0$, we then have $d(Y_p)/d\omega|_{\omega=\omega_r} = j/(2f_r Z_0)$. We equate this to the

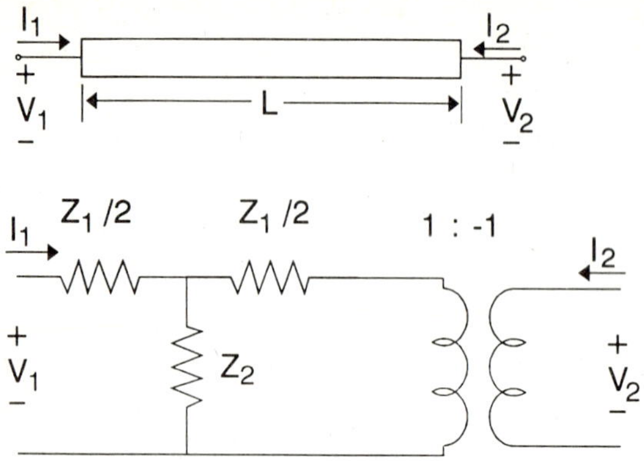

FIGURE 5–22. T-Network Representation of $\frac{1}{2}$ Wave Line.

derivative of the shunt admittance of the inductor-capacitor resonators in Figure 5–21B. The admittance of the shunt LC circuit is $Y = j(\omega C_r - 1/\omega L_r)$ and its derivative with respect to ω evaluated at ω_r is $j\ 2/(\omega_r^2 L_r)$, where $\omega_r^2 = 1/L_r C_r$. Equating these two derivatives, we get the relation $L_r \cong 2Z_0/\pi\omega_r$. Finally, we substitute this into Eqs. (5–47) and (5–48) to obtain Eq. (5–50), which applies for the case of the coupled resonators simulated by half wave lines.

$$Z_0/K_1 = \sqrt{\frac{W\pi}{2g_1}}, \quad Z_0/K_{j+1} = (\pi W/2)\sqrt{\frac{1}{g_j g_{j+1}}}, \quad Z_0/K_{n+1} = \sqrt{\frac{\pi W}{2g_n}} \tag{5–50}$$

Cohn put these formulas into the image impedance formulation.[20] To do this, we consider a coupled resonator circuit consisting of transmission lines connected by impedance inverters. One section of this circuit will have two lines connected by an impedance inverter. From Eqs. (4–5) and (4–6), we get the ABCD matrix of the lines as

$$\begin{pmatrix} \cos\theta & jZ_0 \sin\theta \\ \frac{j\sin\theta}{Z_0} & \cos\theta \end{pmatrix},$$

while that of the impedance inverters is

$$\begin{pmatrix} 0 & -jK \\ \frac{-j}{K} & 0 \end{pmatrix}.$$

The overall ABCD matrix for this section is

$$\begin{pmatrix} \cos\theta & jZ_0\sin\theta \\ \dfrac{j\sin\theta}{Z_0} & \cos\theta \end{pmatrix} \times \begin{pmatrix} 0 & -jK \\ \dfrac{-j}{K} & 0 \end{pmatrix} \times \begin{pmatrix} \cos\theta & jZ_0\sin\theta \\ \dfrac{j\sin\theta}{Z_0} & \cos\theta \end{pmatrix},$$

which results in Eq. (5–51). We can show that the formula for the image impedance is $Z_I = \sqrt{B/C}$.[20] The image impedance of a filter section terminated by its input impedance is discussed in section 4–1. Similarly, the image phase function β is given by $\cos\beta = \sqrt{AD} = A$. Using the ABCD parameters of Eq. (5–51), we arrive at the image impedance formula of Eq. (5–52), where $F = K/Z_0 + (Z_0/K)(1 - \csc^2\theta)$. We also get the image phase relation of Eq. (5–53).

$$\begin{pmatrix} A & B \\ C & D \end{pmatrix} = \begin{pmatrix} \left(\dfrac{Z_0}{K} + \dfrac{K}{Z_0}\right)\cos\theta\sin\theta & jZ_0\left(\dfrac{Z_0}{K}\sin^2\theta - \dfrac{K}{Z_0}\cos^2\theta\right) \\ \dfrac{-j}{Z_0}\left(\dfrac{K}{Z_0}\sin^2\theta - \dfrac{Z_0}{K}\cos^2\theta\right) & \left(\dfrac{Z_0}{K} + \dfrac{K}{Z_0}\right)\cos\theta\sin\theta \end{pmatrix} \quad (5\text{–}51)$$

$$Z_I^2 = Z_0^2 \frac{\dfrac{Z_0}{K}\sin^2\theta - \dfrac{K}{Z_0}\cos^2\theta}{\dfrac{K}{Z_0}\sin^2\theta - \dfrac{Z_0}{K}\cos^2\theta} = \left\{\frac{Z_0^3}{K} - Z_0^2\left(\frac{Z_0}{K} + \frac{K}{Z_0}\right)\cos^2\theta\right\}/(F\sin^2\theta) \quad (5\text{–}52)$$

$$\cos\beta = (Z_0/K + K/Z_0)\sin\theta\cos\theta \quad (5\text{–}53)$$

These relations are intended to characterize a ladder network with coupled resonator circuits as in Figure 5–21B, when these circuits use transmission lines as the resonators.

5.12 NARROW BAND COUPLED LINE FILTERS

Now, we derive the corresponding relations for Z_I and $\cos\beta$ for a resonator circuit that uses coupled lines with its ports 2 and 4 open circuited; port 1 is the input, and port 3 is the output. Using Eq. (5–39) with the assumptions that $k = 1$, $\beta_e L = \beta_o L = \theta$, $Z_{0e}^a = Z_{0e}^b$, and $Z_{0o}^a = Z_{0o}^b$, we have Eq. (5–54), where we have used $Z_+ = Z_{0e} + Z_{0o}$ and $Z_- = Z_{0e} - Z_{0o}$.

$$\begin{pmatrix} V_1 \\ V_3 \end{pmatrix} = -(j/2)\begin{pmatrix} Z_+\cot\theta & Z_-\csc\theta \\ Z_-\csc\theta & Z_+\cot\theta \end{pmatrix}\begin{pmatrix} I_1 \\ I_3 \end{pmatrix} \quad (5\text{–}54)$$

Changing the notation so that the subscript 2 replaces the subscript 3, we can express V_1 as $-(j/2)(Z_+\cot\theta + Z_-\csc\theta)I_1 + (jI_1/2)Z_-\csc\theta - (jI_2/2)Z_-\csc\theta$. Modeling this as the

T network of section 4–1 followed by a 1:–1 transformer (see Figure 5–22), we have $Z_1 = -jZ_+ \cot\theta - jZ_- \csc\theta$ and $Z_2 = (j/2)Z_- \csc\theta$. Using these with the formula for the square of the image impedance determined in section 4–1, $Z_I^2 = Z_1Z_2(1 + Z_1/4Z_2)$, we get Eq. (5–55). Similarly, we use $\cos\beta = 1 + Z_1/2Z_2$ to get Eq. (5–56).

$$Z_I^2 = [(Z_{0e} - Z_{0o})^2 - (Z_{0e} + Z_{0o})^2 \cos^2\theta]/(4\sin^2\theta) \tag{5–55}$$

$$\cos\beta = [(Z_{0e} + Z_{0o})/(Z_{0e} - Z_{0o})] \cos\theta \tag{5–56}$$

Equating the factor of Eq. (5–53) to the similar factor of Eq. (5–56), we get the first term in Eq. (5–57). Equating terms in Eq. (5–52) to the similar ones in Eq. (5–55), we get the remaining terms in Eq. (5–57).

$$\begin{aligned} \frac{Z_{0e} + Z_{0o}}{Z_{0e} - Z_{0o}} &= \left(\frac{Z_0}{K} + \frac{K}{Z_0}\right)\sin\theta \\ (Z_{0e} - Z_{0o})^2 &= \frac{4\,Z_0^3/K}{K/Z_0 + (Z_0/K)(1 - \csc^2\theta)} \\ (Z_{0e} + Z_{0o})^2 &= \frac{4\,Z_0^2\,(Z_0/K + K/Z_0)}{K/Z_0 + (Z_0/K)(1 - \csc^2\theta)} \end{aligned} \tag{5–57}$$

As Cohn notes, these relations are a consistent set. For the narrow band case when $\theta \cong \pi/2$ and $Z_0/K \ll 1$, the first two relations in Eq. (5–57) reduce to the relations

$$\frac{Z_{0e} + Z_{0o}}{Z_{0e} - Z_{0o}} = \left(\frac{Z_0}{K} + \frac{K}{Z_0}\right) \quad \text{and} \quad Z_{0e} - Z_{0o} = 2Z_0^2/K.$$

These relations can be solved for Eq. (5–58).

$$Z_{0e}/Z_0 = 1 + \frac{Z_0}{K} + \left(\frac{Z_0}{K}\right)^2, \qquad Z_{0o}/Z_0 = 1 - \frac{Z_0}{K} + \left(\frac{Z_0}{K}\right)^2 \tag{5–58}$$

Substituting from Eq. (5–50) into Eq. (5–58), we determine the formulas in Eq. (5–59) for calculating the even and odd mode impedances for each section of the coupled lines where we take Z_0 to be the impedance of the input line.[21]

$$\begin{aligned} Z_{0e1}/Z_0 &= 1 + \sqrt{\frac{W\pi}{2g_1}} + \frac{W\pi}{2g_1}, \qquad Z_{0o1} = 1 - \sqrt{\frac{W\pi}{2g_1}} + \frac{W\pi}{2g_1} \\ Z_{0ej+1}/Z_0 &= 1 + (W\pi/2)\sqrt{\frac{1}{g_jg_{j+1}}} + \frac{(W\pi)^2}{4g_jg_{j+1}} \\ Z_{0oj+1}/Z_0 &= 1 - (W\pi/2)\sqrt{\frac{1}{g_jg_{j+1}}} + \frac{(W\pi)^2}{4g_jg_{j+1}} \\ Z_{0en+1}/Z_0 &= 1 + \sqrt{\frac{W\pi}{2g_n}} + \frac{W\pi}{2g_n}, \qquad Z_{0on+1}/Z_0 = 1 - \sqrt{\frac{W\pi}{2g_n}} + \frac{W\pi}{2g_n} \end{aligned} \tag{5–59}$$

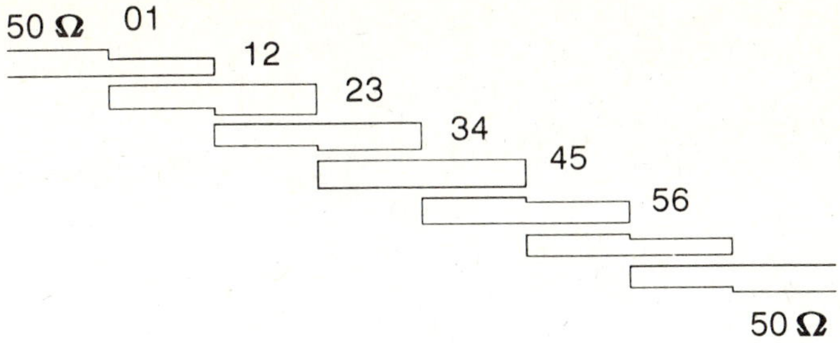

FIGURE 5–23. Coupled Line Filter.

Example: We can design a band pass filter having cut-off frequencies of 1.95 GHz and 2.05 GHz using Eq. (5–59). We choose the Chebyshev 0.2 dB ripple prototype with $n = 5$. The prototype elements are $g_1 = g_5 = 1.3395$, $g_2 = g_4 = 1.337$, and $g_3 = 2.166$. The center frequency is $\sqrt{1.95 \times 2.05} = 1.9994$ GHz. The normalized bandpass is $W = (2.05 - 1.95)/1.9994 = 0.050016$. Using Eq. (5–59), we get the set of even and odd mode impedances for the coupled lines shown in Figure 5–23: $Z_{0e1} = Z_{0e6} = 65.04$, $Z_{0o1} = Z_{0o6} = 40.82$, $Z_{0e2} = Z_{0e5} = 53.11$, $Z_{0o2} = Z_{0o5} = 47.24$, $Z_{0e3} = Z_{0e4} = 52.41$, and $Z_{0o3} = Z_{0o4} = 47.80$. Using calculations for determining the width, spacing, and length of edge coupled lines, we found the dimensions shown in Table 5–3 for these lines. Because the even and odd mode effective dielectric constants were not equal, we initially used line lengths that were the average of the lengths calculated by $L = (\frac{1}{4}) \times 3 \times 10^{11}/(4 \times 10^9\sqrt{\varepsilon_{eff}})$ mm. These are the lengths shown in Table 5–3. They did not work too well, as seen in Figure 5–24. A better filter was obtained by using the length given by the even mode dielectric constant. An even better filter was then obtained by using a CAD optimizing program as shown in Figure 5–24.

With modifications, this method of designing filters can be used to obtain larger bandwidths.[22,23] A report of modifications in this design technique to give better upper stop band rejection and to allow edge coupled lines in the design was given by Chang and Itoh.[24]

TABLE 5–3 Dimensions of Broadside Coupled Lines for a Narrow Band Filter. The substrate for these lines had $\varepsilon_r = 2.23$ and $h = 0.635$ mm.

				Optimized			
Line	Width mm	Length mm	Spacing mm	Length mm	Spacing mm	Z_{0e} Ω	Z_{0o} Ω
01 & 56	.536	14.9	0.35	14.68	.301	65.04	40.82
12 & 45	.607	14.65	1.32	14.62	1.22	53.11	47.24
23 & 34	.609	14.63	1.55	14.60	1.486	52.41	47.80

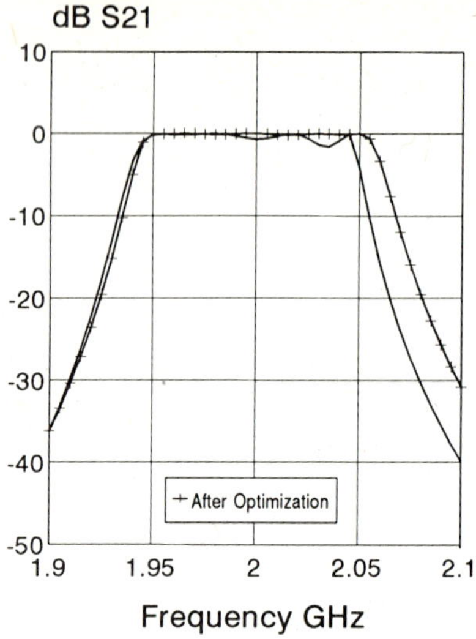

FIGURE 5–24. Coupled Line Filter Response.

5.13 SUSPENDED SUBSTRATE STRIPLINE FILTERS

The main problem associated with edge coupled filters is the inability to achieve high values of coupling without narrow line spacings and attendant high precision fabrication of the coupled line structure. Because of that, an alternative design is to place microstrip lines on two sides of a substrate and obtain broadside coupling of the lines. The amount of coupling can be increased and the accuracy required of the spacing is reduced. For some filters, the substrate is suspended in air and ground planes are located offset from either side of the substrate as shown in Figure 5–25. The result is a line with three layers of different dielectric. The structure is referred to as a *suspended substrate stripline* (*SSS*). Here, we discuss this type of design when used for band pass filters. The innovative step in the design of these filters is that of determining the dimensions for the *broadside coupled suspended substrate striplines* (*BCSSS*) incorporated in these filters. We present two algorithms for designing these filters. In each method, the design of the filter starts with a low pass prototype. Lösche and Malherbe converted the prototype to a high

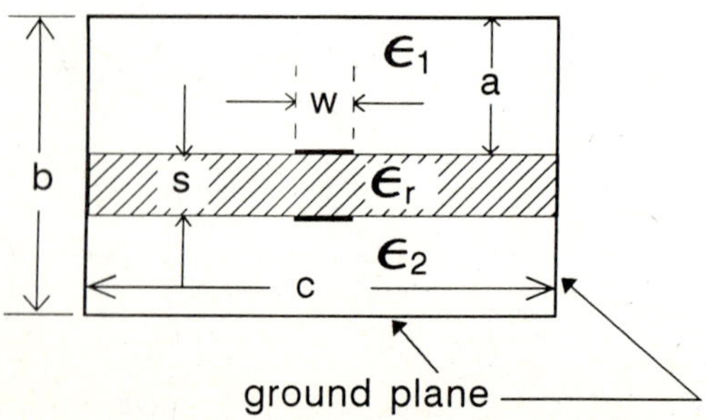

FIGURE 5–25. Broadside Coupled Suspended Substrate Stripline.

pass prototype filter by inversion of the component values.[25] Then, they simulated the lumped elements in the shunt branches with transmission lines. The characteristic impedances of these lines were determined from the lumped element values. The series capacitors were simulated with coupled lines. The dimensions of the broadside coupled lines having the required characteristic impedances were computed. This step in the design required the use of quasi-static design formulas. The other lines in the filter were realized either as suspended substrate striplines or as microstrip. The dimensions for the SSS were calculated using quasi-static formulas.

Sturdivant published a method for calculating SSS filters that is based on the coupled resonator formulas presented in section 5–11.[26] In this method, the coupled lines simulate a coupling capacitor connected between the SSS which are treated as resonators.

5.14 SUSPENDED SUBSTRATE STRIPLINE FILTER DESIGN, METHOD 1

Lösche and Malherbe presented an algorithm for designing a bandpass filter. They started with a low pass prototype, which they converted to a high pass prototype. This became a bandpass filter in the *f* domain with series capacitors connecting the shunt elements. They simulated these capacitors with BCSSS. The basic problem became one of computing the dimensions of the BCSSS so that it performed as the desired capacitance in the pass band The quantities to be determined were the *W*, *b*, and *s* of Figure 5–25, and the length *L* o the BCSSS. To relate these primary quantities to the Z_{0e} and Z_{0o}, they used the eve and odd mode quasi-static impedance formulas of Eq. (5–60), where $s' = s/\sqrt{\varepsilon_r}$ an $b' = b - s(1 - 1/\sqrt{\varepsilon_r})$ are the dimensions adjusted to allow for the effect of the dielectr layer.[27]

$$Z_{0e} = \frac{188.4}{\dfrac{W/b'}{1-s'/b'} + C_{fe}/\varepsilon}, \qquad Z_{0o} = \frac{188.4}{\dfrac{W/b'}{1-s'/b'} + W/s' + C_{fo}/\varepsilon} \qquad \text{(5–6}$$

The capacitor terms in these formulas are given in Eqs. (5–61) and (5–62).

$$C_{fo}/\varepsilon = \left(\frac{1}{b'/s'-1}\ln(b'/s') - \ln(1-s'/b')\right) b'/(s'\pi) \qquad \text{(5–}$$

$$C_{fe}/\varepsilon = 0.4413 + (s'/b')\, C_{fo}/\varepsilon \qquad \text{(5–}$$

The angles θ_e and θ_o, given in Eq. (5–63), are the phase changes of the even and the mode waves propagating over the length of line *L*.

$$\theta_e = 2\pi f L\sqrt{\varepsilon_{effe}}/(3\times10^8),\ \theta_o = 2\pi f L\sqrt{\varepsilon_{effo}}/(3\times10^8) \qquad \text{(5–}$$

The length *L* occurs linearly in these formulas. They assumed that Z_{0e}, Z_{0o}, ε_{effe}, and did not vary with frequency over the filter pass band.

Lösche and Malherbe determined these coupled line dimensions by

1) specifying that the S_{11} into the filter be zero at a frequency near to the folding frequency f_o, and
2) comparing the B parameter of the coupled line to the B parameter of the open circuited line, which represents the prototype series capacitor in the f domain (see Figure 5–27B).

From Eq. (5–46), we have the coupled line parameters of Eq. (5–64).

$$A = D = \frac{Z_{0e}\cot\theta_e + Z_{0o}\cot\theta_o}{Z_{0e}\csc\theta_e - Z_{0o}\cot\theta_o}, \qquad C = \frac{2j}{Z_{0e}\csc\theta_e - Z_{0o}\csc\theta_o}$$
$$B = (j/2)\frac{Z_{0e}^2 + Z_{0o}^2 - 2Z_{0e}Z_{0o}(\cot\theta_e\cot\theta_o + \csc\theta_e\csc\theta_o)}{Z_{0e}\csc\theta_e - Z_{0o}\csc\theta_o} \tag{5–64}$$

The S_{11} into the filter is given in terms of the ABCD parameters as $S_{11} = (B - CZ_0^2)/(2AZ_0 + B + CZ_0^2)$. Setting this to zero at a frequency f_u just above f_o results in Eq. (5–65). Using Eqs. (5–65), (5–60), and (5–63) together, we can calculate a value for L when given b, W, and s.

$$Z_{0e}^2 + Z_{0o}^2 - 2Z_{0e}Z_{0o}(\cot\theta_e\cot\theta_o + \csc\theta_e\csc\theta_o)|_{f=fu} - 4Z_0^2 = 0 \tag{5–65}$$

From Eq. (4–5), we have the ABCD parameters of the open circuited line that represents the capacitor in the f domain as $A = 1$, $B = Z_c/\tan(\theta)$, $C = 0$, and $D = 1$. In this, $\theta = \pi f/2f_o$, where f_o is the center frequency of the pass band. For small angles θ_e and θ_o and for $Z_{0e} \gg Z_{0o}$, the coupled line parameters A and D are approximately unity. The coupled line C parameter approaches zero for small angles θ_e and θ_o. The remaining problem is matching the B parameter of the BCSSS to the f domain equivalent of the prototype capacitor. Lösche and Malherbe adjusted the coupled line dimensions to make this B parameter approximate $Z_C/\tan(\theta)$ over the filter pass band. This required minimizing the difference Δ shown in Eq. (5–66) over the pass band.

$$(j/2)\frac{Z_{0e}^2 + Z_{0o}^2 - 2Z_{0e}Z_{0o}(\cot\theta_e\cot\theta_o + \csc\theta_e\csc\theta_o)}{Z_{0e}\csc\theta_e - Z_{0o}\csc\theta_o} - \frac{Z_C}{\tan(\theta)} = \Delta \tag{5–66}$$

To improve the accuracy of their solution by including a correction for the fringing capacitances at the open ends of the coupled lines, Lösche and Malherbe developed Eq. (5–67) for determining a value of the effective dielectric constants that would compensate for the open end effects on the BCSSS.

$$\varepsilon'_{effi} = (\sqrt{\varepsilon_{effi}} + WZ_{0i}(C_{fi}/\varepsilon)/(376.73\,L))^2 \tag{5–67}$$

In this formula, the subscript i takes values of e and o. They took ε_{effe} to be 1 and ε_{effo} was determined by a method to correct for the effects of the metal cover.[28]

Example: The Design of a 4 to 10 GHz Band Pass Filter. This example follows the procedure explained by Lösch and Malherbe. In this example, the low pass prototype is converted to a high pass prototype, where the folding frequency is the center frequency of the final bandpass filter. These prototypes are shown in Figure 5–26. The series capacitors necessary for the high pass prototype are simulated by BCSSS. The shunt inductors are simulated by shorted transmission lines to ground. The shunt branches consisting of a capacitor and inductor in series are simulated by two series connected quarter wave lines as shown in Figure 5–27A.[29]

The initial prototype element values are those chosen from the generalized Chebyshev polynomial of section 4–7.[30] These correspond to a pass band return loss greater than 20 dB and a minimum stop band insertion loss greater than 40 dB. To derive a high pass prototype filter from these values, we took the inverse of these values and frequency scaled them by the factor $\Omega'_c = \tan[(4/7)(\pi/2)]$ in order to get a lower cut-off fre-

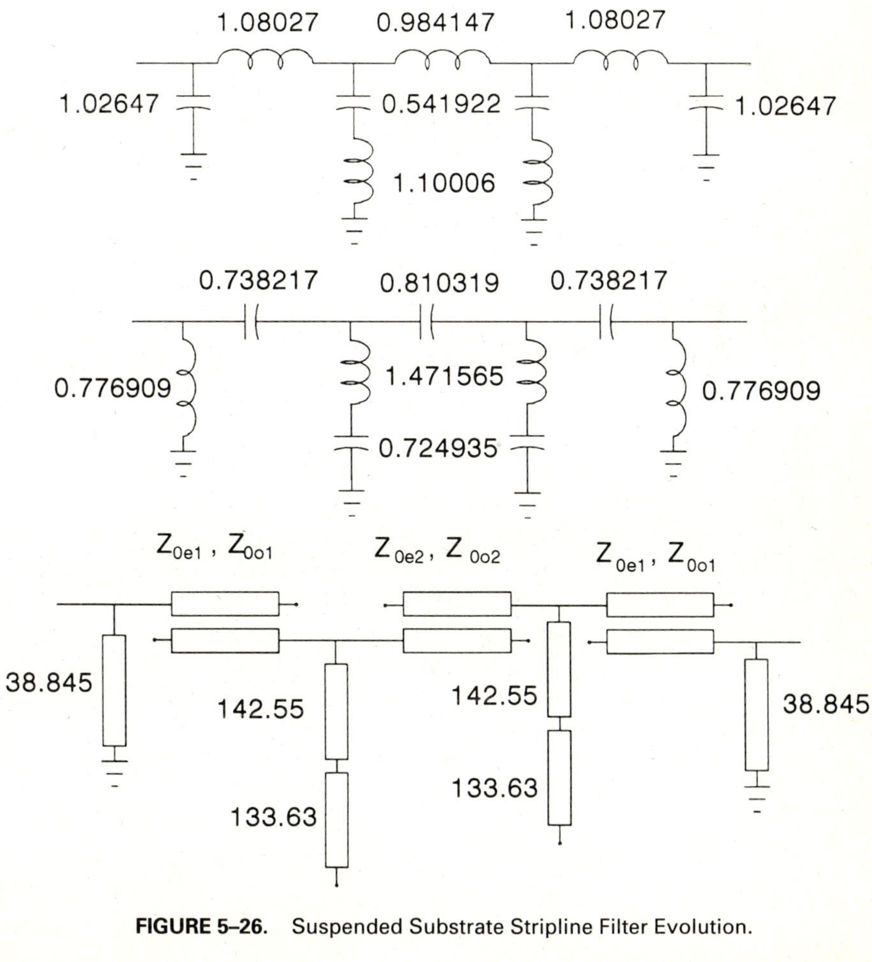

FIGURE 5–26. Suspended Substrate Stripline Filter Evolution.

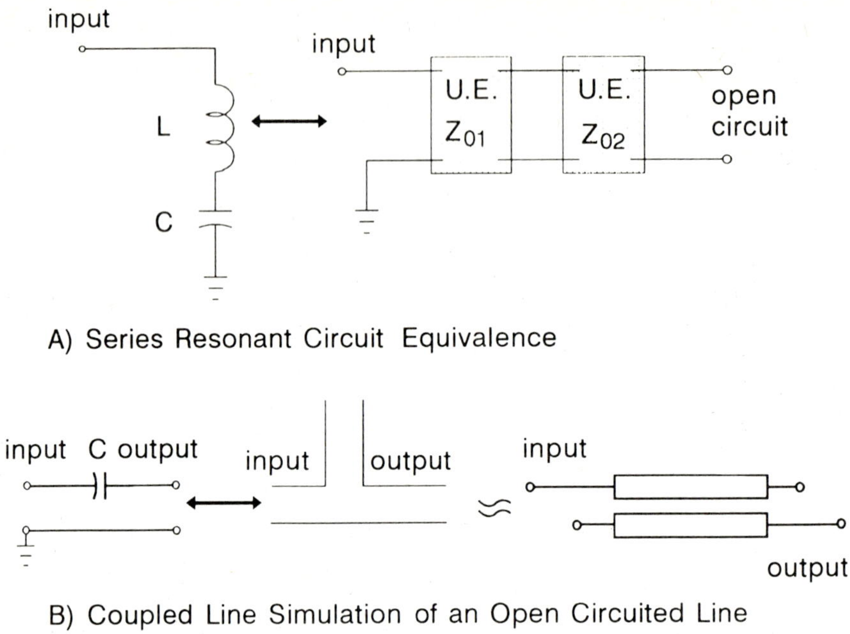

FIGURE 5–27. Equivalent Circuits for a Coupled Line and for Two Series Quarter Wave Lines. The equivalence of the circuits in *A* requires $Z_{01} = L + 1/C$ and $Z_{02} = (L + 1/C)/LC$.

quency of 4 GHz and a center frequency of 7 GHz. For instance, the input element transforms as $g'_1 = (1/g_1)/\tan(\pi f_c/2f_o) = (1/1.02647)/\tan(\pi 4/2 \times 7) = 0.776909$.

The prototype series capacitances are simulated with coupled lines as shown in Figure 5–27B. At the cut-off frequency, the first prototype capacitor impedance scales to 50/0.738217 = 67.73 Ω. For simulating this with BCSSS, we can adjust the dimensions *b, W, L,* and *s*. Because the dimensions *b* and *s* affect the remaining lines, we made this calculation before calculating the dimensions for the other lines. We chose to set s to 0.7112 mm because this is the thickness of a commercially available substrate. Of the remaining dimensions, *L* is determined for any pair of *b* and *W* by the requirement of Eq. (5–65). We wrote a computer program to minimize the squared difference Δ^2 of Eq. (5–66) over the pass band by allowing the dimensions b and W shown in Figure 5–25 to vary and, at the same time, satisfy Eq. (5–65) at the frequency $f_u = 7.2$ GHz. For $s = 0.7112$ mm and $\varepsilon_r = 2.23$, we determined that $b = 5.5$ mm and $W = 2.2$ mm minimized the squared difference. These dimensions required a length of coupled line $L = 4.46$ mm to satisfy Eq. (5–65): $S_{11}(f_u) = 0$. The resulting impedances computed from Eq. (5–60) are $Z_{0e} = 187.06$ and $Z_{0o} = 30.12$ Ω. A similar procedure for the second series capacitor resulted in $W = 2.53$ mm and $L = 4.48$ mm for $b = 5.5$ mm and $s = 0.7112$ mm. The even and odd mode impedances for these were $Z_{0e2} = 175.08$ and $Z_{0o2} = 26.85$ Ω. Figure 5–28A shows

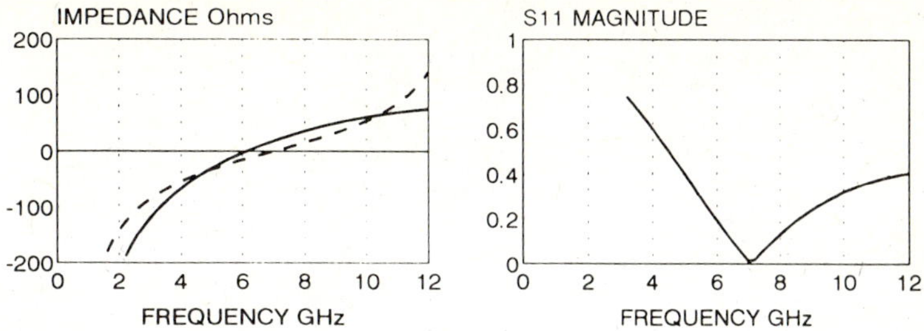

FIGURE 5–28. Coupled Line Performance Graph. The solid line is for the B parameter of the coupled line and the dashed line is for the B parameter of the open circuited stub.

a comparison between the desired B parameter and the B parameter of the coupled line. Figure 5–28B shows the calculated S_{11} into the BCSSS when it is terminated in 50 Ω.

The next step is to simulate the input element (inductor) as a shorted microstrip line. To make this a microstrip line, we required that the substrate under this portion of the filter have a ground plane backing it. This line must have a characteristic impedance given by impedance scaling of $Z_0 = 0.776909 \times 50 = 38.845\ \Omega$. Assuming a substrate with $\varepsilon_r = 2.32$, $h = 0.7112$ mm (28 mils), a top cover 2.39 mm above the line, and using CAD software, we calculated this line to be a microstrip line with a width of 2.01 mm and a quarter wave length at 7 GHz of 7.86 mm. The 2.39 mm dimension was obtained by $(b - s)/2 = (5.5 - 0.7112)/2 = 2.39$.

The shunt LC branches are simulated with back-to-back quarter wave lines as shown in Figure 5–27A. This conversion is given by $Z_{01} = Z_0\,(L + 1/C) = 50(1.471565 + 1/0.724935) = 142.55\ \Omega$ and $Z_{02} = Z_0\,(L + 1/C)/LC = 50(1.471565 + 1/0.724935)/(1.471565 \times 0.724935) = 133.63\ \Omega$ (see problem 5–14). Using CAD software, we determined that these SSS lines must have widths of 0.96 mm and 1.15 mm and quarter wave lengths of 9.51 and 9.56 mm on a substrate of $s = 0.7112$ mm and a ground plane spacing of $b = 5.5$ mm for these two lines.

5.15 SUSPENDED SUBSTRATE STRIPLINE FILTER DESIGN, METHOD 2

Sturdivant discussed another design procedure.[26] In this procedure, the narrow band coupled resonator design equations of section 5–11 were used. The resonators were simulated with half wave SSS. The inversion and coupling required between the resonators were obtained as capacitors between these lines. These capacitors were simulated with BCSSS. The formulas for calculating the capacitance necessary for the required inversion and coupling between the resonators were those given by Matthaei, Young, and Jones.[31] They show that a capacitor, in series with lines of negative length as shown in Figure 5–29, acts as an impedance inverter.

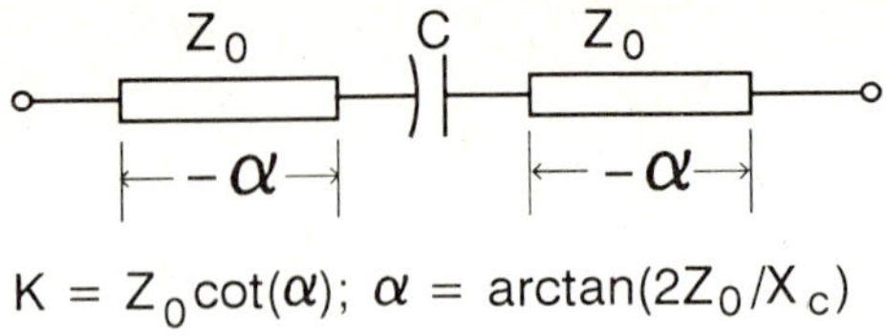

FIGURE 5–29. Impedance Inversion by a Series Capacitor.

First, Sturdivant determined the required Z_0/K_i from the normalized bandwidth W and the prototype element values g_i using Eq. (5–50). Then, the coupling capacitance values required for inversion were found from Eq. (5–68).[31]

$$X_{ci}/Z_0 = K_i/Z_0 - Z_0/K_i \tag{5–68}$$

To complete the inversion circuit shown in Figure 5–29, the electrical length of the resonators was decreased from half wave to the electrical length given in Eq. (5–69).

$$\theta_i = \pi - (\tfrac{1}{2})(\arctan(2Z_0/X_{i+1}) + \arctan(2Z_0/X_i)) \tag{5–69}$$

The needed coupling capacitance C_i determines the length L_{ci} of the overlap in the coupled line section. Sturdivant assumed that the capacitance between the coupled lines per meter was one-half of the odd mode capacitance C_o. With the odd mode capacitance per unit length, Sturdivant calculated the required length of BCSSS. He calculated these lengths by Eq. (5–70), where C_{fend} is the fringing capacitance at the open ends of the coupled line.

$$L_{ci} = 2(C_i - 2C_{\text{fend}})/C_o \tag{5–70}$$

He adjusted L_{ci} to allow for the fringing capacitance at the open ends of the BCSSS by applying the Silvester and Benedek formulas for an open end of a microstrip line as shown in Eq. (5–71).[32]

$$C_{\text{fend}} = 1.4\,(2.2/2.5)\exp\,[(\ln 10)\Sigma A_i\,(\log_{10}(w/s))^{i-1}]\ pF/m. \tag{5–71}$$

Sturdivant introduced the factor of 1.4 in this fringing capacitance calculation to make the calculated capacitances fit measured results. The factor (2.2/2.5) converts from the ε_r of 2.5 used by Silvester and Benedek to that of 2.2 used by Sturdivant. Compensating for this fringing capacitance also can be done by using Eq. (2–12). At least for the example below, Eq. (2–12) yields nearly equal results if the length extension is multiplied by 1.4. These methods of compensating for fringing capacitance do not yield accurate results for all cases. An attempt to use this method with 28 mil substrate and 93 mil line resulted in negative lengths for the overlap.

From the resonator electrical lengths given by Eq. (5–69), Sturdivant calculated the length of SSS needed between the BCSSS by Eq. (5–72).

$$L_{ri} = (\theta_i\, 3 \times 10^8/f_o - (L_{ci} + L_{ci+1})\sqrt{\varepsilon_{reffe}}/2)/\sqrt{\varepsilon_{reff}} \tag{5–72}$$

Sturdivant presented a FORTRAN program for quasi-static determination of the effective relative dielectrics, capacitances per unit length, and impedances when given the SSS and BCSSS dimensions and substrate dielectric constant.

Example: The Design of a 8 to 10 GHz SSS Bandpass Filter. This example uses a 5 element 0.5 dB ripple Chebyshev prototype filter. The suspended substrate has a thickness of s = 5 mils, and the SSS has dimensions of w = 45 mils, b = 35 mils, and c = 90 mils, all chosen to produce a 50 Ω line. The substrate has $\varepsilon_r = 2.2$. The center frequency is $f_o = \sqrt{8 \times 10} = 8.94427$ GHz. The normalized bandwidth is $W = (10 - 8)/8.94427 = 0.22361$. With Sturdivant's FORTRAN program, we calculate the parameters given in Table 5–4.

Using Eq. (5–50), we get $Z_0/K_1 = Z_0/K_6 = \sqrt{(0.22361\ \pi)/(2 \times 1.7058)} = 0.45377$. Using Eq. (5–68), we have $X_{c1}/Z_0 = X_{c6}/Z_0 = 1/0.45377 - 0.45377 = 1.74998$. Using Eq. (5–69), we get the resonator electrical lengths of: $\theta_1 = \theta_5 = 141.96°$, $\theta_2 = \theta_4 = 155.13°$, and $\theta_3 = 157.52°$. The required coupling capacitances are calculated as in $C_1 = (Z_0/X_1)/(\omega_o Z_0) = (1/1.74998)/(2\pi\ 8.94427 \times 10^9 \times 50) = 0.20336 \times 10^{-12}$. To find the lengths of BCSSS that result in these capacitances, we used Eq. (5–71) with $A_1 = 1.295$, $A_2 = -0.2817$, $A_3 = 0.1367$, $A_4 = -0.0133$, and $A_5 = -0.0267$ to find the open end fringing capacitance of $C_{\text{fend}} = 16.128$ pF per meter width of the line (Silvester and Benedek furnished a table of coefficients A_i for use with different ε_r). This is a capacitance of $C = 16.28 \times 0.045/39.37 = 0.018434$ pF, where the factor 39.37 converts from meters to inches. Using Eq. (5–70), we calculate the overlap distance of the BCSSS that couples to the first "half wave" resonator of $L_{c1} = 2(0.20336 - 2 \times 0.018434)/400.6 = 0.00083121$ m or 32.72 mils. Finally, the lengths of SSS between the BCSSS were calculated using Eq. (5–71) as

$$L_{r1} = (141.96 \times 3 \times 10^8/(8.94427 \times 10^9) - (0.00083121 + 0.0002738)\sqrt{1.07519/2})/\sqrt{1.15389} = 0.01178 \text{ m}.$$

A summary of these results is given in Table 5–5.

TABLE 5–4 Line Parameters for Suspended Substrate Stripline Filter.

SSS	BCSSS	BCSSS
$Z_0 = 50.445\ \Omega$	$Z_{0e} = 87.91\ \Omega$	$Z_{0o} = 11.723\ \Omega$
$\varepsilon_{reff} = 1.15389$	$\varepsilon_{reffe} = 1.07519$	$\varepsilon_{reffo} = 1.98465$
$C_0 = 70.981$ pF/m	$C_{0e} = 39.317$ pF/m	$C_{0o} = 400.6$ pF/m

TABLE 5–5 8 to 10 GHz Suspended Substrate Stripline Filter Parameters.

i	Z_0/K_i	X_i/Z_0	C_i (pF)	L_{ri} (mm)	L_{ci} (mm)
1	0.45377	1.74998	0.20336	11.779	0.8312
2	0.24253	3.88074	0.091705	13.225	0.2738
3	0.19871	4.83364	0.073626	13.478	0.1835
4	0.19871	4.83364	0.073626	13.225	0.1835
5	0.24253	3.88074	0.091705	11.779	0.2738
6	0.45377	1.74998	0.20336	—	0.8312

PROBLEMS

5–1. Substitute Eq. (5–16) into Eqs. (5–5) to (5–7) to show that $I_1 = I_2$ for $\gamma_e = j\omega\sqrt{a+b}$ and $I_1 = -I_2$ for $\gamma_o = j\omega\sqrt{a-b}$.

5–2. Verify that the condition $L_1C_{11} = L_2C_{22}$ results in the line currents being related by the inverse of k, where k is given by Eq. (5–13).

5–3. Find the even and odd mode impedances for edge-coupled lines having $W = 1$ mm and $s = 0.3$ mm located on a substrate with $h = 0.7112$ mm and $\varepsilon_r = 2.32$.

5–4. Determine the even and odd mode characteristic impedances for a pair of coupled lines used as a 3 dB directional coupler with $Z_0 = 50\ \Omega$. The center frequency of the coupler is 4 GHz. Use the program in the appendix iteratively to calculate the line widths and spacings required for these lines if they are on a substrate with $\varepsilon_r = 2.32$ and $h = 28$ mils. Your answer should show the impracticality of such close coupling with a single pair of microstrip lines.

5–5. Using a CAD program, modify the design obtained in problem 5–4 to use broadside coupled strip lines by adjusting the line width and line spacing. Use a substrate with $\varepsilon_r = 2.32$.

5–6. Design a Lange coupler with a center frequency of 4 GHz and with $N = 4$, $C = 0.5$, and $Z_{0n} = 30\ \Omega$. Determine the line widths and spacings required if an alumina substrate with $h = 0.635$ mm and $\varepsilon_r = 9.8$ is to be used. Modify your design to determine the widths and spacings if $N = 6$.

5–7. From Eq. (5–35), derive an expression for $|S_{21}|/C$ as a function of θ. Solve this expression for θ and show that the insertion gain of the Lange coupler normalized to its gain at the center frequency is down by 3 dB at a frequency of 0.455 of the center frequency of the coupler.

5–8. From Eq. (5–33), show that the ratio of the odd mode to the even mode impedance of a Lange coupler decreases as the coupling is increased.

5–9. Verify that $Z_{0e} = 2Z_0 + 1/C$ and $Z_{0o} = 1/C$ for circuit C in Figure 5–16.

5–10. Determine the required even and odd mode impedances for a low pass coupled line filter similar to that in the example of section 5–9. Start with a Chebyshev 0.5 dB ripple 5 section inductor input low pass prototype. By bringing two unit elements into the filter from each end, the resulting prototype is symmetrical about the center inductor. Use the equivalent circuit of Figure 5–16A to convert this prototype to the coupled line equivalent. Scale this to 50 Ω.

5–11. A third order Chebyshev 0.5 dB ripple low pass filter is to be made with coupled lines. The filter is to have a 1.6 GHz cut off frequency and a folding frequency f_0 equal to 2.8 GHz. It is to be matched to 50 Ω. Start with two unit elements at the input of the prototype and one at the output. Determine the even mode and odd mode line impedances required for this filter.

5–12. Use a CAD program to simulate the filter design of problem 5–11. The coupled lines are to be broadside coupled strip lines with a dielectric material having $\varepsilon_r = 2.23$, $s = 0.7112$ mm, and $b = 8$ mm. You will not be able to obtain both Z_{0e} and Z_{0o} by adjusting w, so set w to get the correct Z_{0e} for each section. After simulating the resulting broadside coupled stripline filter, optimize it by allowing w and the section line lengths L to vary to get $|S_{21}|$ to less than $\frac{1}{2}$ dB from 0 to 1.5 GHz. Again, simulate the result and compare it with the previous simulation.

5–13. Design a narrow band filter with a pass band from 3.95 to 4.05 GHz. This is to be a

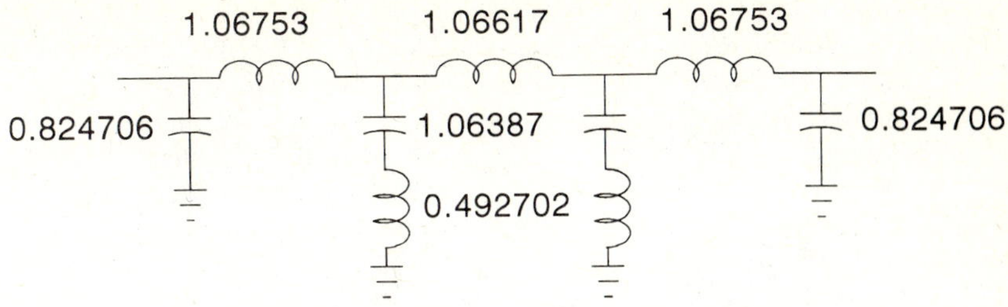

FIGURE 5–30. Seven Element Prototype Filter Circuit

Chebyshev 0.5 dB ripple 5 element prototype. The final filter is to be edge coupled lines on alumina with $h = 0.635$ mm and $\varepsilon_r = 9.8$.

5–14. By equating the input impedances of the two circuits in Figure 5–27A, show that $Z_{01} = (L + 1/C)$ and $Z_{02} = (L + 1/C)/LC$. Note that the input impedance of the cascaded unit elements can be determined from the product of their ABCD parameters.

5–15. Show that the S_{11} into a 2 port is given by $(B - CZ_0^2)/(2AZ_0 + B + CZ_0^2)$ as mentioned in section 5–14.

5–16. Using the method in section 5–14 and using the prototype elements of Figure 5–30, determine the impedances at 4 GHz of the shunt elements for a 4 to 8 GHz bandpass filter. Assume that the series capacitors have equivalent impedances in the f domain of 92.4 Ω at 4 GHz and that the substrate has $\varepsilon_r = 2.23$ and dimensions of $b = 8.5$ mm and $s = 0.7112$ mm. If the equivalent line widths are $W = 2.35$ mm, the equivalent line lengths are 4.29 mm and the even and odd mode effective relative dielectric constants are 1.372 and 2.3117:
- **(a)** Determine the even and odd mode impedances of these coupled lines,
- **(b)** Compute the S_{11} into these coupled lines when they are terminated in 50 Ω, and
- **(c)** Compare the $Z_C/\tan\theta$ with the B parameter of the coupled line at 6 GHz.

ENDNOTES

1. Equations (5–22) through (5–25) were given by Manfred Kirschning and Rolf H. Jansen, "Accurate Wide-Range Design Equations for the Frequency-Dependent Characteristic of Parallel Coupled Microstrip Lines," *IEEE Trans. Microwave Theory Tech., MTT-32, 1* (Jan. 1984), 83–90. Permission to reprint these equations has been given by the IEEE. © 1984 IEEE.
2. M. Kirschning and Rolf H. Jansen, "Corrections to 'Accurate Wide-Range Design Equations for the Frequency-Dependent Characteristics of Parallel Coupled Microstrip Lines,'" *IEEE Trans. Microwave Theory Tech., MTT-33, 3* (March 1985), 288.
3. J. G. Richings and B. Easter, "Measured Odd- and Even-Mode Dispersion of Coupled Microstrip Lines," *IEEE Trans. Microwave Theory Tech., MTT-23, 10* (Oct. 1975), 826–828.
4. Sina Akhtarzad, Thomas R. Rowbotham, and Peter B. Johns, "The Design of Coupled Microstrip Lines," *IEEE Trans. Microwave Theory Tech., MTT-23, 6* (June 1975), 486–492.
5. J. Paul Shelton, "Impedances of Offset Parallel-Coupled Strip Transmission Lines," *IEEE Trans. Microwave Theory Tech., MTT-14, 1* (Jan. 1966), 7–15.

6. Prakash Bhartia and Protap Pramanick, "Computer-Aided Design Models for Broadside-Coupled Striplines and Millimeter-Wave Suspended Substrate Microstrip Lines", *IEEE Trans. Microwave Theory Tech., 36, 11* (Nov. 1988), 1476–1481.
7. P. Pramanick and P. Bhartia, "Corrections to 'Computer-Aided Design Models for Broadside-Coupled Striplines and Millimeter-Wave Suspended Substrate Microstrip Lines,'" *IEEE Trans. Microwave Theory Tech., 37, 10* (Oct. 1989), 1658.
8. Julius Lange, "Interdigitated Stripline Quadrature Hybrid," *IEEE Trans. Microwave Theory Tech., MTT-17, 12* (Dec. 1969), 1150–1151.
9. Raymond Waugh and David LaCombe, "'Unfolding' the Lange Coupler," *IEEE Trans. Microwave Theory Tech., MTT-20, 11* (Nov. 1972), 777–779.
10. Wen Pin Ou, "Design Equations for an Interdigitated Directional Coupler," *IEEE Trans. Microwave Theory Tech., MTT-23, 2* (Feb. 1975), 253–255.
11. Methods for synthesizing lines from the even and odd mode impedances were available before CAD software for synthesizing Lange couplers became available.
12. This approach was proposed in an article by Adolph Presser, "Interdigitated Microstrip Coupler Design," *IEEE Trans. Microwave Theory Tech., MTT-26, 10* (Oct. 1978), 801–805.
13. E. M. T. Jones and J. T. Bolljahn, "Coupled-Strip-Transmission-Line Filters and Directional Couplers", *IRE Trans. Microwave Theory Tech., MTT-4* (April 1956), 75–81.
14. This derivation was first given by George I. Zysman and A. Kent Johnson, "Coupled Transmission-Line Networks in an Inhomogeneous Dielectric Medium," *IEEE Trans. Microwave Theory Tech., MTT-17, 10* (Oct. 1969), 753–759.
15. H. Ozaki and J. Ishii, "Synthesis of a Class of Strip-Line Filters," *IRE Trans. Circuit Theory, CT-5,* (June 1958) 104–109.
16. Zysman and Johnson, "Coupled Transmission Line Networks . . .".
17. Zysman and Johnson, "Coupled Transmission Line Networks . . .".
18. Seymour B. Cohn, "Direct-Coupled-Resonator Filters," *Proc. IRE, 45, 2* (February 1957), 187–196.
19. Seymour B. Cohn, "Parallel-Coupled Transmission-Line-Resonator Filters," *IRE Trans. Microwave Theory Tech., MTT-6, 2* (April 1958), 223–231.
20. A discussion of the image parameters and pertinent formulas is given by George L. Matthaei, Leo Young, and E. M. T. Jones in *Microwave Filters, Impedance-Matching Networks and Coupling Structures,* chapter 3, (Dedham, MA: Artech House, Inc., 1980).
21. A modified form of these formulas is given by George L. Matthaei, Leo Young, and E. M. T. Jones in *Microwave Filters . . .*, pp. 472–473.
22. Edward G. Cristal, "New Design Equations for a Class of Microwave Filters," *IEEE Trans. Microwave Theory Tech., MTT-19, 5* (May 1971), 486–490.
23. Brian J. Minnis, "Printed Circuit Coupled-Line Filters for Bandwidths Up to and Greater than an Octave," *IEEE Trans. Microwave Theory Tech., MTT-29, 3* (March 1981), 215–222.
24. Chi-Yang Chang and Tatsuo Itoh, "A Modified Parallel-Coupled Filter Structure That Improves the Upper Stopband Rejection and Response Symmetry," *IEEE Trans. Microwave Theory Tech., 39, 2* (Feb. 1991), 310–313.
25. Ingo E. Lösch and Johannes A. G. Malherbe, "Design Procedure for Inhomogeneous Coupled Line Sections," *IEEE Trans. Microwave Theory Tech., MTT-36, 7* (July 1988), 1186–1190.
26. Rick Sturdivant, "A Capacitively Coupled BPF Design Using a Suspended Substrate Stripline", *Microwave J., 36, 11* (November 1993), 71–76.
27. The formulas in Eqs. (5–60) through (5–62) are from Seymour B. Cohn, "Characteristic Impedances of Broadside-Coupled Strip Transmission Lines", *IRE Trans. Microwave Theory Tech., MTT-8, 6* (Nov. 1960), 633–637.

28. Lösche and Malherbe suggest using formulas given by Steven March in "Microstrip Packaging: Watch the Last Step", *Microwaves* (Dec. 1981), 51–56.
29. Brian Minnis, "Classes of Sub-Miniature Microwave Printed Circuit Filters with Arbitrary Passband and Stopband Widths", *IEEE Trans. Microwave Theory Tech, 30, 11* (November 1982), 1893–1900.
30. Saod A. Alseyab, "A Novel Class of Generalized Chebyshev Low-Pass Prototype for Suspended Substrate Stripline Filters," *MTT-30, 9* (September 1982), 1341–1347.
31. George L. Matthaei, Leo Young, and E. M. T. Jones, *Microwave Filters . . .*, p. 441.
32. P. Silvester and Peter Benedek, "Equivalent Capacitances of Microstrip Open Circuits," *IEEE Trans. Microwave Theory Tech., MTT-20, 8* (August 1972), 511–516.

APPENDIX

The following formulas for the even and odd mode characteristic impedances are from Kirschning and Jansen and are reprinted with the permission of the IEEE.[1] © 1984 IEEE. The variables are:

W	microstrip line width in mm
h	dielectric thickness in mm
s	microstrip line edge-to-edge spacing in mm
ε_r	relative dielectric constant of the dielectric substrate
f	frequency in GHz
$\varepsilon_{eff}(0)$	quasi-static, single line, effective dielectric constant (The same as is computed using Eq. (2–9))
$Z_0(0)$	quasi-static, single line, characteristic impedance (as calculated using Eq. (2–8))
$\varepsilon_{eff}(f)$	dynamic, single line, effective dielectric constant (as calculated using Eq. (2–15))
$Z_0(f)$	dynamic, single line, characteristic impedance (as calculated using Eq. (2–17))
$\varepsilon_{eff,e,o}(0)$	quasi-static, even- or odd-mode, effective dielectric constant (as calculated using Eq. (5–19) or (5–20))
$Z_{0e,o}(0)$	quasi-static, even- or odd-mode, characteristic impedance
$\varepsilon_{eff,e,o}(f)$	dynamic, even- or odd-mode, effective dielectric constant
$Z_{0e,o}(f)$	dynamic, even- or odd-mode, characteristic impedance
R_{17}	exponent in Eq. (2–17) for $Z_0(f)$

The formulas for this calculation are:

$$Z_{0e}(f) = Z_{0e}(0)[(0.9408(\varepsilon_{eff}(f))^{C_e} - 0.9603)/((0.9408 - D_e)\varepsilon_{eff}^{C_e}(0) - 0.9603)]^{R_{17}}$$

$$Z_{0o}(f) = Z_0(f) + [Z_{0o}(0)(\varepsilon_{eff,o}(f)/\varepsilon_{eff,o}(0))^{Q_{22}} - Z_0(f)Q_{23}]/(1 + Q_{24} + (0.46g)^{2.2}Q_{25})$$

$$P_e(f) = P_1P_2[(P_3P_4 + 0.1844P_7)fh]^{1.5763}$$

$P_1 = 0.27488 + [0.6315 + 0.525/(1 + 0.0157fh)^{20}]u - 0.065683 \exp(-8.7513u)$

$P_2 = -.33622[1 - \exp(-0.03442\varepsilon_r)]$

$P_3 = 0.0363[\exp(-4.6u)][1 - \exp(-(fh/38.7)^{4.97})]$

$P_4 = 1 + 2.751[1 - \exp(-(\varepsilon_r/15.916)^8); P_5 = 0.334 \exp(-3.3(\varepsilon_r/15)^3 + 0.746$

$P_6 = P_5 \exp(-(fh/18)^{0.368})$

$P_7 = 1 + 4.069P_6 g^{0.479} \exp(-1.347g^{0.595} - 0.17g^{2.5})$

$P_8 = 0.7168[1 - 1.076/(1 + 0.0576(\varepsilon_r - 1))]$

$P_9 = P_8 - 0.7913[1 - \exp(-(fh/20)^{1.424}] \arctan[2.481(\varepsilon_r/8)^{0.946}]$

$P_{10} = 0.242(\varepsilon_r - 1)^{0.55}$

$P_{11} = 0.6366[\exp(-0.3401fh) - 1] \arctan(1.263(u/3)^{1.629})$

$P_{12} = P_9 + (1 - P_9)/(1 + 1.183\, u^{1.376}); P_{13} = 1.695\, P_{10}/(0.414 + 1.605\, P_{10})$

$P_{14} = 0.8928 + 0.1072\, [1 - \exp(-0.42(fh/20)^{3.215})]$

$P_{15} = \text{abs}\, [1 - 0.8928\,(1 + P_{11})P_{12}(\exp(-P_{13}g^{1.092}))/P_{14}]$

$P_o(f) = P_1P_2[(P_3P_4 + 0.1844)\, fh\, P_{15})^{1.5763}]; Q_1 = 0.8695\, u^{0.194}$

$Q_2 = 1 + 0.7519\, g + 0.189\, g^{2.31}$

$Q_3 = 0.1975 + (16.6 + (8.4/g)^{-0.3887} + [\ln(g^{10}/(1 + (g/3.4)^{10}))]/241$

$Q_4 = (2Q_1/Q_2)[\exp(-g) + (2 - \exp(-g))/u^{Q_3}]^{-1}$

$Q_5 = 1.794 + 1.14 \ln[1 + 0.638/(g + 0.517\, g^{2.43})]$

$Q_6 = 0.2305 + [\ln(g^{10}/(1 + (g/5.8)^{10}))]/281.3 + [\ln(1 + 0.598\, g^{1.154})]/5.1$

$Q_7 = (10 + 190\, g^2)/(1 + 82.3\, g^3)$

$Q_8 = 1/\exp[6.5 + 0.95 \ln(g) + (g/0.15)^5]$

$Q_9\, (Q_8 + 1/16.5) \ln(Q_7); Q_{10} = [Q_2Q_4 - Q_5 \exp(Q_6\, u^{-Q_9} \ln(u))]/Q_2$

$Z_{0e}(0) = Z_0(0)[\varepsilon_{eff}(0)/\varepsilon_{eff,e}(0)]^{0.5}/[1 - (Z_0(0)/377)(\varepsilon_{eff}(0))^{0.5}Q_{10}]$

$Q_{11} = 0.893(1 - 0.3/(1 + 0.7(\varepsilon_r - 1))); Q_{13} = 1 + 0.038(\varepsilon_r/8)^{5.1}$

$Q_{12} = 2.121[(fh/20)^{4.91}/(1 + Q_{11}\,(fh/20)^{4.91})]/\exp(2.87\, g)\, g^{0.902}$

$Q_{14} = 1 + 1.203(\varepsilon_r/15)^4/(1 + (\varepsilon_r/15)^4)$

$Q_{15} = 1.887[\exp(-1.5\, g^{0.84})]g^{Q_{14}}/[1 + 0.41(fh/15)^3 u^{(2/Q_{13})}/(0.125 + u^{1.626/Q_{13}})]$

$Q_{16} = (1 + 9/(1 + 0.403(\varepsilon_r - 1)^2))Q_{15}$

$Q_{17} = 0.394(1 - \exp(-1.47(u/7)^{0.672}))(1 - \exp(-4.25(fh/20)^{1.87}))$

$Q_{18} = 0.61(1 - \exp(-2.13(u/8)^{1.593}))/(1 + 6.544g^{4.17})$

$Q_{19} = 0.21g^4((1 + 0.18\, g^{4.9})(1 + 0.1\, u^2)(1 + (fh/24)^3))^{-1}$

$Q_{20} = (0.09 + 1/(1 + 0.1(\varepsilon_r - 1)^{2.7}))Q_{19};\ R_e = (fh/28.843)^{12}$

$Q_{21} = abs\,[1 - 42.54\, g^{0.133}(\exp(-0.812g))u^{2.5}/(1 + 0.033\, u^{2.5})]$

$Q_e = 0.016 + (0.0514\, \varepsilon_r Q_{21})^{4.524};\ M_e = 4.866 \exp(-3.228\, u^{0.641})$

$D_e = 5.086(1/(0.3838/Q_e + 0.386))/[\exp(22.2\, u^{1.92})(1/R_e + 1.2992)(1/(\varepsilon_r - 1)^{-6} + 10)]$

$C_e = 1 + 1.275(1 - \exp(-0.004625 M_e \varepsilon_r^{1.674}(fh/18.365)^{2.745})) - Q_{12} + Q_{16} - Q_{17} + Q_{18} + Q_{20}$

$Q_{29} = 15.16/(1 + 0.196(\varepsilon_r - 1)^2);\ Q_{28} = 0.149/(94.5/(\varepsilon_r - 1)^3 + 0.038)$

$Q_{27} = 0.4\, g^{0.84}(1 + 2.5/(5/(\varepsilon_r - 1)^{1.5} + 1))$

$Q_{26} = 30 - 22.2/((13/(\varepsilon_r - 1))^{12} + 3) - Q_{29}$

$Q_{25} = [0.3/(10/(fh)^2 + 1)][1 + 2.333/(5/(\varepsilon_r - 1)^2 + 1)]$

$Q_{24} = 2.506 Q_{28}\, ((1 + 1.3u)fh/99.25)^{4.29}/(2.575/u^{0.894} + 1)$

$Q_{23} = 1 + 0.005 fh\, Q_{27}/[(1 + 0.812(fh/15)^{1.9})(1 + 0.025\, u^2)]$

$Q_{22} = 0.925(fh/Q_{26})^{1.536}/(1 + 0.3(fh/30)^{1.536})$

This program is written in BASIC and computes coupled line parameters according to the formulas given above.

```
10 REM This pgm calculates coupled line parameters, Z0e, Z0o, eeffo, eeffe
20 INPUT "width W in mm";W:INPUT"diel. thickness h in mm";H:C=3*10^8
30 INPUT"separation distance s in mm";S:INPUT"rel. diel. constant";ER
40 INPUT"frequency F in GHz";FF:U=W/H:G=S/H:E=2.718282:PI=3.141593:FH=FF*H
50 REM this calc.s quasi-static diel. const. & CHAR. IMPEDANCE: EQ & ZQ
60 V=U*(20+G^2)/(10+G^2)+G*(1/E)^G:B=.564*((ER-.9)/(ER+3))^.053:ER1=ER-1
70 A=1+LOG((U^4+(U/52)^2)/(U^4+.432))/49+LOG(1+(U/18.1)^3)/18.7
80 AV=1+LOG((V^4+(V/52)^2)/(V^4+.432))/49+LOG(1+(V/18.1)^3)/18.7
90 EQ=(ER+1+ER1*(1+10/U)^(-A*B))/2:AZ=(30.666/U)^.7528:F=FF*10^9:H1=H/10^3
100 ZQ=59.9585*LOG((6+(2*PI-6)*E^(-AZ))/U+SQR(1+(2/U)^2))/SQR(EQ)
110 REM this calc.s single line dynamic char. imped., Z0F & eff. diel. cnst., EF
120 FK=C*ATN(ER*SQR((EQ-1)/(ER-EQ)))/(2*PI*H1*SQR(ER-EQ))
130 F50=FK/(.75+((.75-(.332/ER^1.73))*U)):IF U>.7 THEN GOTO 150
140 MC=1+1.4*(.15-.235/E^(.45*F/F50))/(1+U):GOTO 160
150 MC=1
160 M0=1+1/(1+SQR(U))+.32/(1+SQR(U))^3:M=M0*MC:PF=(F/F50)^M:
    EF=ER-(ER-EQ)/(1+PF)
170 R1=.03891*ER^1.4:R2=.267*U^7:R3=4.766/E^(3.228*U^.641): R5=(FH/28.843)^12
180 R4=.016+(.0514*ER)^4.524:R6=22.2*U^1.92:
    R7=1.206-.3144*(1-(1/E)^R2)*(1/E)^R1
```

```
190 R8=1+1.275*(1-E^(-.004625*R3*(ER^1.674)*(FH/18.365)^2.745)): R8A=(1/E)^R6
200 R9=5.086*R4*R5*ER1^6/((.3838+.386*R4)*(1+1.2992*R5)*(1+10*ER1^6))*R8A
210 R10=.00044*ER^2.136+.0184:R11=1/((19.47/FH)^6-.092): R12=1/(1+.00245*U^2)
220 R13=.9408*EF^R8-.9603:R14=(.9408-R9)*EQ^R8-.9603
230 R15=.707*R10*(FH/12.3)^1.097:R16=1+.0503*ER^2*R11*(1-E^(-(U/15)^6))
240 R17=R7*(1-1.1241*R12/(R16*E^(.026*FH^1.15656-R15))): Z0F=ZQ*(R13/R14)^R17
250 REM this calc.s even and odd mode quasi-static diel. constants: EEQ & EOQ
260 EEQ=(ER+1+ER1/(1+10/V)^(AV*B))/2:BO=.747*ER/(.15+ER)
270 CO=BO-(BO-.207)*E^(-.414*U):DO=.593+.694*E^(-.562*U)
280 AO=.7287*(EF-(ER+1)/2)*(1-E^(-.179*U)): EOQ=((ER+1)/2+AO-EQ)/E^(CO*G^DO)+EQ
290 REM this calc.s even mode effective diel. constant: EER
300 P1=.27488+(.6315+.525*(1/(1+.0157*FH))^20)*U-.065683*(1/E)^(8.7513*U)
310 P2=.33622*(1-E^(-.03442*ER)): P3=.0363*(1/E)^(4.6*U)*(1-E^(-(FH/38.7)^4.97))
320 P4=1+2.751*(1-(1/E)^((ER/15.916)^8))
330 P6=(.334*(1/E)^(3.3*(ER/15)^3)+.746)*(1/E)^((FH/18)^.368)
340 P7=1+4.069*P6*G^.479*(1/E)^(1.347*G^.595+.17*G^2.5)
350 PEF=P1*P2*((P3*P4+.1844*P7)*FH)^1.5763:EER=ER-(ER-EEQ)/(1+PEF)
360 REM this calc.s the odd mode effective diel. constant: EOR
370 P8=.7168*(1+1.076/(1+.0576*ER1)): P13=.41019*ER1^.55/(.414+.3884*ER1^.55)
380 P9=P8-.7913*(1-(1/E)^((FH/20)^1.424))*ATN(2.481*(ER/8)^.946)
390 P11=.6366*((1/E)^(.3401*FH)-1)*ATN(1.263*(U/3)^1.629)
400 P12=P9+(1-P9)/(1+1.183*U^1.376): P14=.8928+.1072*(1-E^(-.42*(FH/20)^3.215))
410 P15=ABS(1-.8928*(1+P11)*P12*(1/E)^(P13*G^1.092)/P14)
420 POF=P1*P2*((P3*P4+.1844)*P15*FH)^1.5763:EOR=ER-(ER-EOQ)/(1+POF)
425 PRINT"quasi-static even mode effective diel. const.=";EEQ
426 PRINT"quasi-static odd mode effective diel. const.=";EOQ
430 PRINT"even mode eff. diel. const. =";EER;"odd mode eff. diel. const.=";EOR
440 REM this calc.s even mode quasi-static char. impedance = ZEQ
450 Q1=.8695*U^.194:Q2=1+.7519*G+.189*G^2.31:
    Q14=1+1.203*(ER/15)^4/(1+(ER/15)^4)
460 Q3=.1975+(16.6+(8.399999/G)^6)^(-.387)+LOG(G^10/(1+(G/3.4)^10))/241
470 Q4=2*(Q1/Q2)/((1/E)^G*U^Q3+(2-(1/E)^G*(1/U)^Q3))
480 ZEQ=ZQ*(EQ/EEQ)^.5/(1-ZQ*EQ^.5*Q4/377)
490 REM this calc.s odd-mode quasi-static char. impedance = ZOQ
500 Q5=1.794+1.14*LOG(1+.638/(G+.5170001*G^2.43)): Q7=(10+190*G^2)/(1+82.3*G^3)
510 Q6=.2305+LOG(1/((1/G)^10+(1/5.8)^10))/281.3+LOG(1+.598*G^1.154)/5.1
520 Q8=(1/E)^(6.5+.95*LOG(G)+(G/.15)^5):Q9=LOG(Q7)*(Q8+1/16.5)
530 Q10=Q4-Q5*E^(LOG(U)*Q6*(1/U)^Q9)/Q2: ZOQ=ZQ*(EQ/EOQ)^.5/(1-ZQ*EQ^.5*Q10/377)
540 REM this calc.s even-mode dynamic char. impedance = Z0E
550 Q11=.893*(1-.3/(1+.7*ER1)):Q13=1+.038*(ER/8)^5.1
560 Q12=2.121/((20/FH)^4.91+Q11)*E^(-2.87*G)*G^.902
570 Q15B=1+.41*(FH/15)^3*U^(2/Q13)/(1/8+U^(1.626/Q13))
580 Q15=1.887*(1/E)^(1.5*G^.84)*G^Q14/Q15B
590 Q16=(1+9/(1+.403*ER1^2))*Q15:PE=4.766*(1/E)^(3.228*U^.641):
    RE=(FH/28.843)^12
600 Q17=.394*(1-E^(-1.47*(U/7)^.672))*(1-(1/E)^(4.25*(FH/20)^1.87))
610 Q18=.61*(1-(1/E)^(2.13*(U/8)^1.593))/(1+6.544*G^4.17)
620 Q19=.21*G^4/((1+.18*G^4.9)*(1+.1*U^2)*(1+(FH/24)^3))
```

```
630 Q21=ABS(1-42.54*G^.133*E^(-.812*G)/(1/U^2.5+.033))
640 Q20=(9.000001E-02+1/(1+.1*ER1^2.7))*Q19:QE=.016+(.0514*ER*Q21)^4.524
650 DE=5.086*(1/E)^(22.2*(U^1.92))/(1/RE+1.2992)/(ER1^(-6)+ 10)/(.3838/QE+.386)
660 CE1=1+1.275*(1-E^(-.004625*PE*ER^1.674*(FH/18.365)^2.745))-Q12+ Q16-Q17+Q18
670 CE=CE1+Q20:Z0E=ZEQ*((.9408*EF^CE-.9603)/((.9408-DE)*EQ^CE-.9603))^R17
680 REM this calc.s odd-mode dynamic char. impedance = Z0O
690 Q29=15.16/(1+.196*ER1^2):Q28=.149*ER1^3/(94.5+.038*ER1^3)
700 Q27=.4*G^.84*(1+2.5/(5/ER1^1.5+1)):Q26=30-22.2/((13/ER1)^12+3)-Q29
710 Q25=.3/(10/FH^2+1)*(1+2.333/(5/ER1^2+1))
720 Q24=2.506*Q28*U^.894*((1+1.3*U)*FH/99.25)^4.29/(3.575+U^.894)
730 Q23=1+.005*FH*Q27/((1+.812*(FH/15)^1.9)*(1+.025*U^2))
740 Q22=.925*(FH/Q26)^1.536/(1+.3*(FH/30)^1.536)
750 Z0O=Z0F+(ZOQ*(EOR/EOQ)^Q22-Z0F*Q23)/(1+Q24+(.46*G)^2.2*Q25)
760 PRINT"quasi-static even mode char. impedance =";ZEQ
770 PRINT "quasi-static odd mode char. impedance = ";ZOQ
780 PRINT"dynamic even-mode char. impedance=";Z0E
790 PRINT"dynamic odd mode char. impedance=";Z0O
800 END
```

Chapter

6

Linear Amplifier Design Using S Parameters

6.0 INTRODUCTION

Transistors are inherently nonlinear devices in that their characteristic curves of voltage versus current are not straight lines over the entire range. A sample set is shown in Figure 6–1. The collector current corresponding to a given base current varies nonlinearly with collector voltage V_{CE}. The collector current corresponding to a given collector voltage varies nonlinearly with base current I_B. However, these characteristics are nearly straight lines over part of the region. We can perform a linear analysis of these devices by expressing the collector current I_C as the first order terms of Taylor's series in these two variables as in Eq. (6–1).

$$I_C = f(V_{CE}, I_B) \cong I_{CO} + \frac{\partial I_C}{\partial V_{CE}} \Delta V_{CE} + \frac{\partial I_C}{\partial I_B} \Delta I_B = I_{CO} + h_{oe}\, \Delta V_{CE} + h_{fe}\, \Delta I_B \quad (6\text{–}1)$$

The increments ΔV_{CE} and ΔI_B are small changes in these quantities about some operating point or Q (for quiescent) point. The quantity I_{CO} is called the *dc bias current* for the collector. The assumption implicit in Eq. (6–1) is that the higher order terms in the series are negligible when describing a small (signal) variation about the quiescent point. This requires the selection of the Q point on the characteristics so that the deviation from a linear variation is small over the operating range. The ΔV_{CE} and ΔI_B must remain within the linear region. This type of operation is called *class A* operation. We let the increments ΔI_C, ΔI_B, and ΔV_{CE} in Eq. (6–1) be the signal currents and voltages. These are the voltages and currents used subsequently in this chapter. The partial derivatives are equated to the device parameters h_{oe}, h_{fe}, and h_{ie} as shown in Eq. (6–1).

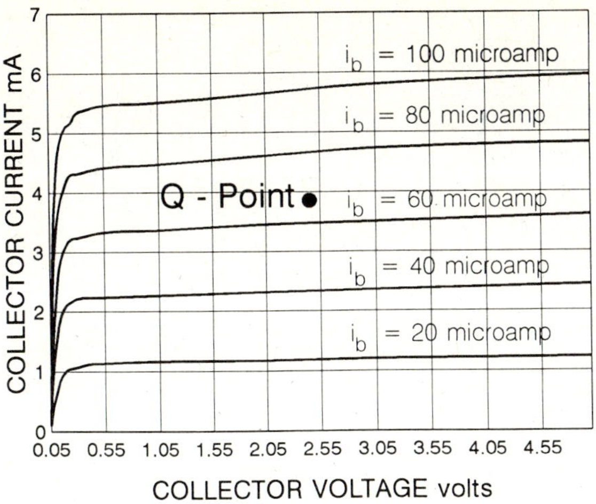

FIGURE 6–1. Characteristic Curves of a Transistor.

If the Q point is near to transistor cut-off such that the transistor does not conduct for the negative one-half of a sinusoidal base-to-emitter signal voltage swing, then the transistor is said to be operating in *class B.* In class B, the transistor is operating nonlinearly. If the Q point is below cut-off, so the transistor conducts only during a small portion of the peak of the sinusoidal base-to-emitter voltage, then it is said to be operating *class C.*

6.1 BASIC DESIGN FORMULAS

The design of microwave amplifiers differs from amplifier design at lower frequencies because of the transmission line effects for the lines connecting components. Measurements of microstrip circuits and microwave transistors usually are made using network analyzers that often present the results in terms of the S parameters or in terms of reflection coefficients and transmission coefficients as functions of frequency. The design equations are usually expressed in terms of these S parameters.

The basic circuit where we have a source with a source impedance, a 2 port with a transistor, and a load impedance is shown in Figure 6–2. The reference planes for all the reflection coefficients are at the ports of the 2 port. The reflection coefficient Γ_S is measured looking back from the input of the 2 port into the source with the ideal voltage generator of the source circuit shorted out. The reflection coefficient Γ_L is measured at the 2 port looking into the load. Both of these parameters depend upon the internal impedances of these portions of the circuit and also upon the characteristic impedance of the transmission line connecting to these parts of the circuit. For this discussion, we take the 2 port to be the linear model of a transistor characterized by its S parameters. The formulas in Eq. (6–2) for the reflection coefficient Γ_{in} looking into the 2 port and the reflection coefficient Γ_{out} looking back into the output terminals of the 2 port form the core from which

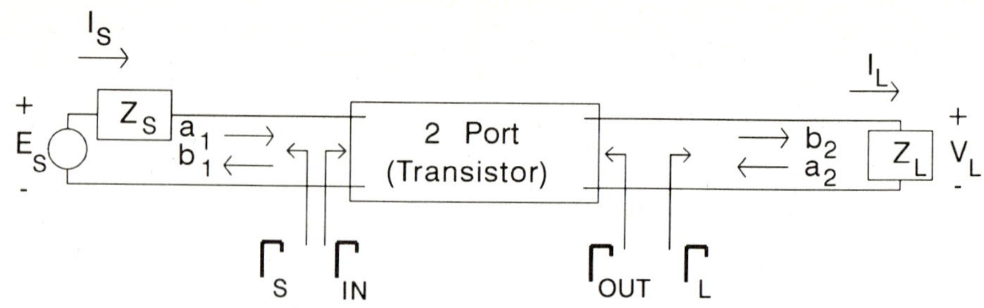

FIGURE 6–2. Linear Transistor Amplifier Circuit.

we derive our analysis procedures. We also use power gain formulas, one of which is shown in Eq. (6–3). The derivation of these is in the next section. From Eq. (6–2), we see that Γ_{in} depends upon the match at the load and Γ_{out} depends upon the match at the source.

$$\Gamma_{in} = S_{11} + \frac{S_{12}S_{21}\Gamma_L}{1 - S_{22}\Gamma_L}, \qquad \Gamma_{out} = S_{22} + \frac{S_{12}S_{21}\Gamma_S}{1 - S_{11}\Gamma_S} \tag{6–2}$$

$$G_T = \frac{1 - |\Gamma_S|^2}{|1 - \Gamma_{in}\Gamma_S|^2}|S_{21}|^2 \frac{1 - |\Gamma_L|^2}{|1 - S_{22}\Gamma_L|^2} \tag{6–3}$$

A simplified form of these formulas occurs in the case of a unilateral transistor. To be useful, a transistor should have a transfer gain $|S_{21}|$ larger than unity, and the reverse parameter $|S_{12}|$ should be much smaller than unity. Often we can assume $|S_{12}|$ to be zero. This is called the *unilateral transistor* case. These formulas then become simpler. In the unilateral case, Γ_{in} is given by S_{11} and Γ_{out} is S_{22}, so that there is no interaction of the output on the input circuit. This assumption should be used cautiously, because it ignores the possibility that the amplifier might oscillate at some frequency. If we can make this assumption, Eq. (6–3) also becomes symmetrical and more easily evaluated. In evaluating Eq. (6–3), sometimes we separate it into three terms, as in Eq. (6–4), and refer to the terms as though they described the gain caused by three separate blocks in the circuit.

$$G_1 = \frac{1 - |\Gamma_S|^2}{|1 - S_{11}\Gamma_S|^2}, \qquad G_0 = |S_{21}|^2, \qquad G_2 = \frac{1 - |\Gamma_L|^2}{|1 - S_{22}\Gamma_L|^2} \tag{6–4}$$

In the unilateral case, these individual gain terms separately describe the effects of the input mismatch, the internal gain of the transistor, and the output mismatch.

Example: Unilateral Transistor. We are given a transistor with the S parameters $S_{11} = 0.4 \angle 170°$, $S_{21} = 4 \angle 50°$, $S_{22} = 0.5 \angle -40°$, and $S_{12} \cong 0$. These parameters are referenced to 50 Ω. The signal source has an internal impedance of $Z_S = 19.81 + j20.24\ \Omega$, and the source reflection coefficient is

$$\Gamma_S = \frac{19.81 + j20.24 - 50}{19.81 + j20.24 + 50} = 0.5\angle 130°.$$

Using Eq. (6–4), we compute G_1 as

$$G_1 = \frac{1 - |0.5\angle 130°|^2}{|1 - (0.4\angle 170°)(0.5\angle 130°)|^2} = 0.893, \qquad \text{or a } G_1 \text{ of } -0.5 \text{ dB}.$$

We are given the load reflection coefficient of $\Gamma_L = 0.5\angle{-130°}$. From Eq. (6–4), we have G_2 equal to

$$G_2 = \frac{1 - |0.5\angle{-130°}|^2}{|1 - (0.5\angle{-40°})(0.5\angle{-130°})|^2} = 0.482, \qquad \text{or a } G_2 \text{ of } -3.166 \text{ dB}.$$

Then, we have $|S_{21}|^2 = 16$ or 12.04 dB. The overall power gain in dB is the sum of these numbers, and is equal to 8.4 dB.

6.2 INPUT AND OUTPUT REFLECTION COEFFICIENTS

To derive Eq. (6–2) and (6–3), we first determine equations relating the input and output wave functions a_1, b_1, a_2, and b_2. These wave functions will not necessarily have the same value as those determined when measuring the S parameters, because the source and load impedances are not set to the reference impedance (usually 50 Ω). We relate the input circuit variables E_S and I_S of Figure 6–2 to the wave functions a_1 and b_1 by writing the equation for the sum of voltages around the input circuit $E_S = I_S Z_S + V_{in}$, and then expressing this in terms of the wave functions as $E_S = (a_1/\sqrt{Z_0} - b_1/\sqrt{Z_0})Z_S + a_1\sqrt{Z_0} + b_1\sqrt{Z_0}$. We define b_S to be $E_S\sqrt{Z_0}/(Z_S + Z_0)$ then this relation becomes $b_S = a_1 - b_1\Gamma_S$. In a similar way, we write Kirchhoff's voltage equation for the output loop $V_L = I_L Z_L$ as the relation $a_2 - b_2\Gamma_L = 0$ between the output line wave functions a_2 and b_2. These relations, along with the linear relations between the input and output wave functions for the 2 port in terms of the S parameters, are four equations in the variables a_1, b_1, a_2, and b_2 shown as Eq. (6–5). These are four equations in the parameters a_i and b_i.

$$\begin{pmatrix} 1 & 0 & -\Gamma_S & 0 \\ S_{11} & S_{12} & -1 & 0 \\ S_{21} & S_{22} & 0 & -1 \\ 0 & 1 & 0 & -\Gamma_L \end{pmatrix} \begin{pmatrix} a_1 \\ a_2 \\ b_1 \\ b_2 \end{pmatrix} = \begin{pmatrix} b_2 \\ 0 \\ 0 \\ 0 \end{pmatrix} \qquad (6\text{–}5)$$

Because there are four of these parameters and four independent equations, we can solve for any one of these parameters in terms of the coefficients and b_S. However, doing so implies that the parameter value found is the value that will hold when the circuit is energized by E_S and terminated by Z_L. The formula for the reflection coefficient looking into the amplifier is independent of the source parameters b_S and Γ_S. This formula is Γ_{in}.

To find $\Gamma_{\text{in}} = b_1/a_1$, we can use the last three of the equations in Eq. (6–5) and rewrite them as

$$\begin{pmatrix} S_{11} & S_{12} & 0 \\ S_{21} & S_{22} & -1 \\ 0 & 1 & -\Gamma_L \end{pmatrix} \begin{pmatrix} a_1 \\ a_2 \\ b_2 \end{pmatrix} = \begin{pmatrix} b_1 \\ 0 \\ 0 \end{pmatrix}.$$

Expanding by minors,[1] we arrive at

$$a_1/b_1 = \frac{1 - \Gamma_L S_{22}}{S_{11}(1 - S_{22}\Gamma_L) + S_{21}S_{12}\Gamma_L} = 1/\Gamma_{\text{in}}.$$

Inverting this, we have Eq. (6–2).

To find the formula for Γ_{out}, we proceed in a similar manner. We use the three equations given in Eq. (6–5) that do not involve the load parameters. However, we must determine this reflection coefficient under the condition that the input signal source is shorted out ($E_S = 0$). Again, we have three equations that can be solved for the ratio of a_2/b_2. This solution will yield Γ_{out} as given in Eq. (6–2).

6.3 POWER GAIN FORMULAS

To discuss power gain formulas, we recognize that the amplifier circuit of Figure 6–2 can have a mismatch at the input ($\Gamma_S \neq \Gamma_{\text{in}}^*$), at the output ($\Gamma_L \neq \Gamma_{\text{out}}^*$), or at both the input and the output. In any of these cases, the power delivered to the load will be reduced from that which would be delivered if the input and the output circuits were both matched. If we were to match the input, then the power into the amplifier would be the available power from the source P_A. If we were to match the output (but not necessarily the input), then the power from the amplifier would be the available power from the network called P_{AVO}. When the transistor is stable, matching either the input or the output increases the power delivered to the load relative to the unmatched cases.

First, we discuss the *operating power gain* G_p. This is the ratio of the power delivered to the load to the power into the amplifier. We write the power into the amplifier as $P_{\text{in}} = |a_1|^2 - |b_1|^2$, and the power to the load as $P_L = |b_2|^2 - |a_2|^2$. The term $|b_1|^2$ represents power reflected back from the amplifier input, and $|a_2|^2$ represents power reflected back from the load. Substituting Γ_L for the ratio a_2/b_2 and Γ_{in} for the ratio b_1/a_1, we have

$$G_P = \frac{|b_2|^2(1 - |\Gamma_L|^2)}{|a_1|^2(1 - |\Gamma_{\text{in}}|^2)}.$$

Now, we solve Eq. (6–5) for $b_2 = b_S S_{21}/D$ and $a_1 = b_S(1 - S_{22}\Gamma_L)/D$, where D is the determinant of the coefficient matrix in Eq. (6–5). Substituting these into the foregoing expression for G_P, we get Eq. (6–6) as a formula for calculating the operating power gain.

$$G_P = \frac{1}{1-|\Gamma_{\text{in}}|^2} |S_{21}|^2 \frac{1-|\Gamma_L|^2}{|1-S_{22}\Gamma_L|^2} \tag{6–6}$$

The *transducer power gain* G_T is the ratio of the power delivered to the load P_L to the power available from the source P_A. Reflection at the input might cause the input power to be less than P_A. In that case, G_T will be a number less than G_P. The transducer gain will equal G_P only if the input of the amplifier is matched to the source. The transducer gain gives a measure of the change in power to the load realized when the (possibly mismatched) amplifier is inserted into the circuit between a source and load, if the source was previously matched to the load. For this reason, it is sometimes called the *insertion gain.*

To determine a formula for calculating G_T, we start with the power delivered to the load $P_L = |b_2|^2(1 - |\Gamma_L|^2)$. We solve for b_2 using Eq. (6–5) as shown in Eq. (6–7).

$$b_2 = \frac{b_s\, S_{21}}{(1-S_{22}\,\Gamma_L)(1-S_{11}\Gamma_S) - S_{12}S_{21}\Gamma_S\Gamma_L} = \frac{b_S S_{21}}{(1-S_{22}\Gamma_L)(1-\Gamma_S\Gamma_{\text{in}})} \tag{6–7}$$

We can factor the term $(1 - S_{22}\Gamma_L)$ in the denominator of Eq. (6–7) and rewrite it by substituting Γ_{in} into the other denominator factor as shown. Substituting from Eq. (6–7) into the expression for P_L, we get Eq. (6–8). Next, we solve for P_A.

$$P_L = \frac{|b_s|^2\, |S_{21}|^2}{|1-\Gamma_S\Gamma_{\text{in}}|^2} \frac{1-|\Gamma_L|^2}{|1-S_{22}\Gamma_L|^2} \tag{6–8}$$

We start with $P_{\text{in}} = |a_1|^2(1 - |\Gamma_{\text{in}}|^2)$ and note that P_A is the input power when $\Gamma_{\text{in}} = \Gamma_S^*$, so we have $P_A = |a_1|^2(1 - |\Gamma_S|^2)$. From Eq. (6–5), we have $b_S = a_1 - b_1\Gamma_S$. This can be rewritten as $b_S = a_1(1 - \Gamma_S b_1/a_1) = a_1(1 - \Gamma_S\Gamma_{\text{in}})$. For a matched input, this becomes $a_1 = b_S/(1 - |\Gamma_S|^2)$. Using this for a_1 in the expression for P_A, we get $P_A = |b_S|^2/(1 - |\Gamma_S|^2)$. Using this and Eq. (6–8) in the ratio P_L/P_A, we obtain Eq. (6–9) for the transducer power gain.

$$G_T = \frac{1-|\Gamma_S|^2}{|1-\Gamma_S\,\Gamma_{\text{in}}|^2} |S_{21}|^2 \frac{1-|\Gamma_L|^2}{|1-S_{22}\,\Gamma_L|^2} \tag{6–9}$$

An alternate expression can be found by factoring the term $(1 - S_{11}\Gamma_S)$ in the denominator of Eq. (6–7) to get

$$b_2 \frac{b_s S_{21}}{(1-S_{11}\Gamma_S)(1-\Gamma_L\Gamma_{\text{out}})}.$$

Then, we substitute this into the expression $P_L = |b_2|^2(1 - |\Gamma_L|^2)$ and proceed as before. The result is Eq. (6–10) for the transducer gain.

$$G_T = \frac{1-|\Gamma_S|^2}{|1-S_{11}\Gamma_S|^2}|S_{21}|^2\frac{1-|\Gamma_L|^2}{|1-\Gamma_L\Gamma_{\text{out}}|^2} \tag{6–10}$$

Some comment about G_T is necessary. Frequently, we will refer to the "transducer gain" of a network, where the S_{ij} refer to the S parameters of the network and not just to an active device embedded in the network. When Γ_S and Γ_L are zero, the transducer gain of such a network becomes $|S_{21}|^2$.

Finally, we define the *available power gain* G_A. This is the available power output from the amplifier P_{AVO} divided by the available power from the source P_A. The available power out of the amplifier is $|b_2|^2$, which is the power output from the amplifier if it were to have a matched load (Γ_L set equal to Γ^*_{out}). However, the input is not necessarily matched to the source. If the input is not matched, then P_{AVO} is less than the maximum power that we would get with a matched input. The power P_A is $|a_1|^2$ and P_{AVO} is $|b_2|^2$. So, the available gain is the ratio of two hypothetical powers $|b_2|^2/|a_1|^2$. To find an expression for this, we recognize that the definition given above agrees with the definition of the transducer gain G_T, if it were evaluated for the case where the output is matched to the load. Substituting $\Gamma_L = \Gamma^*_{\text{out}}$ into Eq. (6–10), we arrive at Eq. (6–11) for the available gain.

$$G_A = \frac{1-|\Gamma_S|^2}{|1-S_{11}\Gamma_S|^2}|S_{21}|^2\frac{1}{1-|\Gamma_{\text{out}}|^2} \tag{6–11}$$

Example: Transistor Amplifier Power Gain. Consider Figure 6–3 where initially we have a source $E_S = 0.1$ volts, which is matched to the load and is delivering its maximum power to the load. A transistor is then inserted between the source and the load. The transistor is not matched either to the load or to the source. The transistor S parameters are $S_{11} = 0.4\angle 170°$, $S_{21} = 4\angle 50°$, $S_{12} = 0.07\angle 70°$, and $S_{22} = 0.5\angle -40°$. We determine the operating power gain and the transducer power gain for this case. First, we calculate P_A for the Figure 6–3A. When the source is matched to the load, the load resistance is equal to the resistive portion of the source impedance. Then one-half of the voltage E_S is across the load resistance. For a 50 Ω reference and $\Gamma_S = 0.5\angle 130°$, we get $Z_S = 19.81 + j20.24$. So, $P_A = |E_S/2|^2/R_L = (0.1/2)^2/19.81 = 126.19$ mW. Now, we find the amplifier input reflection coefficient. We use Eq. (6–2) as

$$\Gamma_{\text{in}} = 0.4\angle 170 + \frac{(4\angle 50°)(0.07\angle 70°)(0.5\angle -130°)}{1-(0.5\angle -40°)(0.5\angle -130°)} = 0.2878\angle 170.778°.$$

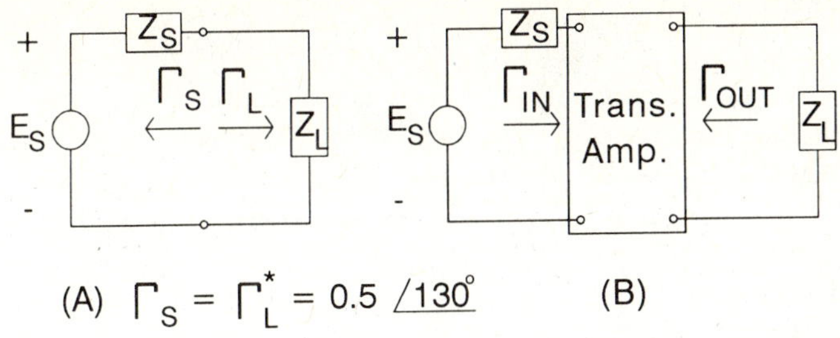

FIGURE 6–3. Transistor Amplifier Insertion between a Source and a Load. (A) Source is matched to a Load: Z_S = 19.81 + j20.24, Z_L = 19.81 −j20.24. These correspond to $\Gamma_S = 0.5\angle 130° = \Gamma_L^*$. (B) Transistor amplifier S parameters are $S_{11} = 0.4\angle 170°$, $S_{21} = 4\angle 50°$, $S_{12} = 0.07\angle 70°$, and $S_{22} = 0.5\angle -40°$. All S parameters are referenced to 50 Ω. $E_S = 0.1$.

Using this value of Γ_{in} in Eq. (6–6), we get

$$G_P = \frac{1}{1-0.2878^2}(4^2)\frac{1-0.5^2}{|1-(0.5\angle -40°)(0.5\angle -130°)|^2} = 8.414 \qquad \text{or a } G_P \text{ of 9.25 dB.}$$

We determine the power to the load and the power to the transistor with the mismatched amplifier of Figure 6–3B using Eq. (6–8) with $b_S = E_S\sqrt{Z_0}/|Z_S + Z_0| = 0.0097283$. The load power is

$$\frac{(0.0097283)^2(4^2)}{|1-(0.5\angle -40°)(0.5\angle -130°)|^2}\frac{1-0.5^2}{|1-(0.5\angle 130°)(0.2878\angle 170.8°)|^2} = 0.836 \text{ mW.}$$

We calculate P_{in} from $|a_1|^2(1 - |\Gamma_{\text{in}}|^2)$, where a_1 is given by $a_1 = b_S/(1 - \Gamma_S\Gamma_{\text{in}})$. Using these formulas, we get P_{in} = 99.38 mW. Because the transistor amplifier is not matched to the source, we see that the power into the transistor amplifier P_{in} is less than P_A. We also note that $P_L/P_{\text{in}} = 0.836/0.09938 = 8.41$ agrees with the operating power gain calculated using Eq. (6–6). We can find G_T directly from the P_L/P_A or we can use Eq. (6–9). Using the calculated values of power, we get $G_T = 0.836/0.12619$, and we get a G_T of 6.63 that is 8.2 dB. This is less than the operating power gain of 9.25 dB because there is a mismatch at the input of the amplifier that reflects some power back to the source. The transducer power gain is equal to the operating power gain only when Γ_{in} is matched to Γ_S.

Although there are other power gains that have been defined, these are the most commonly used power gain expressions. Of these three power gains, the transducer power gain is probably the most important. One reason for this is that the transducer power gain of a 2 port with $Z_S = Z_0$ and $Z_L = Z_0$ is given by the overall S_{21} of the 2 port as $G_T = |S_{21}|^2$. Another reason is that that G_T is the insertion gain.

6.4 BILATERAL MATCH

The maximum power output from a stable amplifier occurs when the amplifier is conjugately matched at both the input and output ports with $\Gamma_S = \Gamma^*_{\text{in}}$ and $\Gamma_L = \Gamma^*_{\text{out}}$. From the formulas in Eq. (6–2), we see that to calculate Γ_{in}, we must know Γ_L and to calculate Γ_{out}, we must know Γ_S. A simultaneous solution of these two formulas is

$$\Gamma_S^* = \Gamma_{in} = S_{11} + \frac{S_{12}S_{21}\Gamma_L}{1 - S_{22}\Gamma_L} \qquad \text{and} \qquad \Gamma_L^* = \Gamma_{\text{out}} = S_{22} + \frac{S_{12}S_{21}\Gamma_S}{1 - S_{11}\Gamma_S}.$$

This results in the Γ_S and Γ_L required for a conjugate match at both the input and the output. The reflection coefficients that simultaneously satisfy the pair of equations are referred to as Γ_{MS} and Γ_{ML}. The formulas resulting from this solution are given below as Eqs. (6–12), (6–13), and (6–14) where the quantity Δ is $S_{11}S_{22} - S_{12}S_{21}$.

$$\Gamma_{\text{MS}} = \frac{B_1 \pm \sqrt{B_1^2 - 4|C_1|^2}}{2C_1}, \qquad \Gamma_{\text{ML}} = \frac{B_2 \pm \sqrt{B_2^2 - 4|C_2|^2}}{2C_2} \tag{6–12}$$

$$B_1 = 1 + |S_{11}|^2 - |S_{22}|^2 - |\Delta|^2, \quad C_1 = S_{11} - \Delta S_{22}^* \tag{6–13}$$

$$B_2 = 1 + |S_{22}|^2 - |S_{11}|^2 - |\Delta|^2, \quad C_2 = S_{22} - \Delta S_{11}^* \tag{6–14}$$

The physically permissible solutions for an unconditionally stable transistor are those that result in reflection coefficients whose magnitudes are not greater than unity. Usually, these will be the solutions with the negative sign in Eq. (6–12).

A bilaterally matched transistor amplifier will have the maximum gain possible without incorporating some feedback into the amplifier circuit external to the transistor. This gain is sometimes called the *maximum available gain (MAG).* A formula for calculating this value of gain can be obtained by substituting Eqs. (6–12), (6–13), and (6–14) into Eq. (6–9). The result is Eq. (6–15), where K is given in Eq. (6–16).

$$G_{\text{MAX}} = \frac{|S_{21}|}{|S_{12}|}(K - \sqrt{K^2 - 1}) \tag{6–15}$$

$$K = \frac{1 - |S_{11}|^2 - |S_{22}|^2 + |\Delta|^2}{2|S_{21}S_{12}|} \tag{6–16}$$

This formula applies only to transistors that are stable (do not oscillate) under any passive load or source impedance terminations. A bilaterally matched amplifier has $G_{\text{MAX}} = G_T = G_P = G_A$.

Example: Bilateral Match. Using the unconditionally stable transistor of the previous example in the circuit shown in Figure 6–4, we can achieve a match at both ports by adjusting the positions and lengths of the input and output stubs. The transistor S parameters are $S_{11} = 0.4\angle 170°$, $S_{21} = 4\angle 50°$, $S_{12} = 0.07\angle 70°$, and $S_{22} = 0.5\angle{-40°}$. The re-

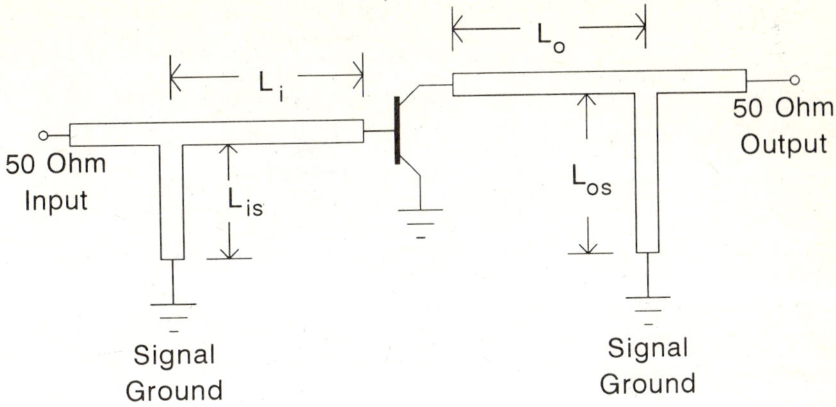

FIGURE 6–4. Bilaterally Matched Amplifier.

quired Γ_{MS} and Γ_{ML} are calculated as $0.7888\angle{-166.8°}$ and $0.8215\angle 42.1°$, where $\Delta = (0.4\angle 170°)(0.5\angle{-40°}) - (0.07\angle 70°)(4\angle 50°) = 0.09\angle{-82.7°}$,

$$B_1 = 1 + 0.4^2 - 0.5^2 - 0.09^2 = 0.9019,$$

$$B_2 = 1 + 0.5^2 - 0.4^2 - 0.09^2 = 1.0819,$$

$$C_1 = 0.4\angle 170° - (0.09\angle{-82.7°})(0.5\angle 40°) = 0.43854\angle 166.822°, \quad \text{and}$$

$$C_2 = 0.5\angle{-40°} - (0.09\angle{-82.7°})(0.4\angle{-170°}) = 0.53065\angle{-42.1°}.$$

Figure 6–5 is the Smith® chart showing the matching steps.

Using $\Delta = 0.09\angle 82.7°$ in Eq. (6–16), we calculate

$$K = \frac{1 - 0.4^2 - 0.5^2 + 0.09^2}{2 \times 4 \times 0.07} = 1.068.$$

From Eq. (6–15), we calculate

$$G_{MAX} = (4/0.07)(1.068 - \sqrt{1.068^2 - 1}) = 39.6 \text{ or } 16 \text{ dB}.$$

6.5 STABILITY

The fundamental criterion for stability is that the network not have poles in the right half of the complex frequency plane. From the pole locations for a network we can find the solution of the homogeneous differential equation for the network as a sum of terms, each incorporating an exponential factor exp $(S_i t)$, where the S_i is the complex frequency of the pole location. If the pole is located in the left half of the S plane, then the exponential term is a decaying exponent by virtue of the negative real part of S_i. If the pole location is in the right half plane, then the term has an increasing exponential factor in the time variation. The latter case represents an unstable circuit. A detailed discussion of this topic can be found in many books.

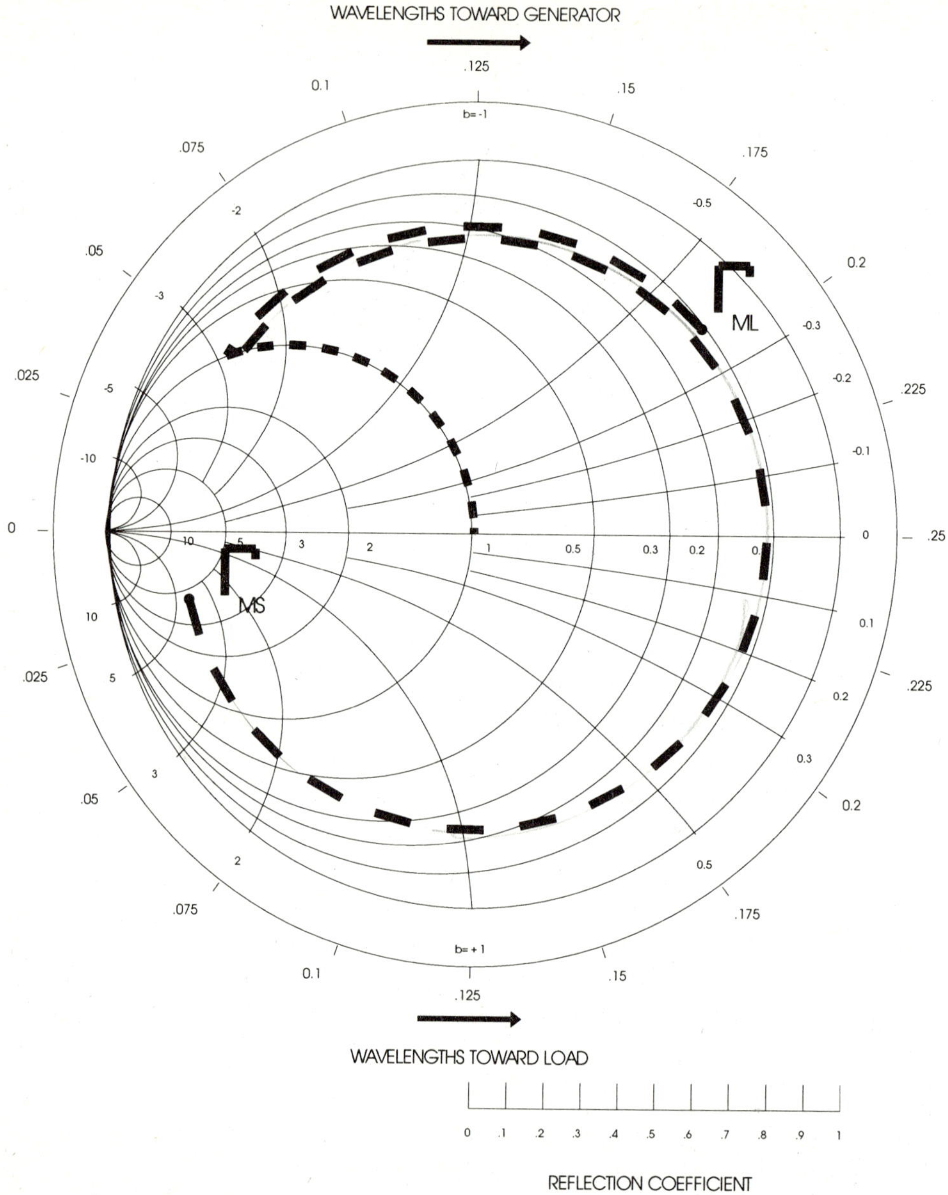

FIGURE 6–5. Smith® Chart for Matching the Input and Output of the Transistor Amplifier of Fig. 6–4. Permission to produce the Smith charts in this book has been granted by Analog Instruments Co., New Providence, NJ under renewal copyright issued in 1976.

6.6 STABILITY CIRCLES

The usual case encountered when designing microwave networks is one where the S parameters of the active device are known or can be measured over only a limited range of frequencies. Given an accurate model of the device, one can translate this information into pole locations. Usually that is not a practical alternative. Instead of trying to determine the pole locations of a network, the microwave engineer commonly employs the criterion that the impedance looking into the network must have a positive real part. In essence, a negative resistor is equivalent to a signal generator. So, the negative input resistance leads to oscillations. The reflection coefficient looking into the amplifier from a line with the characteristic impedance Z_0 is given by Eq. (6–17) (if Z_0 is real and positive).

$$|\Gamma_{in}| = \left|\frac{Z_{in} - Z_0}{Z_{in} + Z_0}\right| = \frac{\sqrt{(R_{in} - Z_0)^2 + X_{in}^2}}{\sqrt{(R_{in} + Z_0)^2 + X_{in}^2}} \tag{6–17}$$

When used in Eq. (6–17), the condition $R_{in} > 0$ implies $|\Gamma_{in}| < 1$. So, the stability criterion becomes $|\Gamma_{in}| < 1$. This criterion is applied at a single frequency (or maybe a small number of frequencies) and to both the input and the output ports of a single stage amplifier or transistor. Although this criterion does not guarantee that the circuit will not oscillate at some other frequency, it usually helps in the detection of instabilities if used wisely. For multiple stage amplifiers, the stability should be determined by other means.[2]

This stability criterion is used in a way that is compatible with the use of S parameters and with Smith® charts. The input reflection coefficient is a function of the transistor S parameters and Γ_L. A transistor might have $|\Gamma_{in}|$ greater than 1 for some Γ_L and not for other Γ_L. The locus of points, Γ_L where $|\Gamma_{in}| = 1$, form a circle on the plane of the Smith® chart. This circle is called the *output stability circle* and it is the perimeter of the unstable region. The choice of Γ_L outside of this region will produce a stable design (assuming that the output circuit is passive). If the circuit is stable for all passive load impedances, then it is said to be *unconditionally stable.* If it is stable only for some values of passive load impedance but not for others, then it is said to be *conditionally stable.*

Example: Output Stability Circles. As examples of this technique, we choose two transistors with the S parameters:

TRANSISTOR A: $S_{11} = 0.8\angle 60°$, $S_{21} = 2.5\angle 38°$, $S_{12} = 0.05\angle 75°$, $S_{22} = 0.44\angle -60°$

TRANSISTOR B: $S_{11} = 0.68\angle -109°$, $S_{21} = 2.17\angle 88°$, $S_{12} = 0.10\angle 41°$, $S_{22} = 0.51\angle -52°$

Using these S parameters, we calculated (from formulas developed subsequently) the output stability circles shown in Figure 6–6. These are plotted on the same coordinates as the Smith® chart for the load reflection coefficient. For any passive Z_L, we know that $|\Gamma_L| \le 1$. This region is that of the usual Smith® chart.

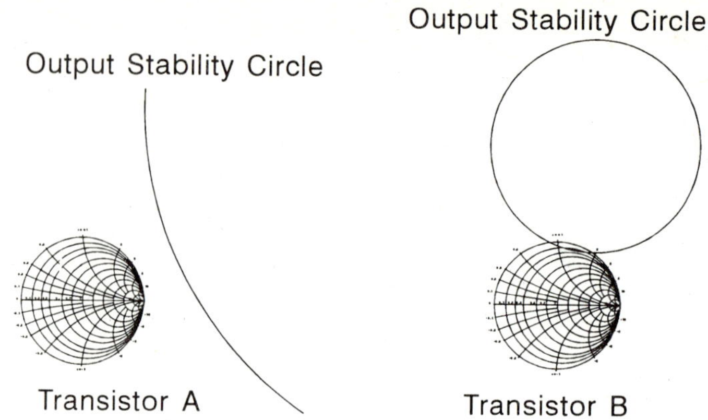

FIGURE 6–6. Output Stability Circles for Two Transistors.

For transistor A, the stable region is outside of the output stability circle. The stability circle does not enclose any of that part of the Smith® chart where $|\Gamma_L| \leq 1$. Any choice of a passive load reflection coefficient, including an open circuit or a short circuit, will not cause this transistor to oscillate at this frequency. The transistor is unconditionally stable. This does not preclude an oscillation occurring at some other frequency.

For transistor B, we see that the stability circle divides the Smith® chart into two regions. For this transistor, the region inside the output stability circle is one that represents values of Γ_L that will cause this transistor to oscillate. So, this transistor will not be stable for these loads; however, it will not oscillate for other passive Γ_L outside of this region. This transistor is conditionally stable.

The discussion of stability can also be put in terms of the input circuit and Γ_S. We use then an *input stability circle,* which is the locus of points Γ_S on the Smith® chart where $|\Gamma_{out}| = 1$. The comments for the output stability circle apply for the input stability circle if Γ_L is replaced with Γ_S and Γ_{in} is replaced with Γ_{out}.

The intuitive notion that the "inside" of the stability circles represents the unstable region is not always correct. One must determine which is the stable region. To explain this process, we refer to Eq. (6–2). We see that for $\Gamma_L = 0$, Γ_{in} is equal to S_{11}. To determine which region has $|\Gamma_{in}| < 1$, we check the center of the Smith® chart where $\Gamma_L = 0$. If $|S_{11}| > 1$ and the center of the Smith® chart is "inside" the output stability circle, then the inside of this stability circle is the unstable region (see Figure 6–7A). If $|S_{11}| < 1$ and the center of the Smith® chart is inside the output stability circle, then the "inside" of this stability circle is the stable region (see Figure 6–7B). Conversely, if the center of the Smith® chart is not contained in the output stability circle and $|S_{11}| > 1$, then the inside of the stability circle is the stable region.

Similarly, $|S_{22}|$ is used to determine the stability region for the input stability circles (see Figure 6–7C).

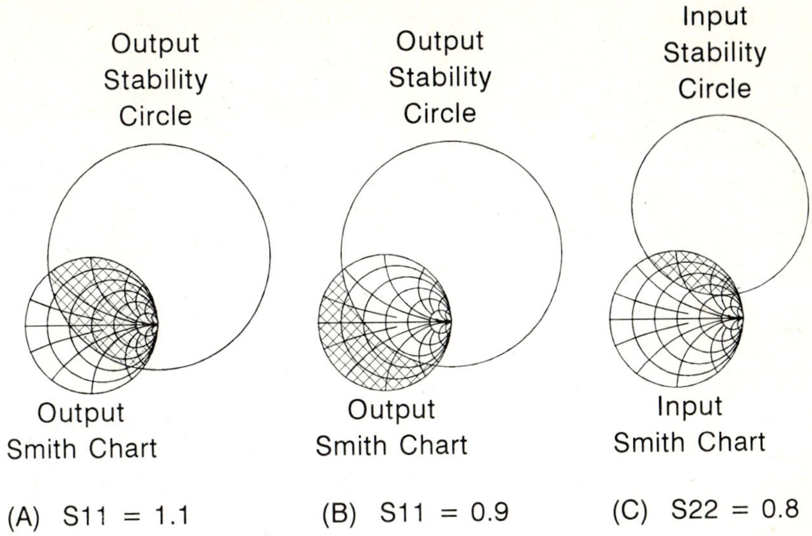

FIGURE 6–7. Determination of Stable Regions.

6.7 DERIVATION OF STABILITY CIRCLE PARAMETERS

We show that the stable region is bounded by a circle by substituting Eq. (6–2) into the formula $|\Gamma_{in}|^2 = \Gamma_{in}\,\Gamma_{in}^* = 1$. This results in Eq. (6–18).

$$[S_{11}(1 - S_{22}\Gamma_L) + S_{21}S_{12}\Gamma_L][S_{11}\,(1 - S_{22}\Gamma_L) + S_{21}S_{12}\Gamma_L]^* = (1 - S_{22}\Gamma_L)(1 - S_{22}\Gamma_L)^* \qquad (6\text{–}18)$$

Using $\Delta = S_{11}S_{22} - S_{21}S_{12}$ in Eq. (6–18) and rearranging terms, we get the expression

$$|\Gamma_L|^2 - 2\frac{\mathcal{Re}\,\{\Gamma_L(S_{11}^*\Delta - S_{22})\}}{|\Delta|^2 - |S_{22}|^2} = \frac{1 - |S_{11}|^2}{|\Delta|^2 - |S_{22}|^2}.$$

We let

$$X = \frac{\mathcal{Re}\,(S_{11}^*\Delta - S_{22})}{|\Delta|^2 - |S_{22}|^2}, \qquad Y = \frac{\mathcal{Im}\,(S_{11}^*\Delta - S_{22})}{|\Delta|^2 - |S_{22}|^2}$$

and $\Gamma_L = U + jV$ in this expression, then we complete the square to obtain Eq. (6–19).

$$(U - X)^2 + (V - Y)^2 = \frac{1 - |S_{11}|^2}{|\Delta|^2 - |S_{22}|^2} + X^2 + Y^2 \qquad (6–19)$$

This is an expression for a circle in the U, V plane with its center at $U = X$ and $V = Y$. From this, we arrive at the formulas in Eq. (6–20) for C_L and R_L the center and radius of the output stability circle.

$$C_L = \frac{S_{22}^* - \Delta^* S_{11}}{|S_{22}|^2 - |\Delta|^2}, \qquad R_L = \frac{|S_{21}S_{12}|}{\left||S_{22}|^2 - |\Delta|^2\right|} \qquad (6–20)$$

Similarly, we can arrive at the formulas in Eq. (6–21) for the center C_S and the radius R_S of the input stability circle.

$$C_S = \frac{S_{11}^* - \Delta^* S_{22}}{|S_{11}|^2 - |\Delta|^2}, \qquad R_S = \frac{|S_{21}S_{12}|}{\left||S_{11}|^2 - |\Delta|^2\right|} \qquad (6–21)$$

6.8 STABILIZING A TRANSISTOR

The technique illustrated in section 3–5 of adding a shunt resistor to the input of a transistor can be used to make a conditionally stable transistor unconditionally stable. We can stabilize transistors by adding a series resistor or a shunt resistor to the input or the output. The resistance required for stability is determined from a plot of the stability circle on a Smith® chart. The conductance of the largest constant conductance circle, which does not include any of the unstable regions, is the minimum normalized shunt conductance required for stability. Similarly, the resistance of the largest constant resistance circle that does not include any unstable regions is the series resistance required for stability. Generally, the shunt resistor at the input is the best option, because it dissipates less power and is not susceptible to an open circuit in the signal source.

Example: Input Damping Resistor. Consider an NE70000 GaAs MESFET with the S parameters shown in Table 6–1. These parameters are supplied by the manufacturer and are available from the data libraries in some of the microwave CAD software. Using the S parameters for 5 GHz in Eq. (6–21), we obtain the results

$$\Delta = S_{11}S_{22} - S_{12}S_{21} = (0.86\angle{-63°})(0.60\angle{-31°}) - (0.07\angle 56°)(3.08\angle 127°) = 0.534\angle 70.4°,$$

$$C_S = \frac{0.86\angle 63° - (0.534\angle{-70.4°})(0.6\angle{-31°})}{0.86^2 - 0.534^2} = 1.279\angle 75.78°, \qquad \text{and}$$

$$R_S = \frac{(3.08)(0.07)}{0.86^2 - 0.534^2} = 0.475.$$

TABLE 6–1 NE70000 Scattering Parameters and Stability Circles

Freq GHz	Scattering Parameters S_{11}	S_{21}	S_{12}	S_{22}	Input Circle R_S	C_S
2	$0.95\angle{-26}$	$3.57\angle 157$	$0.04\angle 76$	$0.66\angle{-14}$	0.287	$1.1\angle 35.9$
5	$0.86\angle{-63}$	$3.08\angle 127$	$0.07\angle 56$	$0.60\angle{-31}$	0.475	$1.3\angle 75.8$
9	$0.75\angle{-95}$	$2.39\angle 101$	$0.09\angle 44$	$0.56\angle{-46}$	0.552	$1.44\angle 107$
15	$.68\angle{-129}$	$1.71\angle 68$	$0.11\angle 36$	$0.52\angle{-74}$	0.492	$1.47\angle 137$
18	$.65\angle{-136}$	$1.51\angle 57$	$0.11\angle 40$	$0.54\angle{-77}$	—	—

Performing these calculations for all frequencies in Table 6–1, we determine the R_S and C_S for the input stability circles listed in Table 6–1. The instability regions for these circles are shown as shaded regions on the Smith® chart in Figure 6–8. The admittance y_s of the circuit connected to the input of the transistor can be plotted on this Smith® chart. By choosing this admittance outside of the shaded regions, we avoid causing oscillations in an amplifier with this transistor. Note that the normalized source admittance $y_s = 0.2 - j0.8$ is quite close to the unstable region for 5 GHz. Calculating the Γ_{out} for this admittance using Eq. (6–2), we arrive at a value of $0.97\angle{-47°}$, while the source admittance of $y_s = 0.1 - j0.8$ lies just inside the shaded region and yields $\Gamma_{out} = 1.14\angle{-46.9°}$. The latter value of $|\Gamma_{out}|$ is greater than 1, which clearly indicates oscillations would occur for that choice of source admittance. A shunt resistor of $5Z_0$ (corresponding to $g = 0.2$) across the input will cause the transistor to be unconditionally stable, because the circle for this conductance on the Smith® chart does not enclose any of the unstable regions shown, and any passive source circuit admittance in parallel with this shunt resistor will result in a point on the Smith® chart with a conductance equal to or greater than 0.2. Of course, we have no data on the transistor at very low frequencies, so we are not sure what will happen at those frequencies.

6.9 STABILITY PARAMETERS

The conditions for stability at a given frequency are $|\Gamma_{in}| < 1$ and $|\Gamma_{out}| < 1$, which must hold for all values of Γ_L and Γ_S obtainable with passive input and output circuits. Passive output and input circuit terminations correspond to $|\Gamma_L| \le 1$ and $|\Gamma_S| \le 1$. Using these conditions with Eq. (6–2) for Γ_{in} and Γ_{out}, we can obtain a set of inequalities in the S parameters for the transistor (or any 2 port). These inequalities can be solved for the relations shown in Eq. (6–22).[3]

$$K = \frac{1 - |S_{11}|^2 - |S_{22}|^2 + |\Delta|^2}{2|S_{12}\ S_{21}|} > 1, \qquad |\Delta| = |S_{11}S_{22} - S_{12}S_{21}| < 1 \tag{6–22}$$

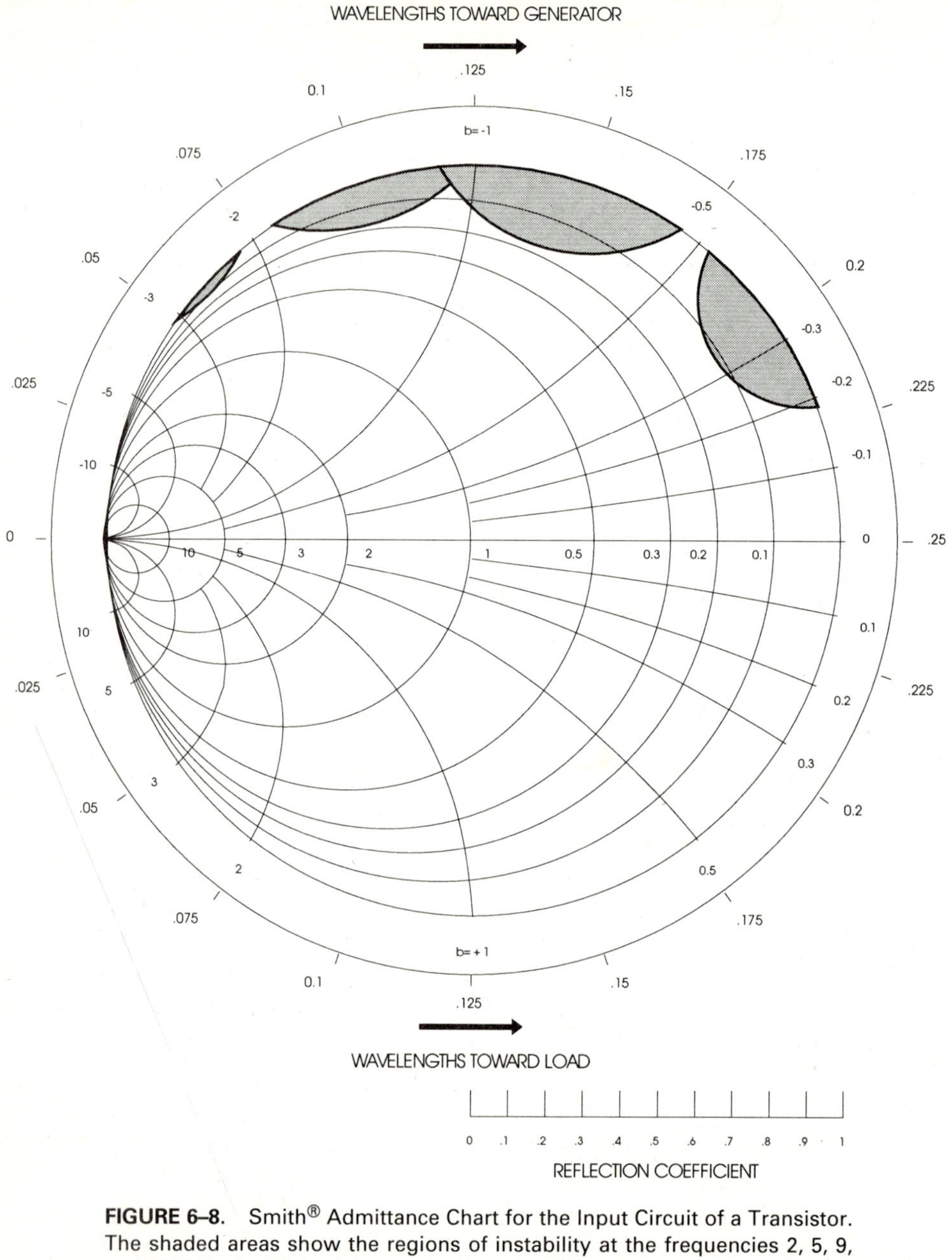

FIGURE 6–8. Smith® Admittance Chart for the Input Circuit of a Transistor. The shaded areas show the regions of instability at the frequencies 2, 5, 9, and 15 GHz determined from the input stability circles. Permission to produce the Smith charts in this book has been granted by Analog Instruments Co., New Providence, NJ under renewal copyright issued in 1976.

So, K and $|\Delta|$ are known as *stability parameters.* These parameters must satisfy $K > 1$ and $|\Delta| < 1$ for a transistor to be unconditionally stable. We note that Eq. (6–15) takes on the form of $|S_{21}|/|S_{12}|$ when K is at the limit of unity. We refer to this quantity as the *maximum stable gain (MSG).* Other stability parameters are used as well (see problem 6–11).

Example: Given the GaAs MESFET parameters at 5 GHz in Table 6–1, we calculate the stability parameters for this transistor at 5 GHz as

$$|\Delta| = |(0.86\angle{-63°})(0.60\angle{-31°}) - (0.07\angle 56°)(3.08\angle 127°)| = 0.534 < 1, \qquad \text{and}$$

$$K = \frac{1 - 0.86^2 - 0.6^2 + 0.534^2}{2 \times 0.07 \times 3.08} = 0.431 < 1.$$

We see that the stability criterion $K > 1$ is not met, so the transistor is not unconditionally stable at this frequency. This agrees with the result that the input stability circle encloses part of the input circuit Smith® chart. However, $K > 1$ and $|\Delta| < 1$ are both met at the frequency of 18 GHz as is shown in these calculations:

$$|\Delta| = |(0.65\angle{-136°})(0.54\angle{-77°}) - (0.11\angle 40°)(1.51\angle 57°)| = 0.275 < 1 \qquad \text{and}$$

$$K = \frac{1 - 0.65^2 - 0.54^2 + 0.275^2}{2 \times 0.11 \times 1.51} = 1.089 > 1.$$

The unconditional stability criteria are met at 18 GHz, but not at 5 GHz.

6.10 CONSTANT GAIN CIRCLES

One of the more useful design tools is the constant gain circle. The basic idea is that, given an unconditionally stable transistor (or any linear 2 port) with known S parameters and a desired value of G_P less than the maximum, there is a range of load impedances that will provide that operating power gain. These impedance values form a circle on the Smith® chart. One can select the load impedances (or Γ_L) along this circle to satisfy some other criterion, such as minimum noise.

This is not restricted to the operating power gain. The locus of Γ_S, which yields a constant G_A, is a circle on the Smith® chart. Also, in the unilateral case, there are constant gain circles based on the gain expressions for both the input (G_1) and output (G_2) gains. In all of these cases, the transistor or 2 port should be stable. If it is not stable, it can be stabilized by resistors in the input or the output circuits so that the S parameters used in these expressions are for a stable 2 port.

6.11 CONSTANT GAIN CIRCLE FORMULAS

To derive the formulas for determining the constant operating gain circles, we first define the parameter $g_p = G_p/|S_{21}|^2$.

$$g_P = \frac{1-|\Gamma_L|^2}{\left(1-\left|\frac{S_{11}-\Delta\Gamma_L}{1-S_{22}\Gamma_L}\right|^2\right)|1-S_{22}\Gamma_L|^2} \tag{6–23}$$

Then, we substitute this and the expression for Γ_{in} from Eq. (6–2) into Eq. (6–6) to result in Eq. (6–23). Equation (6–23) can be rewritten as

$$g_P\big((1-S_{22}\Gamma_L)(1-S_{22}^*\Gamma_L^*)-(S_{11}-\Delta\Gamma_L)(S_{11}^*-\Delta^*\Gamma_L^*)\big)=1-\Gamma_L\Gamma_L^*.$$

That expression in turn can be rewritten as

$$|\Gamma_L|^2 - 2\frac{\mathcal{Re}\left\{\Gamma_L(S_{22}-\Delta S_{11}^*)g_P\right\}}{1+g_P(|S_{22}|^2-|\Delta|^2)} = \frac{1-g_P(1-|S_{11}|^2)}{1+g_P(|S_{22}|^2-|\Delta|^2)}.$$

Letting $\Gamma_L = U + jV$,

$$X=\frac{g_P\,\mathcal{Re}\left\{S_{22}-\Delta S_{11}^*\right\}}{1+g_P(|S_{22}|^2-|\Delta|^2)} \quad \text{and} \quad Y=\frac{g_P\,\mathcal{Im}\left\{S_{22}-\Delta S_{11}^*\right\}}{1+g_P(|S_{22}|^2-|\Delta|^2)}$$

and, completing the square in this formula results in

$$(U-X)^2+(V-Y)^2=\frac{1-g_P(1-|S_{11}|^2)}{1+g_P(|S_{22}|^2-|\Delta|^2)}+X^2+Y^2.$$

This is the formula for a circle with its center C_L at $U = X$, $V = Y$, and its radius R_L equal to the square root of the right-hand side. The center and the radius of the *constant operating gain circle* are given by Eq. (6–24).

$$C_L=\frac{g_P(S_{22}^*-\Delta^*S_{11})}{1+g_P(|S_{22}|^2-|\Delta|^2)} \quad \text{and} \quad R_L=\frac{\sqrt{1-2g_PK|S_{12}S_{21}|+g_P^2|S_{12}S_{21}|^2}}{1+g_P(|S_{22}|^2-|\Delta|^2)} \tag{6–24}$$

In Eq. (6–24), we have used the subscript L to remind the reader that this constant gain circle is a plot of possible values for Γ_L. Any load with a Γ_L that lies on this circle will result in the same operating gain as all other Γ_L that lie on this circle. However, this does not mean that the power output is the same for all Γ_L located on a constant gain circle. Note that Eq. (6–24) was derived without requiring that the input be matched. So, the amplifier power output could be varied by changing Γ_S if desired. Doing so will not change the operating power gain, nor will it change the constant operating gain circle location or will it change the radius on the Smith® chart. If Γ_S is kept constant for different values of Γ_L on the circle, the power to the load will vary with Γ_L, even though the operating gain has not changed. This occurs because Γ_{in} changes with Γ_L while Γ_S is kept constant. The result is a change in the power transferred from the source to the amplifier.

Example: Operating Power Gain Circles. A transistor with the S parameters $S_{11} = 0.48\angle{-148°}$, $S_{21} = 3.51\angle 52°$, $S_{12} = 0.063\angle 18°$, and $S_{22} = 0.23\angle{-67°}$ is stable with the stability factors $|\Delta|$ and K:

$$\Delta = S_{11}S_{22} - S_{21}S_{12} = (0.48\angle{-148})(0.23\angle{-67}) - (0.063\angle 18)(3.51\angle 52) = 0.22\angle 139°$$

and

$$K = \frac{\left(1 - |S_{11}|^2 - |S_{22}|^2 + |\Delta|^2\right)}{\left(2|S_{21}S_{12}|\right)} = \frac{\left(1 - 0.48^2 - 0.23^2 + 0.22^2\right)}{(2 \times 3.51 \times 0.063)} = 1.73.$$

The MAG given by Eq. (6–15) is

$$G_{MAX} = \frac{|S_{21}|}{|S_{12}|}\left(K - \sqrt{K^2 - 1}\right) = \left(\frac{3.51}{0.063}\right)\left(1.73 - \sqrt{1.73^2 - 1}\right) = 17.73$$

or 12.49 dB. This gain will occur only if the input and the output are bilaterally matched as calculated from Eqs. (6–12), (6–13), and (6–14). Suppose we wish to reduce the operating gain by 5 dB below this maximum for some purpose, such as increasing the bandwidth. We can mismatch the output by choosing Γ_L to be some value other than Γ^*_{out}. Because we choose to mismatch the output, we must work with the operating power gain circle. Reducing the operating power gain by 5 dB requires g_P to be equal to $10^{-0.5}(17.73/3.51^2) = 0.455$. From Eq. (6–24), we calculate the center and radius of the constant gain circle to be

$$C_L = \frac{0.455(0.23\angle 67° - (0.22\angle{-139°})(0.48\angle{-148°}))}{1 + 0.455(0.23^2 - 0.22^2)} = 0.057\angle 61.93° \qquad \text{and}$$

$$R_L = \frac{\sqrt{1 - 2 \times 0.455 \times 1.73 \times 0.063 \times 3.51 + (0.455 \times 0.063 \times 3.51)^2}}{1 + 0.455(0.23^2 - 0.22^2)} = 0.8119.$$

This locus is shown in Figure 6–9 as the dashed circle. One possible choice of load is given by $\Gamma_L = 0.716\angle{-62°}$ and is shown on Figure 6–9 as a straight line from the origin of the Smith® chart out to the constant gain circle. With this choice in Eq. (6–2), we get

$$\Gamma_{in} = 0.48\angle 148 + \frac{(0.063\angle 18°)(3.51\angle 52°)(0.716\angle{-62°})}{1 - (0.23\angle{-67°})(0.716\angle{-62°})} = 0.365\angle 223.5°.$$

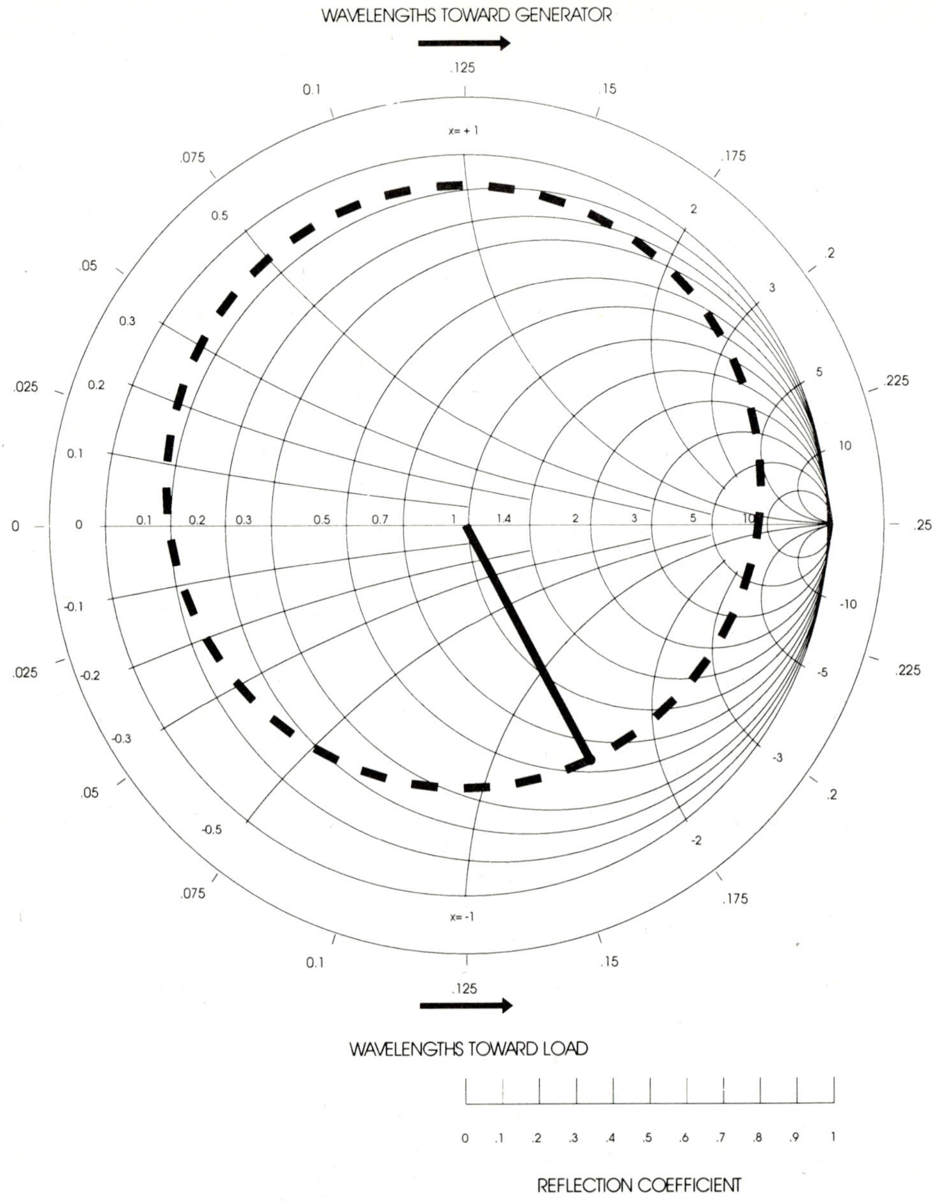

FIGURE 6–9. A Constant Operating Power Gain Circle. The circle shown is a locus of Γ_L for which the gain is 5 dB below the maximum gain. Permission to produce the Smith charts in this book has been granted by Analog Instruments Co., New Providence, NJ under renewal copyright issued in 1976.

Using Eq. (6–6), we calculate the operating power gain to be

$$G_P = \frac{1}{1-0.365^2}\left(3.51^2\right)\frac{1-0.716^2}{\left|1-(0.23\angle{-67^\circ})(0.716\angle{-62^\circ})\right|^2} = 5.61.$$

This is 5 dB below the maximum power gain of 12.49 dB. If we match Γ_S to the new value of Γ^*_{in}, the output power will also be reduced by 5 dB.

To derive a formula for the circle of constant available power gain, we observe that Eq. (6–11) for the available gain is identical in form to Eq. (6–6) for the operating power gain, with the exception that subscripts have changed. Therefore, the derivation of the circle formulas for the available gain case is the same as for the operating gain case. We simply exchange the subscripts: $_A$ for $_P$, $_S$ for $_L$, and interchange the subscripts $_1$ and $_2$ in Eq. (6–24) to arrive at the Eq. (6–25) for the center and the radius of the *constant available gain circle,* where $g_A = G_A/|S_{21}|^2$.

$$C_S = \frac{g_A(S_{11}^* - \Delta^* S_{22})}{1+g_A\left(|S_{11}|^2 - |\Delta|^2\right)}, \qquad R_S = \frac{\sqrt{1-2g_A K|S_{12}S_{21}| + g_A^2|S_{12}S_{21}|^2}}{1+g_A\left(|S_{11}|^2 - |\Delta|^2\right)} \tag{6–25}$$

We use the subscript $_S$ to remind the reader that this is a locus of values for Γ_S. Rather than matching the input, we deliberately mismatch it to arrive at the decrease in gain from the maximum gain to the desired gain. In deriving Eq. (6–11), we required that the output circuit be matched to output of the transistor $\Gamma_L = \Gamma^*_{\text{out}}$. Because of that requirement, the output circuit must be matched if we are to achieve the calculated decrease in available gain. So different Γ_S, all on the same constant available gain circle, will require different Γ_L for G_A to be the same. The constant available gain circle formulas are different from the constant operating gain circle formulas in that respect. As mentioned, the S parameters must be for an unconditionally stable 2 port for Eqs. (6–24) and (6–25) to be valid. If we wish to adjust the gain by mismatching the output, then we would use constant operating gain circles. If we wish to adjust the gain by mismatching the input, then we would use constant available gain circles.

Example: Available Gain Circles. This is an example showing the process of designing an amplifier for use at 5 GHz. The transistor selected is an NE70000 and the S parameters are those given in Table 6–1. In the example of section 6–8, we determined that this transistor would be stable at all frequencies of interest if a 250 Ω shunt resistor were placed across the input. This resistor can be incorporated with the transistor to make a 2 port as shown in Figure 6–10. The S parameters of this 2 port can be determined from the S parameters of the transistor and the conductance of the resistor as shown in section 3–5. These parameters are:

GaAs MESFET NE70000

$S_{11} = 0.86\angle{-63^\circ}$, $S_{21} = 3.08\angle{127^\circ}$, $S_{12} = 0.07\angle{56^\circ}$, and $S_{22} = 0.6\angle{-31^\circ}$

2 PORT (Transistor and shunt resistor)

$S_{11} = 0.643\angle{-66.1^\circ}$, $S_{21} = 2.698\angle{130.8^\circ}$, $S_{12} = 0.0613\angle{59.8^\circ}$, and $S_{22} = 0.615\angle{-29.9^\circ}$

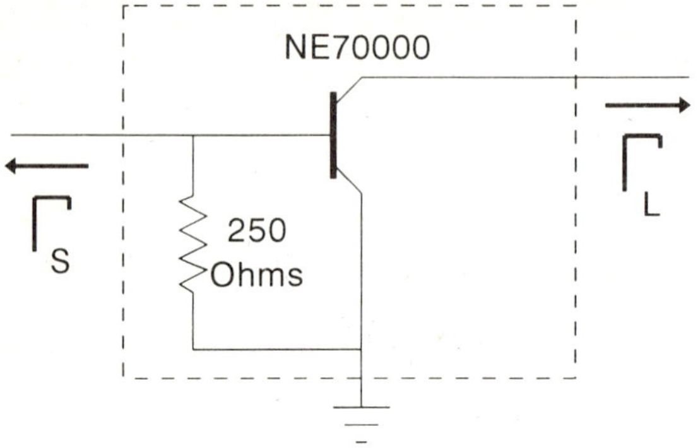

FIGURE 6–10. Stabilized Transistor.

Using these 2-port S parameters and Eqs. (6–15) and (6–16), we determine the stability factors and the MAG as

$$\Delta = (0.643\angle{-66.1°})(0.615\angle 29.9°) - (2.698\angle 130.8°)\,(0.0613\angle 59.8°) = 0.382\angle{-71.6°},$$

$$K = \frac{1 - 0.643^2 - 0.615^2 + 0.382^2}{2(2.698)(0.0613)} = 1.0711, \qquad \text{and}$$

$$G_{\max} = \frac{2.698}{0.0613}\left(1.0711 - \sqrt{1.0711^2 - 1}\right) = 30.2.$$

Because K is greater than 1 and $|\Delta|$ is less than 1, the 2 port is unconditionally stable. The maximum gain of the 2 port is 30.2; however, to achieve this gain, one must conjugately match both the input and the output of the 2 port. Using these 2 port S parameters in Eq. (6–12), we get the values $\Gamma_{MS} = 0.866\angle 78.9°$ and $\Gamma_{ML} = 0.855\angle 44.5°$. In Figure 6–11, we see that this Γ_{MS} lies near, or possibly in, the unstable region for the transistor when operated without the shunt resistor. If we were to use this gain and the shunt resistor were to open, the amplifier might start to oscillate. If we specify a gain that is less than this, then there will be a range where we can choose a Γ_S outside of the transistor unstable region. Then, a failure of the shunt resistor will not cause oscillations. With $G_A = 26$, we get $g_A = 26/(2.6979)^2 = 3.572$. We use Eq. (6–25) to calculate the center and radius of the constant gain circle as

$$C_S = \frac{3.57(0.643\angle 66.1° - (0.382\angle 71.6°)(0.615\angle{-29.9°}))}{1 + 3.57(0.643^2 - 0.382^2)} = 0.804\angle 78.9° \qquad \text{and}$$

$$R_S = \frac{\sqrt{1 - 2 \times 3.57 \times 1.07 \times 0.0613 \times 2.698 + (3.57 \times 0.0613 \times 2.698)^2}}{1 + 3.57(0.643^2 - 0.382^2)} = 0.1476.$$

This circle is shown on the Smith® chart of Figure 6–11. Any point on this circle represents a Γ_S that results in an available gain of 26. For instance, choosing the point labeled

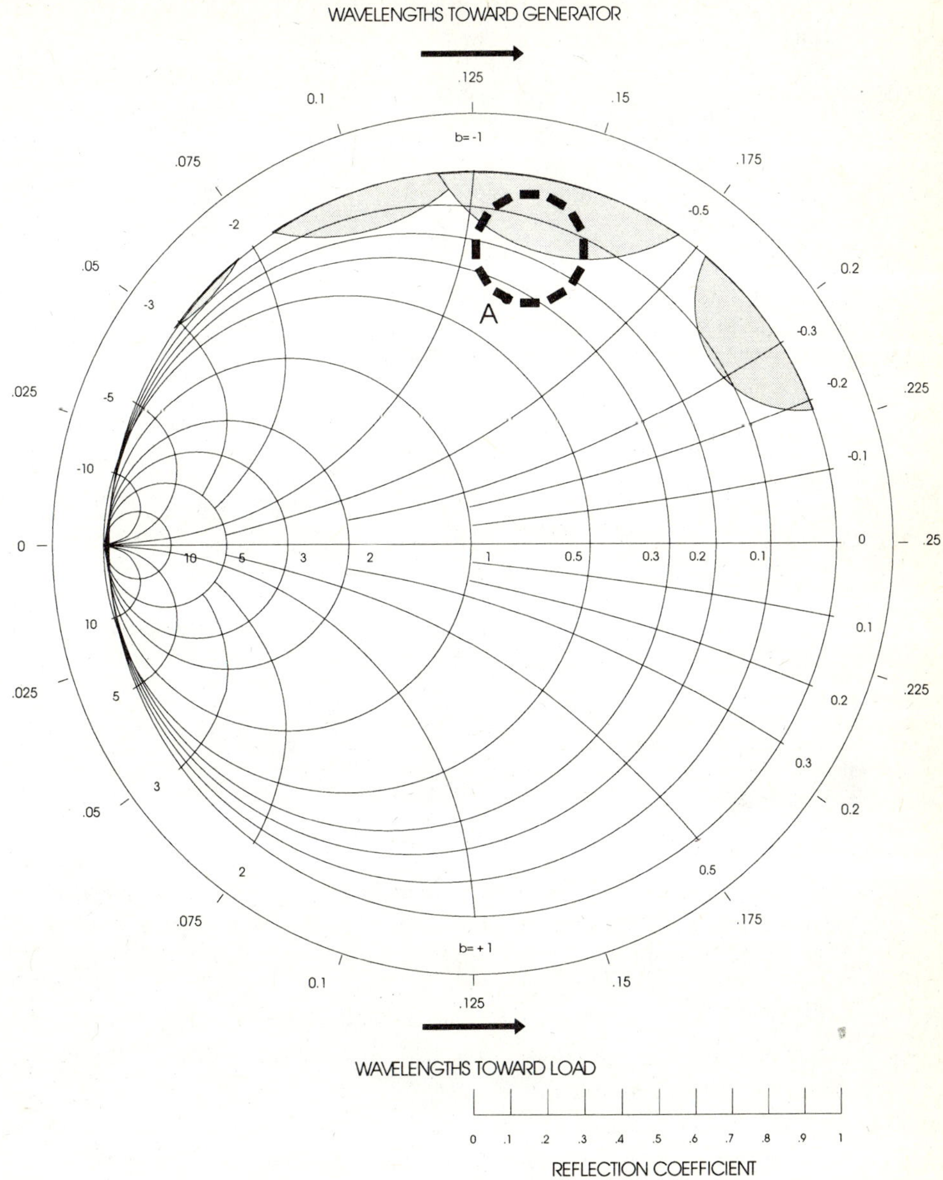

FIGURE 6–11. Constant Available Gain Circle. The dashed circle is for G_A = 26. Permission to produce the Smith charts in this book has been granted by Analog Instruments Co., New Providence, NJ under renewal copyright issued in 1976.

A on this circle, we have $\Gamma_S = 0.662\angle 82°$. Then, we calculate $\Gamma_{out} = 0.7523\angle 39.97°$. Using Eq. (6–11), we check G_A:

$$G_A = \frac{1-0.662^2}{|1-(0.643\angle -66.1°)(0.662\angle 82°)|^2}(2.70^2)\frac{1}{1-0.7523^2} = 26.$$

To finish the design, we must design circuits for the input and the output of the 2 port that yield $\Gamma_S = 0.662\angle 82°$ and $\Gamma_L = 0.752\angle -39.97°$ i.e., the output port must be conjugately matched). This is left as problem 6–12.

6.12 UNILATERAL CONSTANT GAIN CIRCLES

To derive constant gain circles in the case when the assumption can be made that the transistor is unilateral, one can start with the unilateral gain formulas from Eq. (6–4):

$$G_1 = \frac{1-|\Gamma_S|^2}{|1-S_{11}\Gamma_S|^2} \quad \text{and} \quad G_2 = \frac{1-|\Gamma_L|^2}{|1-S_{22}\Gamma_L|^2}.$$

Then, constant gain circle formulas can be derived for both the G_1 term and the G_2 term. They are Eq. (6–26). Their derivation is left as problem 6–13.

$$C_i = \frac{g_i\, S_{ii}^*}{1-|S_{ii}|^2(1-g_i)}, \qquad R_i = \frac{\sqrt{1-g_i}\,(1-|S_{ii}|^2)}{1-|S_{ii}|^2(1-g_i)} \tag{6–26}$$

In these formulas, the subscript i is 1 for the gain circle corresponding to G_1 (the input gain term), and 2 for the gain circle corresponding to G_2 (the output gain term). The factor g_i is the gain normalized to the maximum gain as in Eq. (6–27). The maximum gain used in Eq. (6–27) follows from the unilateral gain formulas in Eq. (6–4) by substituting $\Gamma_S = \Gamma_{in}^* = S_{11}^*$ and $\Gamma_L = \Gamma_{out}^* = S_{22}^*$ into Eq. (6–4).

$$g_i = G_i/G_{i\max} = G_i(1-|S_{ii}|^2) \tag{6–27}$$

The principal advantage obtained in using these formulas is their simplicity. However, the unilateral assumption also allows us to work with either or both the input and the output circuits to achieve the desired result. In the case of the available gain circles, we worked with the input circuit, and in the case of the operating gain circles, we worked with the output circuit.

The hazard is using these formulas is the possibility of oscillations. If the stability criteria $K > 1$ and $|\Delta| < 1$ are met, then one need not worry about oscillations.

Example: Unilateral Constant Gain Circles. Consider the transistor used in calculating the operating power gain circle in Figure 6–9. The S parameters of that transistor are $S_{11} = 0.48\angle -148°$, $S_{21} = 3.51\angle 52°$, $S_{12} = 0.063\angle 18°$, and $S_{22} = 0.23\angle -67°$. If we assume that S_{12} is zero, then we can use Eqs. (6–26) and (6–27) to calculate constant gain circles for the output gain G_2. Using this assumption, we have the following maximum gains

$$G_{1MAX} = \frac{1}{1-0.48^2} = 1.299, \qquad G_{2MAX} = \frac{1}{1-0.23^2} = 1.056, \qquad \text{and}$$

$$G_{MAX} = G_{1MAX}|S_{21}|^2 G_{2MAX} = 1.299 \times 3.51^2 \times 1.056 = 16.9.$$

When we included S_{12} in the calculations for the example of Figure 6–9, we obtained a maximum gain of 17.73 as compared with the G_{MAX} here of 16.9. This change is due to assuming $S_{12} = 0$. So, the assumption results in an 0.2 dB error. To decrease the gain by 5 dB as in the previous example, we have $g_2 = G_2/G_{2MAX} = 10^{-0.5} = 0.31623$ from Eq. (6–27). Then, Eq. (6–26) results in

$$C_2 = \frac{0.31623 \times 0.23\angle 67}{1-0.23^2(1-0.31623)} = 0.0755\angle 67^\circ \qquad \text{and}$$

$$R_2 = \frac{\sqrt{1-0.31623(1-0.23^2)}}{1-0.23^2(1-0.31623)} = 0.8126.$$

This circle is plotted on Figure 6–12 as a solid black circle. For comparison, we have plotted the 5 dB operating power gain circle on this figure as a dashed circle. The difference between these two is caused by assuming that S_{12} is zero; this difference becomes smaller for smaller values of $|S_{12}|$. A choice of Γ_L on the constant gain circle results in a G_2 that is reduced from G_{2MAX} by $10^{-0.5}$

6.13 SINGLE STAGE BAND PASS AMPLIFIER

A single stage bandpass amplifier requires input and output circuits that cause the amplifier gain to be constant throughout the desired band. The transistor $|S_{21}|$ rolls off with frequency, so these circuits must offset this decrease in gain. Because of this roll-off, the maximum gain is set by one that can be achieved at the high frequency end of the band. The design is best accomplished by first designing for a bilateral match at the high end of the frequency band f_2. Then, the match at both the input and the output is degraded to increase the gain at the low frequency end of the band f_1 until a constant gain throughout the band is achieved.

The process of varying the input and output matching circuits to achieve a flat gain throughout the band is best done using the optimizing routine from one of the microwave CAD programs. Because these programs can get lost in local minima or otherwise obtain undesirable results, computer optimization must be done with care. It is best to start by varying only a few components at a time. To determine which components to vary, we can use Smith® chart plots of both the input circuit and the output circuit at f_1, similar to that in Figure 3–16. The effect of each of the components (lines and lumped circuit elements) is shown on these charts. Constant operating gain circles determined at f_1 are compared with these Smith® chart plots of the output circuit. Constant available gain cir-

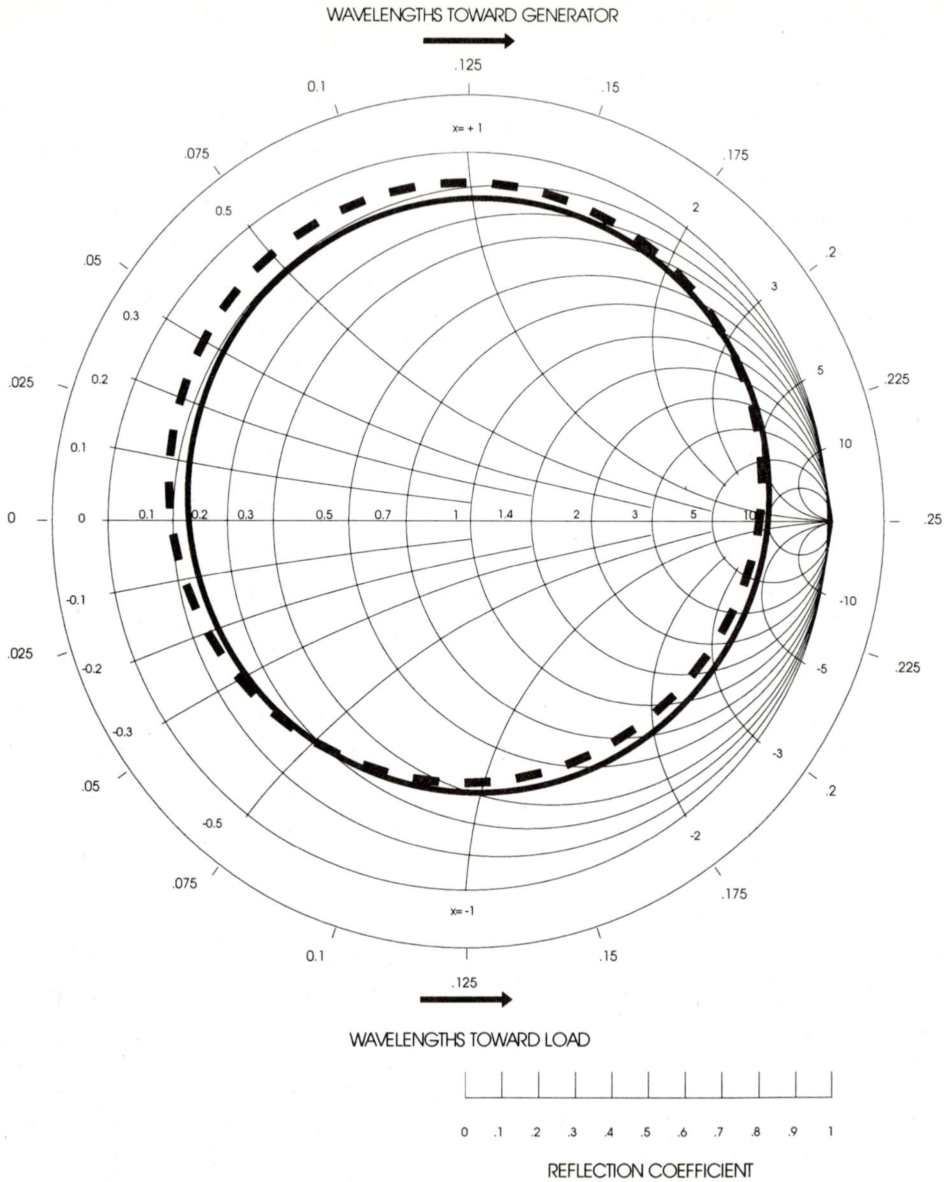

FIGURE 6–12. Unilateral Constant Gain Circle. The solid circle is that for a gain 5 dB below the maximum gain. The corresponding 5 dB constant operating gain circle is shown as a dashed circle. Permission to produce the Smith charts in this book has been granted by Analog Instruments Co., New Providence, NJ under renewal copyright issued in 1976.

cles (again determined at f_1) are compared with the Smith® chart for the input circuit at f_1. From these comparisons, we can estimate which components can be varied to most efficiently increase the gain at f_1. After achieving a relatively flat gain curve, we can optimize all components simultaneously to produce a final design.

The gain is decreased outside of the pass band by the proper choice of input and output circuit topology. A coupling capacitor followed by a shorted stub can be used to match inside the pass band. If the stub is considerably less than a quarter wavelength at f_1, the stub impedance will decrease rapidly with decreasing frequency below the pass band. This decrease, coupled with the increase in the capacitive impedance, results in a voltage divider action that causes the circuit gain to decrease below the pass band. At the high frequency end of the band, a shorted stub that is between one-quarter and one-half wavelength at f_2 will cause the circuit gain to fall off rapidly above the pass band. This stub length can be chosen to be exactly one-half wavelength at some selected frequency above the band. At this frequency, the stub will present zero impedance to ground, causing the overall circuit gain to be zero. Similarly, an open circuited stub that is less than one-quarter wavelength in the pass band will become an apparent short circuit at its quarter wave frequency above the pass band. The designer can adjust the rate of gain roll-off above the pass band by properly choosing this frequency and using the stub length to make the open circuited stub a quarter wavelength long at this frequency.

Example: Band Pass Amplifier. The specification for this amplifier is a flat gain from 1.4 GHz to 2 GHz and a roll-off outside this band. The input and output are to be capacitively coupled to prevent dc bias voltages appearing in other parts of the circuit. Stub lines with their ends "shorted" at signal frequencies with bypass capacitors will be used in both the input and output circuits to allow for bias voltage insertion. The circuit to be used for this amplifier is the one shown in Figure 6–13. The transistor used for this amplifier is a NE02135. The S parameters of this transistor are shown in Table 6–2.

Using Eq. (6–12), we calculate $\Gamma_{MS} = 0.790\angle{-154.8°}$ and $\Gamma_{ML} = 0.602\angle 84°$ at 2 GHz. These are points D and 4 in Figure 6–14. To match the transistor input to 50 Ω, we start with a 1 pF coupling capacitor at the input and then choose the line lengths to

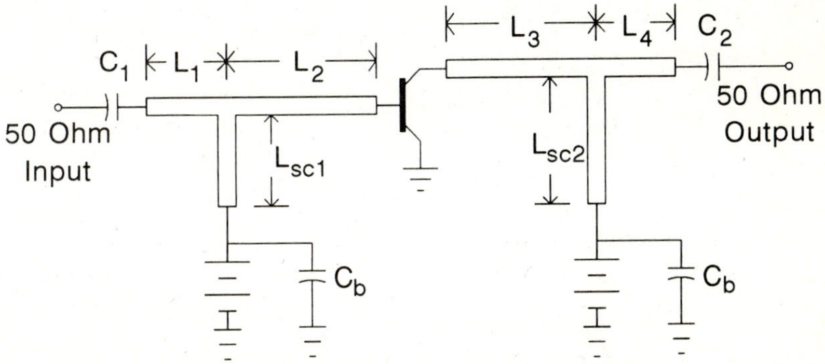

FIGURE 6–13. Band Pass Transistor Amplifier.

cause a match as shown in Figure 6–14. Similarly, we obtain the collector match after assuming a 1.5 pF coupling capacitor at the output of the amplifier. This match is also shown in Figure 6–14. The transmission line lengths for this match are shown in Table 6–3. With these circuits at the input and output of the transistor, we obtain the transducer gain curve shown in Figure 6–15. This gain curve does not meet the specifications. It must be increased at the low frequency end in such a manner as not to degrade it too severely at the high frequency end. We determine which output circuit components to vary to achieve an increased gain at the low frequency from Figure 6–16. This figure is a plot of the 1.4 GHz constant operating power gain circles for gains of 1 and 3 dB below the maximum, superimposed on a plot of the effect of the components in the output circuit. The point for Γ_L at 1.4 GHz is also plotted as point 4. From the position of Γ_L relative to the gain circles, we see that operating power gain realized at 1.4 GHz is 3 dB below the maximum operating power gain at that frequency. The arc lengths 1–2 and 3–4 in Figure 6–16 correspond to the transmission lines L_4 and L_3 of Figure 6–13. Inspection of Figure 6–16 reveals that changing these arc lengths will not materially increase the gain at 1.4 GHz. Varying C_2 or L_{sc2} varies the arc lengths 0–1 and 2–3 in Figure 6–16. From the constant gain circles, it appears that changing these arc lengths can change the position of Γ_L into a region of higher gain. Because this change in gain will be small, the output circuit changes are left to a later step in the design.

The effect of the input mismatch at 1.4 GHz is found by comparing Γ_S to the constant available gain circles determined from Eq. (6–25). The gain circles are for 3, 6, and 9 dB below the maximum gain. These are plotted in Figure 6–17, along with the arcs showing the effect of the input circuit elements at 1.4 GHz. Point D is Γ_S for the input circuit. The available gain at D is at least 6 dB below the maximum gain. We note that the Eq. (6–25) assumes that the output is matched. Although this is not so at 1.4 GHz, the output is not greatly mismatched as shown in Figure 6–16. So, we can use Figure 6–17 to determine which of the input circuit components can best be varied to increase the gain at 1.4 GHz. Arc 0-A corresponds to the input coupling capacitor. That arc length is decreased by increasing this capacitance. This causes point D to move toward a higher gain region. Increasing *C–B* and *D–C* simultaneously also moves point D to a higher gain region. These changes are made by decreasing the stub length L_{sc1} and increasing the transmission line length L_2. For the arc length *0-A* shown, one can see that changing the arc length *A–B* will not materially change the gain. Of these alternatives, we choose to increase the input coupling capacitor to achieve an increased gain.

We take two more steps to finish the design. The first step is to calculate the transducer gain for the input capacitor set to 2.5 pF. The result is shown in Figure 6–15 as the

TABLE 6–2 NE02135 Transistor S Parameters.

Frequency	S_{11}	S_{21}	S_{12}	S_{22}
1.4 GHz	$0.6\angle 166$	$3.5\angle 65$	$.068\angle 40$	$.256\angle -85$
2.0 GHz	$0.6\angle 152$	$2.6\angle 46$	$0.11\angle 36$	$0.19\angle -104$

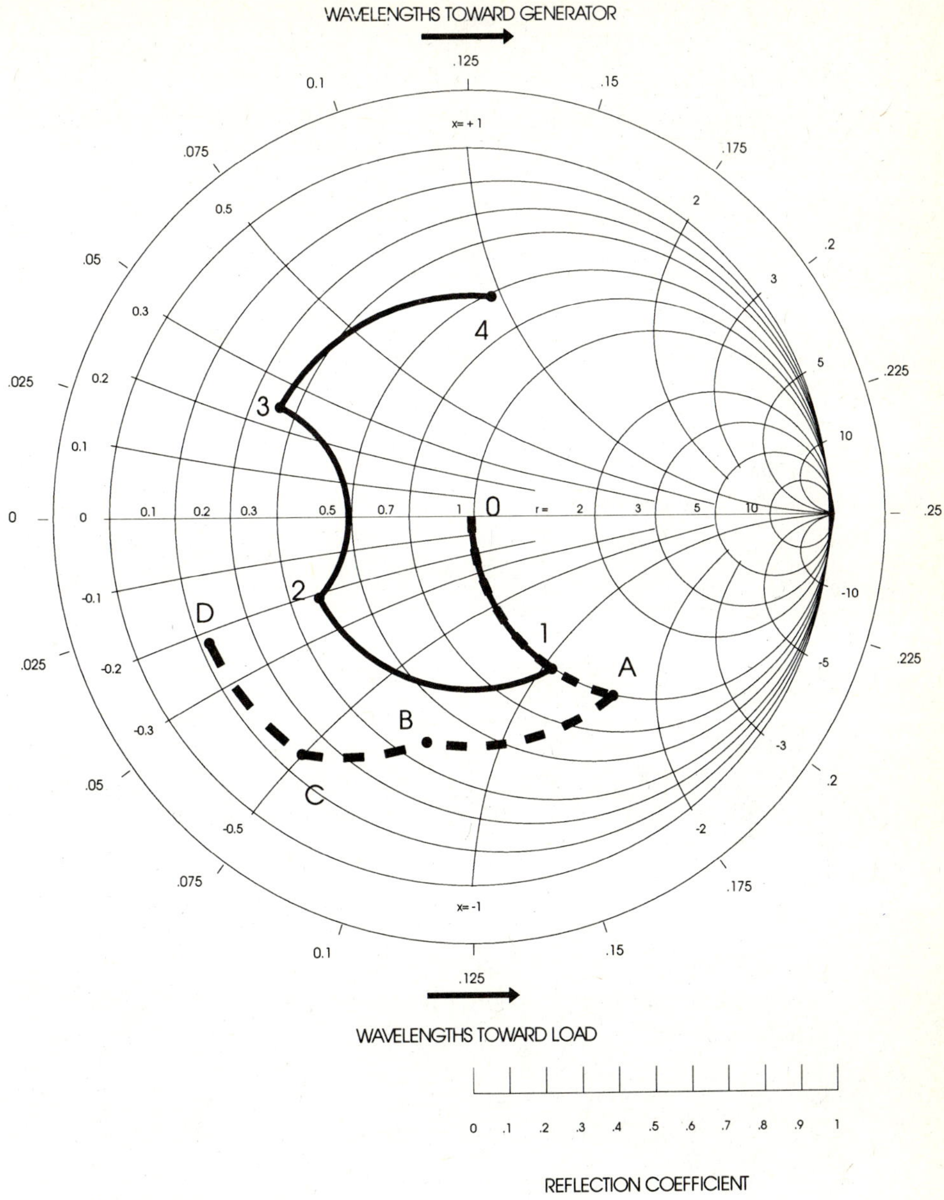

FIGURE 6–14. Amplifier Input and Output Match at 2 GHz shown on a Smith® chart. Arcs 0A, AB, BC, and CD show the effect of x_{c1}, L_1, L_{sc1}, and L_2. Arcs 01, 12, 23, and 34 show the effect of x_{c2}, L_4, L_{sc2}, and L_3. Point D is Γ_{MS} and point 4 is Γ_{ML}. Permission to produce the Smith charts in this book has been granted by Analog Instruments Co., New Providence, NJ under renewal copyright issued in 1976.

TABLE 6–3 Circuit Line Lengths and Reactances for the Smith Charts shown in Figure 6–14, 6–16, and 6–17. The line lengths are in fractions of a wavelength λ. The reactance is normalized to 50 Ω.

Frequency Input Circuit	2 GHz	1.4 GHz	Output Circuit	2 GHz	1.4 GHz
x_{c1}	1.59	2.27	x_{c2}	1.06	1.52
L_1	0.0694λ	0.0486λ	L_3	0.0914λ	0.0639λ
L_{sc1}	0.354λ	0.248λ	L_{sc2}	0.0516λ	0.0361λ
L_2	0.04λ	0.028λ	L_4	0.125λ	0.0877λ

gain curve for $C_1 = 2.5$ pF. This clearly improves the gain at the low end and decreases the gain at the high end by less than 1 dB. This circuit is now close enough to a final design so that a CAD optimizing routine is able to function well. The second step is to fix this capacitance at 2.2 pF (which is a commercially available value) and use an optimizing routine of a microwave CAD program. We allow the program to vary the lengths of all transmission lines, but we keep the capacitors fixed at 2.2 pF and 1.5 pF. The transducer gain curve and the input and output reflection coefficients resulting after several computer optimization runs are shown in Figure 6–18. Because the output reflection coefficient is below −10 dB over the band, the reflected power at the output is relatively small and the transducer gain will be close to the available gain for the amplifier. The circuit elements for this final design are:

Input Circuit $C_1 = 2.2$ pF, $L_1 = 0.0462\lambda$, $L_{sc1} = 0.3465\lambda$, and $L_2 = 0.03731\lambda$ and

Output Circuit $C_2 = 1.5$ pF, $L_3 = 0.1079\lambda$, $L_{sc2} = 0.09225\lambda$, and $L_4 = 0.1562\lambda$

All the line lengths are given in fractions of a wavelength at 2 GHz.

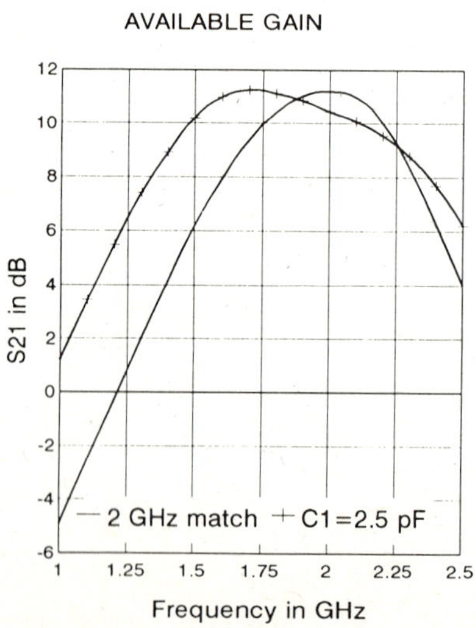

FIGURE 6–15. Transducer Gain versus Frequency. The curves are for the amplifier when matched at 2 GHz and when mismatched at the input with a 2.5 pF capacitor.

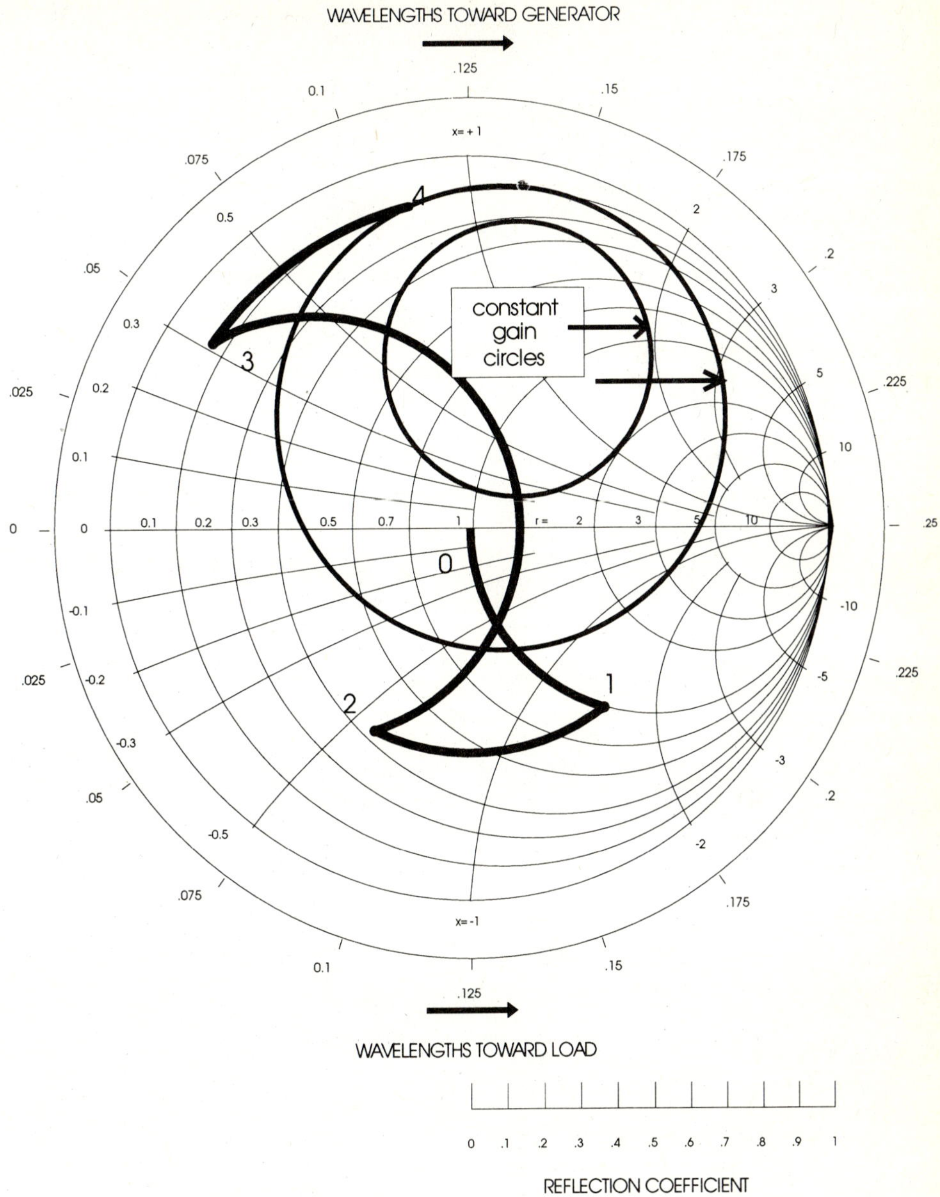

FIGURE 6–16. Location of Γ_L Relative to the Constant Operating Gain Circles at 1.4 GHz. The circles are for 1 dB and 3 dB below the maximum gain. Permission to produce the Smith charts in this book has been granted by Analog Instruments Co., New Providence, NJ under renewal copyright issued in 1976.

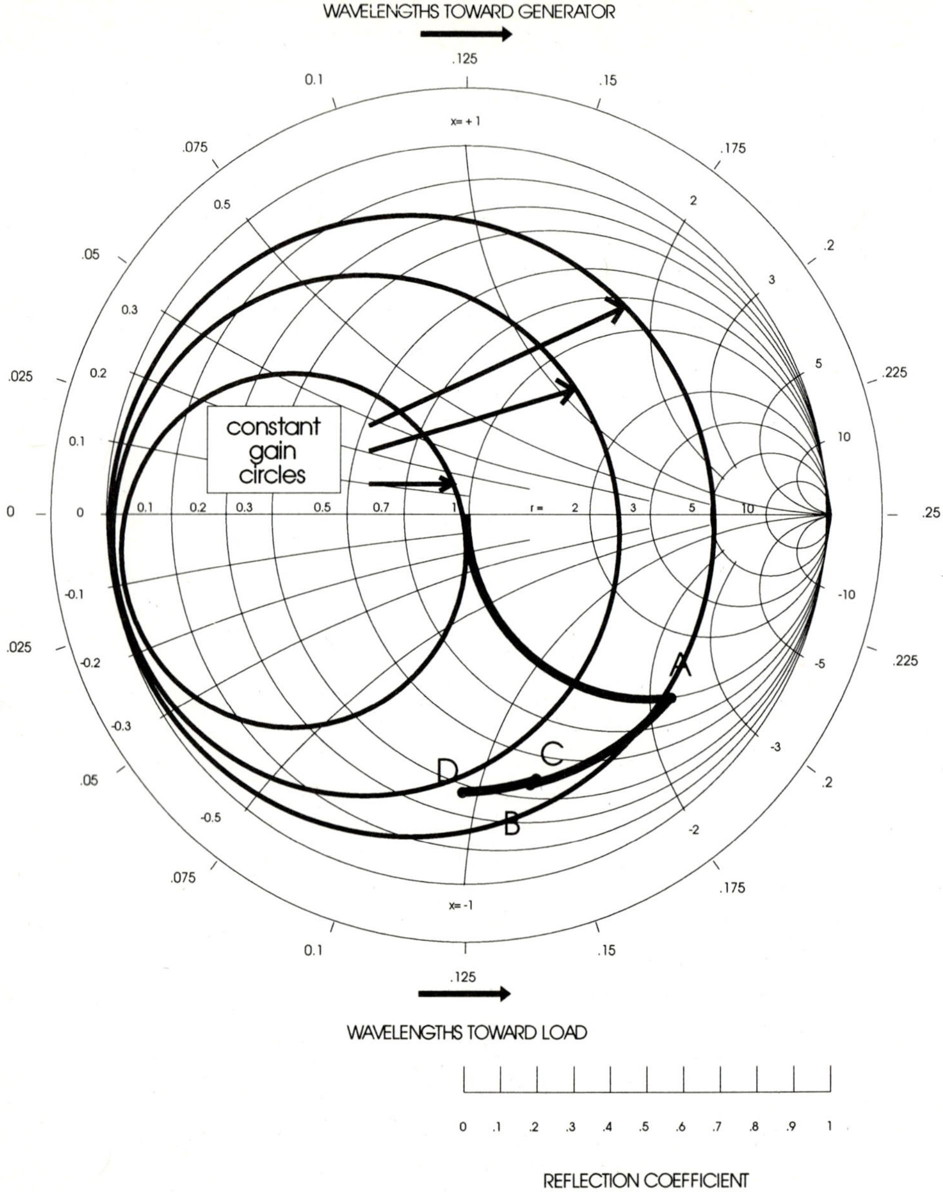

FIGURE 6–17. Location of Γ_S Relative to the Constant Available Gain Circles at 1.4 GHz. The gain circles are for 3, 6, and 9 dB below the maximum gain. Permission to produce the Smith charts in this book has been granted by Analog Instruments Co., New Providence, NJ under renewal copyright issued in 1976.

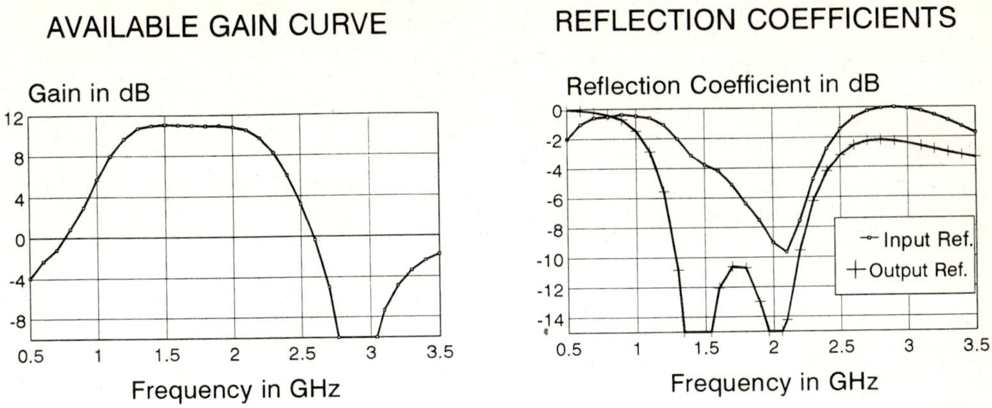

FIGURE 6–18. Amplifier Available Gain and Reflection Curves after Changing the Input Capacitor to 2.2 pF and after Final Optimization.

6.14 AMPLIFIER NOISE

Random fluctuation of charges in electronic devices causes spurious voltages to appear at the terminals of these devices. These voltages are referred to as *noise voltages.* When the signal is at a sufficiently low level, these voltages can degrade or completely mask the signal voltage. In a resistor, these fluctuations are due to the thermal energy of the charges, and they represent electrical power in the circuit. The resulting noise voltage is sometimes called Johnson noise or Nyquist noise. The average amount of power per unit Hertz due to such fluctuations in a resistor is given in Eq. (6–28). This quantity is called a *power spectral density* η.

$$\eta = 2KT \quad \text{Watts/Hz} \tag{6–28}$$

The constant K is Boltzmann's constant, which is equal to 1.38×10^{-23}, and T is the temperature of the resistor in degrees Kelvin. The conversion between Kelvin scale and the Celsius scale is $T(K) = T(C) + 273.15$. To find the total average noise power present in a resistor between the radian frequencies ω_1 and ω_2, one can integrate as

$$\text{Noise Power} = (1/2\pi)\int_{\omega_1}^{\omega_2} 2KT\, d\omega + (1/2\pi)\int_{-\omega_2}^{-\omega_1} 2KT\, d\omega.$$

Because η is not a function of frequency, this results in Noise Power = 4KTB, where $B = (\omega_2 - \omega_1)/2\pi$. This noise power density is a statistical average power density. The noise voltage corresponding to this noise power is considered to be a random variable. Here, we will use the notation $V(t)$ to represent one possible sample function of the random variable for the voltage in question. The statistical average denoted by $\langle V(t)^2 \rangle$ is called the mean square voltage and will be referred to simply as V^2. Then, we have $V^2 = \langle V(t)^2 \rangle = 4KTBR$. The noise voltage is assumed to have the special property that the time average of its squared sample function is equal to the statistical average of squared value.

Because these voltages are random fluctuations rather than sine waves, one cannot characterize them as phasors to add their phasor values directly when two such resistors are connected in series as in Figure 6–19. Rather, the mean square voltage across the series resistor combination is given by the $<(V_1(t) + V_2(t))^2> = <V_1(t)^2> + 2<V_1(t)><V_2(t)> + <V_2(t)^2>$, where the cross product terms separate into a product of two mean values because the noise voltages in the two resistors are assumed to be independent (uncorrelated). The noise voltages have a zero mean value, so the mean square voltage across the series resistor combination is the sum of the mean square voltages across the individual resistors, as in $V_t^2 = V_1^2 + V_2^2$. From this, we have the relation for the total voltage in terms of the individual resistors as $V_t = 2\sqrt{KTB(R_1 + R_2)}$. The principle of adding mean square values also applies to noise currents through resistors. The currents in Figure 6–19 add, as in $I_t^2 = I_1^2 + I_2^2$.

The *noise bandwidth* of a circuit is defined as the noise power passed by the circuit divided by the noise power spectral density passed by the circuit at midband. This is expressed in Eq. (6–29), where ω_o is the radian frequency at the midband of the circuit.

$$\text{Noise Bandwidth} = B_n = \frac{1}{2\pi}\frac{\int_{-\infty}^{+\infty}\eta|H(\omega)|^2\,d\omega}{\eta|H(\omega_o)|^2} \tag{6–29}$$

Example: To compute the noise bandwidth of a third order low pass Butterworth filter with a transfer characteristic $|H(\omega)| = 1/\sqrt{1 + (\omega/\omega_u)^6}$, we use Eq. (6–29) to form the integral

$$B_n = (1/2\pi)\int_{-\infty}^{\infty} KT/\left(1+(\omega/\omega_u)^6\right)d\omega/\left(2KT\,|H(0)|\right).$$

This can be evaluated exactly using a table of integrals to be $B_n = \omega_u/6$. Notice that the 3 dB radian frequency where $|H(\omega)|$ is down to $1/\sqrt{2}$ of $|H(0)|$ is $\omega = \omega_u$. The ratio of the noise bandwidth to the 3 dB bandwidth is $B_n/f_u = 2\pi/6$. So, the noise bandwidth is larger than the 3 dB bandwidth by the ratio of $2\pi/6 = 1.047$.

When a noisy resistor is connected to the input of an amplifier, the noise power available for amplification, is equal to KTB rather than 4KTB. This is equivalent to the argument that a matched generator supplies $(\frac{1}{4})(E^2/R_L)$ power to its matched load resistor.

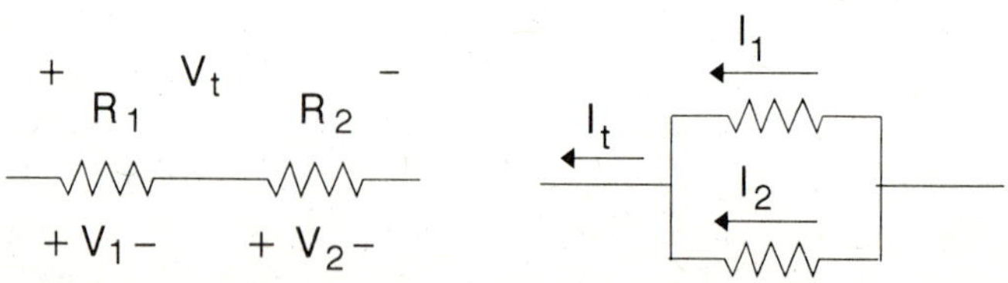

FIGURE 6–19. Resistor Noise Addition.

For this reason, the component of output power from an amplifier due to thermally generated noise in the resistive input is given by GKTB, where G is the power gain of the amplifier and B is its noise bandwidth.

Internal noise sources such as the diffusion noise present in the high field region of the channel in a FET are present in electronic amplifiers. The power from these internal noise sources also appears at the output. So, the total output power in the absence of signal is larger than GKTB and this increase is characterized by an *equivalent temperature* T_e of the amplifier. The equivalent temperature is defined for an amplifier at 290° K (room temperature) from the difference between the actual output noise power and the amplified input noise power due to thermal resistor noise at the input. The equivalent noise temperature is given by Eq. (6–30), where G is the power gain of the amplifier (assumed to be constant over the band) and P_o is its output noise power.

$$P_o = \mathrm{GK290B} + \mathrm{GKT_eB}. \tag{6–30}$$

Another parameter, also giving a measure of these internal noise sources in an amplifier, is the *noise figure* F_{dB}. This is given in the formula in Eq. (6–31), where T is taken to be 290° K (room temperature) and P_o is the power output given in Eq. (6–30).

$$F_{dB} = 10\log_{10}(P_o/\mathrm{GKTB}) \tag{6–31}$$

The ratio P_o/GKTB is frequently called the noise figure; however, it is more properly termed the *noise factor* F. The relation between the noise factor and the equivalent noise temperature is given in Eq. (6–32), where F is the noise factor at 290° K.

$$F = 1 + \frac{T_e}{290} \tag{6–32}$$

The *signal-to-noise ratio* is a ratio of signal power to noise power for a given amplifier and signal level. Usually, one is interested in the input signal level for a 3 dB signal-to-noise ratio at the output. This figure gives a measure of the noise figure of the amplifier.

Example: A receiver has a noise bandwidth of 500 KHz, a noise figure of 4 dB, a gain of 60 dB, and a source impedance of 50 Ω. Using Eq. (6–32), we calculate the equivalent temperature of the receiver to be $T_e = 290(10^{0.4} - 1) = 438°$ K. At room temperature (290° K), the output noise power in the absence of signal can be calculated using Eq. (6–30) as $P_o = \mathrm{GK}(T + T_e)B = 10^6 \times 1.38 \times 10^{-23}(290 + 438)500 \times 10^3 = 5.02$ nW. This corresponds to 0.5 mV across 50 Ω. An input signal of $10.04/10^6$ nW would be required for a 3 dB signal-to-noise level ratio at room temperature.

When cascading two amplifiers, the overall noise factor of the two amplifiers is given by Eq. (6–33), where the subscript 1 is for the first amplifier and G_1 is the gain of the first amplifier. For the general case of n cascaded amplifiers, see problem 6–18.

$$F = F_1 + \frac{F_2 - 1}{G_1} \tag{6–33}$$

Example: Low noise amplifiers are usually constructed with a very low noise preamplifier with moderate gain cascaded with a relatively noisy amplifier with high gain. The overall combination yields good results. Assume that we are given a preamplifier with an equivalent noise temperature of 170° K and a gain of 14 dB, followed by an amplifier with a noise figure of 12 dB and a gain of 40 dB. To find the overall noise figure, we determine the amplifier noise factor and the preamp noise factor with Eq. (6–32) and substitute those into Eq. (6–33): $F_1 = 1 + 170/290 = 1.586$, $F_2 = 10^{1.2} = 15.849$, and $F = 1.586 + (15.849 - 1)/10^{1.4} = 2.217$ or 3.458 dB.

6.15 2 PORT NOISE PARAMETERS

Although noise figure and noise factor are defined in terms of power rather than power density, we often use them as though they were power density functions of frequency. In that case, we consider the noise power in 1 Hz band, so that the transfer function of the circuit involved is essentially constant over the band. The noise factor is calculated using three parameters: F_{min}, R_n, and Y_{opt}, which depend only on the transistor and not on the circuit surrounding it.[4] Then, the noise factor determined from these parameters depends upon Y_s of the source driving the 2 port. The noise voltage $v(t)$ corresponding to the narrow band power is a random variable that is assumed to be *stationary.* This means that the statistics of it do not depend upon time. It is also assumed to be *ergodic,* so that time averages can be used to determine its statistical properties.[5] This noise voltage is not periodic, but one can define the Fourier transform $(1/2\pi)\int_{-T/2}^{T/2} v(t) \exp(-j\omega t)\, dt$, which has the same statistical properties in the limit as T goes to infinity and where $v(t)$ is a sample function of the random variable. We will use $V(\omega) = \text{Lim}_{T\to\infty} (1/2\pi)\int_{-T/2}^{T/2} v(t) \exp(-j\omega t)\, dt$ with a complex admittance Y determined at the center frequency of the narrow band to find a current $I(\omega) = Y\, V(\omega)$. This allows us to use phasors and ac circuit analysis techniques with random variables.

To define the parameters F_{min}, R_n, and Y_{opt}, and to relate them to the noise figure of a 2 port, we start with a linear 2 port that has internal noise sources as shown in Figure 6–20. In Figure 6–20, the resistor is shown as a noise current source I_s and a noise free shunt admittance Y_s. The resistor noise supplies the current $I_s(\omega)$ that drives the 2 port. The internal noise sources are separated using a generalization of Thevenin's theorem to appear connected to the input of the noiseless linear 2 port,[6] as in Figure 6–21.

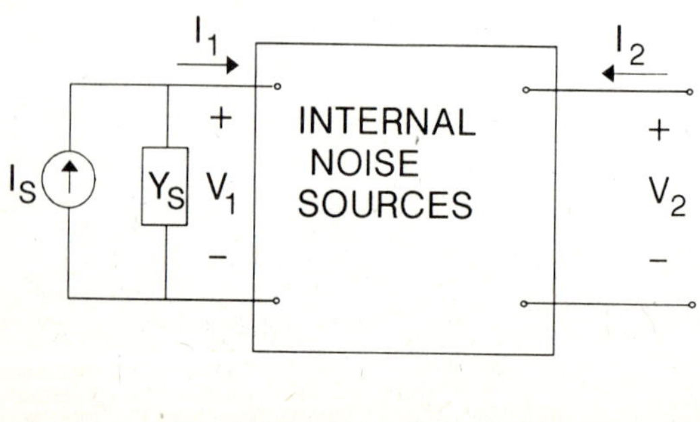

FIGURE 6–20. Transistor Noise Model.

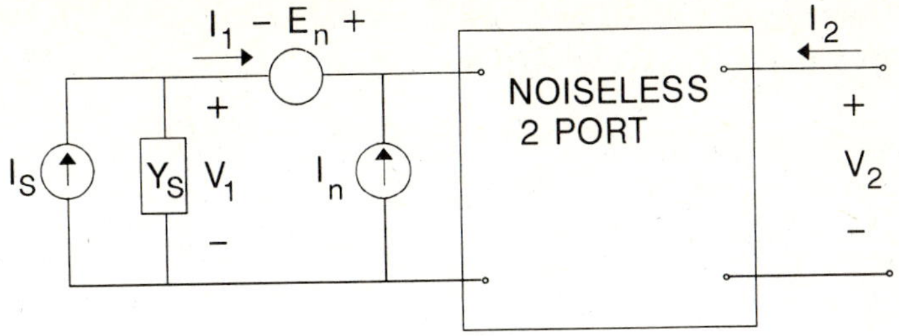

FIGURE 6–21. Equivalent Noise Sources with a Noise-Free 2 Port.

We now make a Norton's equivalent circuit of that part of this circuit, having only the resistor noise signal source (consisting of $I_s(\omega)$ and Y_s) and the internal noise sources (consisting of $E_n(\omega)$ and $I_n(\omega)$). Because these are random variables, the mean square short circuit noise current is obtained by adding the mean square currents as in $<|I_{eq}(\omega)|^2> = <|I_s(\omega)|^2> + <|I_n(\omega) + Y_sE_n(\omega)|^2>$. Here, we have assumed that neither $I_n(\omega)$ nor $E_n(\omega)$ are "correlated" with $I_s(\omega)$, so the cross product terms $<|I_s(\omega)\ I_n(\omega)|>$ and $<|I_s(\omega)\ E_n(\omega)|>$ are the product of the mean values, which are assumed to be zero. The noise factor is the output power divided by the amplified input resistor noise. This ratio is equal to the ratio of the equivalent mean square noise current at the input to the mean square current from the source resistor noise as in Eq. (6–34).

$$F = <I^2_{eq}(\omega)>/<I^2_s(\omega)> = 1 + <|I_n(\omega) + E_n(\omega)Y_s|^2>/<I^2_s(\omega)> \tag{6–34}$$

Let $I_n(\omega)$ be equal to $I'_n(\omega) + I_u(\omega)$, where $I'_n(\omega)$ is the component of $I_n(\omega)$ correlated with $E_n(\omega)$, and $I'_u(\omega)$ is the component uncorrelated with $E_n(\omega)$. Then, the numerator term in Eq. (6–34) becomes $<|I'_n(\omega) + E_n(\omega)Y_s + I_u(\omega)|^2> = <|I'_n(\omega) + E_n(\omega)Y_s|^2> + <|I_u(\omega)|^2>$. We define a *correlation admittance* Y_γ by requiring the statistical average $<I'^*_n(\omega)E_n(\omega)>$ to be equal to $Y^*_\gamma<|E_n(\omega)|^2>$. Substituting these into Eq. (6–34) results in Eq. (6–35).

$$F = 1 + <|I_u(\omega)|^2>/<I^2_s(\omega)> + \left(<|E_n(\omega)|^2>|Y_\gamma + Y_s|^2\right)/<I^2_s(\omega)> \tag{6–35}$$

Thermal noise power in a resistor with conductance G is related to $I(\omega)$ by $<|I(\omega)|^2> =$ 4KTBG, and so we define $G_u = <|I_u(\omega)|^2>/4KTB$ and $R_n = <|E_n(\omega)|^2>/4KTB$. The input conductance is related to the input noise current by $G_s = <|I_s(\omega)|^2>/4KTB$. Using these in Eq. (6–35), we obtain Eq. (6–36).

$$F = 1 + G_u/G_s + (R_n/G_s)\left((G_\gamma + G_s)^2 + (B_\gamma + B_s)^2\right) \tag{6–36}$$

The noise factor F in Eq. (6–36) can be minimized with respect to G_s and B_s by evaluating $\partial F/\partial G_s = 0$ and $\partial F/\partial B_s = 0$. Doing this results in an optimum source admittance $Y_s = Y_{opt} = G_{opt} + jB_{opt}$, given in Eq. (6–37). When the source admittance is equal to Y_{opt}, then the circuit will have the minimum noise factor F_{min}, given in Eq. (6–38).

Substituting these into Eq. (6–38) yields Eq. (6–39), which is a relation for the noise factor in terms of the parameters defined above and the actual source admittance Y_s.

$$G_{opt} = \left(\frac{G_u + R_n\, G_\gamma^2}{R_n} \right), \qquad B_{opt} = -B_\gamma \tag{6–37}$$

$$F_{\min} = 1 + 2R_n(G_\gamma + G_{\text{opt}})^2 \tag{6–38}$$

$$F = F_{\min} + R_n/G_s|Y_s - Y_{\text{opt}}|^2 \tag{6–39}$$

The parameters $F_{\min}$, R_n, and Y_{opt} are supplied by the manufacturer of the transistor. These are called the minimum noise factor, the noise resistance, and the source admittance for minimum noise factor. They are used in the design of low noise amplifiers.

The formula given as Eq. (6–39) is more useful with S parameters if given in terms of the reflection coefficient at the input rather than the input admittance. To do this, we normalize the parameters as $r_n = R_n/Z_0$, $g_s = G_sZ_0$, $y_{\text{opt}} = Y_{\text{opt}}Z_0$, and $y_s = Y_sZ_0$. Then, we substitute $y_s = (1 - \Gamma_s)/(1 + \Gamma_s)$ and $y_{\text{opt}} = (1 - \Gamma_{\text{opt}})/(1 + \Gamma_{\text{opt}})$ into Eq. (6–39) to obtain:

$$F = F_{\min} + (r_n/g_s)\frac{\left|(1-\Gamma_s)(1+\Gamma_{\text{opt}}) - (1+\Gamma_s)(1-\Gamma_{\text{opt}})\right|^2}{\left|1+\Gamma_{\text{opt}}\right|^2\left|1+\Gamma_s\right|^2}.$$

This can be rearranged as:

$$F = F_{\min} + (4r_n/g_s)\frac{\left|\Gamma_{\text{opt}} - \Gamma_s\right|^2}{\left|1+\Gamma_{\text{opt}}\right|^2\left|1+\Gamma_s\right|^2}. \tag{6–40}$$

Using $2g_s = (y_s + y_s^*)$ and $y_s = (1 - \Gamma_s)/(1 + \Gamma_s)$, we get

$$2g_s = (1 - \Gamma_s + \Gamma_s^* + 1 - \Gamma_s^* + \Gamma_s - 2|\Gamma_s|^2)/|1 + \Gamma_s|^2 = (1 - |\Gamma_s|^2)/|1 + \Gamma_s|^2.$$

When this is substituted into Eq. (6–40), we have Eq. (6–41).

$$F = F_{\min} + 4r_n\frac{\left|\Gamma_{\text{opt}} - \Gamma_s\right|^2}{(1-|\Gamma_s|^2)\left|1+\Gamma_{\text{opt}}\right|^2} \tag{6–41}$$

6.16 NOISE CIRCLES

The loci of Γ_s for constant values of F in Eq. (6–41) are circles in the Γ_s plane whose locations and radii are functions of these parameters. These circles are called *constant noise figure circles.* They are used in the design of amplifiers, along with other criteria such as stability and gain circles. The noise circle location and radius depends on the parameters of $F_{\min}$, r_n, and Γ_{opt}, which are dependent upon the physical construction of the transistor and do not vary with the source circuit parameters. To derive the equations of the constant noise figure circles, we define the parameter N_i as in Eq. (6–42), and rewrite Eq. (6–41) as $N_i = |\Gamma_{opt} - \Gamma_s|^2/(1 - |\Gamma_s|^2)$ or $N_i\,(1 - |\Gamma_s|^2) = (\Gamma_s - \Gamma_{opt})(\Gamma_s^* - \Gamma_{opt}^*)$.

$$N_i = (F_i - F_{min})\,|1 + \Gamma_{opt}|^2/(4r_n) \tag{6–42}$$

Now, we let $\Gamma_s = U + jV$ and complete the square in the result to get a formula for a circle in the U, V plane (Γ_s plane). The center of the constant N_i circle in the Γ_s plane is given by Eq. (6–43) and the radius by Eq. (6–44).

$$C_i = \Gamma_{opt}/(1 + N_i) \tag{6–43}$$

$$R_i = \sqrt{N_i^2 + N_i(1 - |\Gamma_{opt}|^2)}\;/\;(1 + N_i) \tag{6–44}$$

Example: To design the input circuit for a moderately low noise 4 GHz amplifier using an NE76100 GaAs MESFET, we referred to the transistor parameters given in Table 6–4. We saw that the stability factor K is less than unity at 4 GHz so, we calculated the input stability circle using Eq. (6–21) to avoid the unstable region for Γ_s. We got $C_s = 1.493\ \angle 103.2°$ and $R_s = 0.8613$. The stability circle is shown in Figure 6–22. Next, we calculated the noise figure circles for $F_{dB} = 2$ and 3 dB using Eqs. (6–42), (6–43), and (6–44). For 2 dB, we have $F_2 = 10^{0.2} = 1.5849$, $F_{min} = 10^{0.08} = 1.2023$,

$$N = (1.5849 - 1.2023)|1 + 0.7\angle 65°|^2/(4 \times 0.45) = 0.4425,$$

$$C_2 = 0.7\angle 65°/(1 + 0.4425) = 0.4853\angle 65°, \qquad \text{and}$$

$$R_2 = \sqrt{0.4425^2 + 0.4425(1 - 0.7^2)}/(1 + 0.4425) = 0.4501.$$

The noise circles for 2 dB and 3 dB are shown in Figure 6–22 as solid lines. We calculated the maximum stable gain from $G_A = |S_{21}/S_{12}| = 26.259$ or 14.19 dB. From this, we determined that the gain would be less than 14 dB. We calculated the gain circles for 12 dB and 10 dB from Eq. (6–25). These are also shown in Figure 6–22 as dashed lines. In this figure, we see that the 2 dB noise figure circle and the 12 dB gain circle intersect near the origin of the Smith® chart. It is apparent that using the 50 Ω input will yield a gain of 12 dB and a noise figure of 2 dB. Thus, we selected this for our input circuit.

TABLE 6–4 S Parameters and Noise Parameters for NE76100.

Freq.	S11		S21		S12		S22	
2	0.951	–43.9	3.505	144.7	0.068	65.0	0.731	–22.0
4	0.859	–81.4	2.941	114.6	0.112	45.5	0.657	–39.7
6	0.784	–111.4	2.411	90.3	0.135	31.5	0.598	–53.1

	K	F_{min} (dB)	$\|\Gamma_{opt}\|$	$\angle\Gamma_{opt}$	r_n
2	0.145	0.60	0.82	34	0.48
4	0.281	0.80	0.70	65	0.45
6	0.410	1.15	0.60	95	0.39

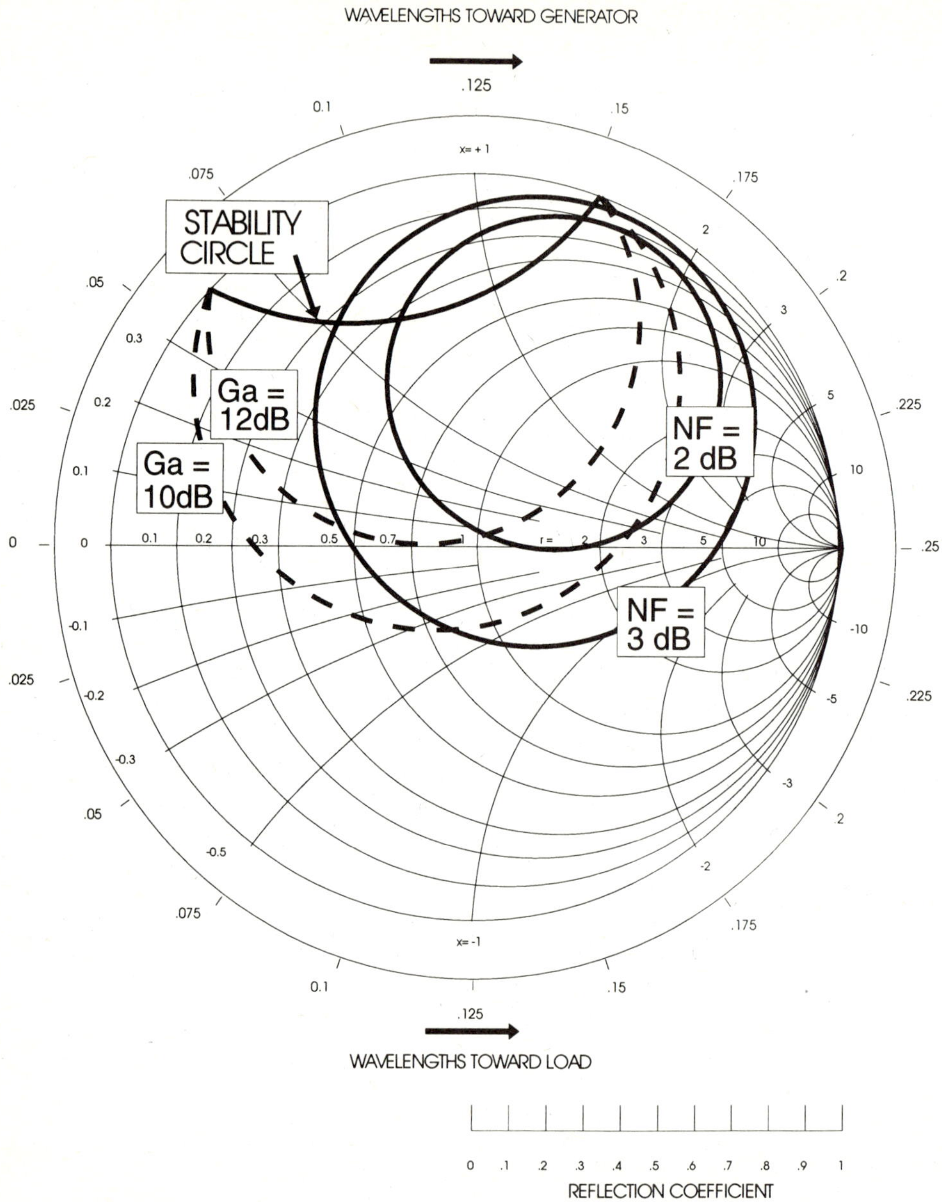

FIGURE 6–22. Noise Circles and Gain Circles Used for Amplifier Design. Permission to produce the Smith charts in this book has been granted by Analog Instruments Co., New Providence, NJ under renewal copyright issued in 1976.

6.17 NOISE MODELS

Internal noise sources in FETs are thermal noise in the resistive paths and diffusion noise in the channel. Theoretical descriptions of these noise sources have been published.[7,8,9] The problem of modeling these noise sources with accurate but usable equivalent circuits has been addressed.[10,11,12]

One simple noise model for the intrinsic part of a FET is shown in Figure 6–23A, where R_m and R_i are resistors in which thermal noise originates and I_g and I_d are noise current sources that model the channel diffusion noise.[9] The thermal noise sources in the resistors are shown explicitly as ideal voltage generators E_m and E_i whose mean square voltages are 4KTR. Then, the resistors are shown as noiseless resistors in series with these voltage generators. The source I_s is the noise current occurring in the resistive part of the source admittance.

To express the noise factor of this circuit in terms of these resistors and voltages, we solve for the short circuit drain current I_{sc} caused by the internal and external noise sources, and the short circuit drain current I_o caused by the external noise source acting alone. In this discussion, we will require that the noise currents and voltages used in the analysis be the Fourier transform of the time functions as in section 6–15. We then use these currents in the expression: $F = \langle|I_{sc}|^2\rangle / \langle|I_o|^2\rangle$. From a sequence of Norton's to Thevenin's to Norton's transformations, we arrive at the source circuit of Figure 6–23B, where $Y_x = 1/(Z_s + R_m + Z_{gs}) = 1/Z_x$, $Z_{gs} = R_{gs} + 1/j\omega C_{gs}$, $Z_s = 1/Y_s$, and $I_x = (I_s Z_s + E_m + I_g Z_{gs})Y_x$. From Figure 6–23B, we obtain Eq. (6–45). A further use of Norton's equivalent circuit results in Figure 6–23C.

$$I_{sc} = I_d - g_m E_g \tag{6–45}$$

From this figure, we get $I_{in} = I_x - E_x Y_x$ and $E_x = (I_x - I_i + g_m E_g - I_d)/(Y_x + G_i)$. A simultaneous solution of these equations results in Eq. (6–46).

$$E_g = [I_s Z_s + E_m + E_i - I_g(Z_x + R_i) + I_d R_i]/[g_m R_i + j\omega C_{gs}(R_i + Z_x)] \tag{6–46}$$

Substitution of Eq. (6–46) into Eq. (6–45) allows us to write the short circuit noise current as

$$I_{sc} = I_d - g_m \frac{I_s Z_s + E_m + E_i - I_g(Z_x + R_i) + I_d R_i}{g_m(R_i + j\omega C_{gs}(R_i + Z_x)/g_m)}$$

For the case when E_m, E_i, I_g, and I_d all are zero, we have $I_o = -g_m E_g$. Solving for E_g and substituting into this we get

$$I_o = -\frac{I_s Z_s}{R_i + j\omega C_{gs}(Z_s + R_m + R_i + Z_{gs})/g_m}.$$

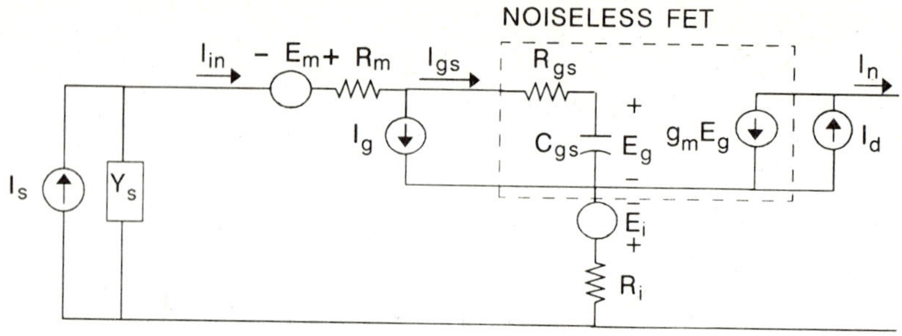

A NOISE EQUIVALENT CIRCUIT OF A FET

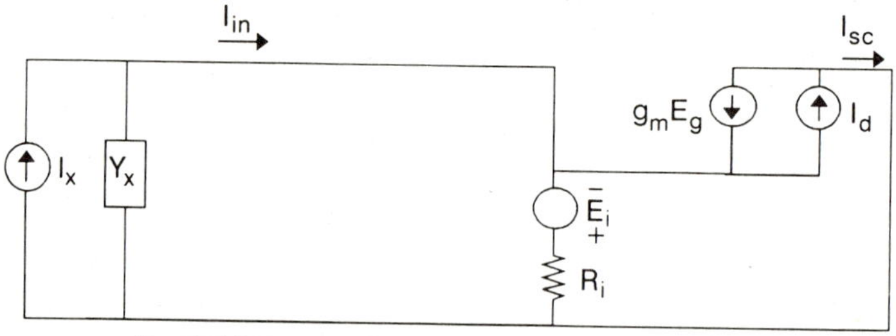

B SIMPLIFIED NOISE CIRCUIT OF A FET

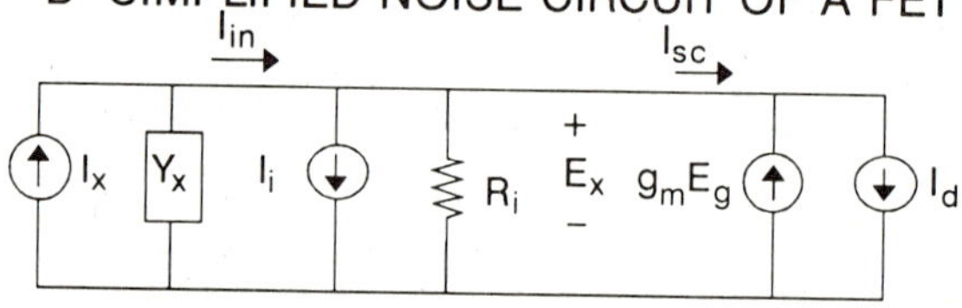

C FINAL SIMPLIFICATION

FIGURE 6–23. Transistor Noise Model.

Substituting these into the expression for the noise factor, we get the noise factor as

$$F = \frac{< \left| I_s Z_s + E_m + E_i - I_g (Z_s + R_m + R_i) - I_d j\omega C_{gs} (Z_s + Z_{gs} + R_m + R_i)/g_m \right|^2 >}{< \left| I_s Z_s \right|^2 >}.$$

Recognizing that the noise voltages E_m and E_i are uncorrelated with each other or with the noise currents, and using E = 4KTR for these voltages, we can rewrite this as Eq. (6–47), where R_n' is $R_m + R_i$.

$$F = 1 + R_n'/(|Z_s|^2 G_s) + \frac{R_n'}{|Z_s|^2 G_s} \frac{< \left| I_g (1 + Z_s/R_n') + I_d j\omega C_{gs} (1 + (Z_s + Z_{gs})/R_n')/g_m \right|^2 >}{4KT} \tag{6–47}$$

A comparison of Eq. (6–47) with Eq. (6–35) shows that they have the same form. Pucel, Haus, and Statz derived relations for these parameters in terms the transistor bias currents, gate lengths, and thicknesses, and showed how the noise factor parameters varied with them.[9]

Other noise models have been proposed as well. Fukui proposed an empirical model based on the assumed relations between the equivalent circuit elements of a transistor and its noise parameters given in Eq. (6–48).[13] The resistors R_g and R_d are series resistors in the gate and drain circuits, and the k_i are fitting factors that can be adjusted to fit a given transistor.

$$F_{\min} = 1 + k_1 f C_{gs}\sqrt{(R_g + R_d)/g_m}, \qquad R_n = k_2/g_m^2$$
$$Z_{\text{opt}} = k_3(R_g + R_d + 1/4g_m) + j\,k_4/(fC_{gs}) \tag{6–48}$$

To be useful, the noise model must

1. have components that can be determined by a simple measurement procedure,
2. be easily related to the noise parameters $F_{\min}$, R_n, and Y_{opt}, and
3. produce results capable of extrapolation to different frequency ranges and different biases.

A discussion of the use of various noise models in CAD programs has been given by Pitzalis.[14]

PROBLEMS

6–1. The NE68035C transistor has the S parameters $S_{11} = 0.28\angle 177°$, $S_{21} = 4.26\angle 65°$, $S_{12} = 0.04\angle 57°$, and $S_{22} = 0.56\angle{-35°}$ at 2 GHz. Determine the input reflection coefficient when $\Gamma_L = 0.6\angle 110°$.

6–2. Assume the transistor in problem 6–1 is unilateral and has a $\Gamma_s = 0.3\ \angle 50°$, then calculate the unilateral power gains G_1, G_2, and G_0. What is the overall power gain?

6–3. The 2 port shown in the Figure 6–24, consists of a series inductor. For this 2 port, determine an expression for the available power from the input network shown in terms of Z_0 and Z_s, when Z_0 is assumed to be purely resistive. Find the expressions for the operating gain, the transducer gain, and the available gain in terms of the circuit parameters.

6–4. If the source and load impedances for a 2 port satisfy $Z_S = Z_L = Z_0$, show that the transducer gain is $|S_{21}|^2$, where S_{21} is that of the 2 port referenced to Z_0.

6–5. Find G_T, G_P, and G_A for the transistor in problem 6–1 if $Z_S = 2Z_0$.

6–6. The NE68135 transistor has S parameters at 4 GHz of $S_{11} = 0.68\angle 167°$, $S_{21} = 2.8\angle 65°$, $S_{12} = 0.06\angle 65°$, and $S_{22} = 0.39\angle{-33°}$. Find the Γ_{ML}, Γ_{MS}, and G_{MAX} for a bilateral match of this transistor at this frequency.

6–7. Design input and output matching circuits of the form shown in Figure 6–4 to bilaterally match the NE68135 transistor of problem 6–6 at 4 GHz. Use a substrate with $h = 0.635$ mm and $\varepsilon_r = 10$ in your design and determine the microstrip line widths.

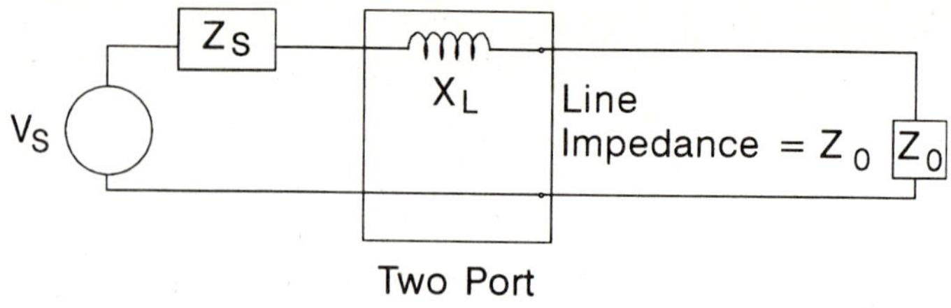

FIGURE 6–24. Circuit for Problem 6–3.

6–8. The NE710 transistor has the S parameters at 3 GHz of $S_{11} = 0.92\angle{-50°}$, $S_{21} = 4.14\angle{142°}$, $S_{12} = 0.07\angle{63°}$, and $S_{22} = 0.56\angle{-26°}$ and at 6 GHz of $S_{11} = 0.87\angle{-90°}$, $S_{21} = 3.36\angle{114°}$, $S_{12} = 0.12\angle{42°}$, and $S_{22} = 0.5\angle{-47°}$. Calculate and plot the input stability circles for these frequencies. Which would be a better design choice for Γ_S with this transistor at these frequencies, $\Gamma_S = 0.7\angle{90°}$ or $\Gamma_S = 0.7\angle{-90°}$?

6–9. For the NE710 of problem 6–8, determine which of the stability relations of Eq. (6–22) are not met at both 3 GHz and 6 GHz.

6–10. The criterion for stability that the input resistance not be negative is the same as the criterion that the input conductance of a circuit not be negative. For the 2 port of Figure 6–25, show that $\mathcal{Re}\,\{Y_{in}\} = 0$ corresponds to the equation: $(G_L + G_{22} - X)^2 + (B_L + B_{22} - Y)^2 = X^2 + Y^2$, where $X = \mathcal{Re}\,\{Y_{21}Y_{12}\}/2(G_{11} + G_S\}$ and $Y = \mathcal{Im}\,\{Y_{21}Y_{12}\}/2(G_{11} + G_S)$. This defines a stability circle in the load admittance plane whose center is at $G_L = X - G_{22}$, $B_L = Y - B_{22}$, and whose radius is $\sqrt{X^2 + Y^2}$.

6–11. For the circuit of Figure 6–25 and the stability circles of problem 6–10, the least stable case is when G_S is zero. In that case, the circuit is said to be inherently stable if the stability circle lies entirely in the left half of the Y_L plane. If $C = |Y_{12}Y_{21}|/(2G_{11}G_{22} - \mathcal{Re}\,\{Y_{12}Y_{21}\})$ is the Linville stability factor, show that $0 \le C \le 1$ corresponds to the stability circle in the left half of the Y plane.

6–12. Finish the design of the 2 port in the second example of section 6–11 to achieve the $\Gamma_S = 0.662\angle{82°}$ and $\Gamma_L = 0.75\angle{-39.97°}$.

6–13. Derive the circle formulas for the unilateral gain given in Eq. (6–26).

6–14. For the NE68135 transistor of problem 6–6, find the operating gain circle corresponding to G_P of 3 dB below the G_{MAX}.

6–15. Find the constant available gain circle that is 3 dB below the maximum gain for the NE68035C transistor of problem 6–1.

6–16. Find the MSG (see section 6–9) at 3 GHz for the NE710 transistor of problem 6–8. Then, calculate the available gain circle for 3 dB below this MSG and find the Γ_S on that circle that is furthest from the 3 GHz unstable region. Find the corresponding Γ_L required for this

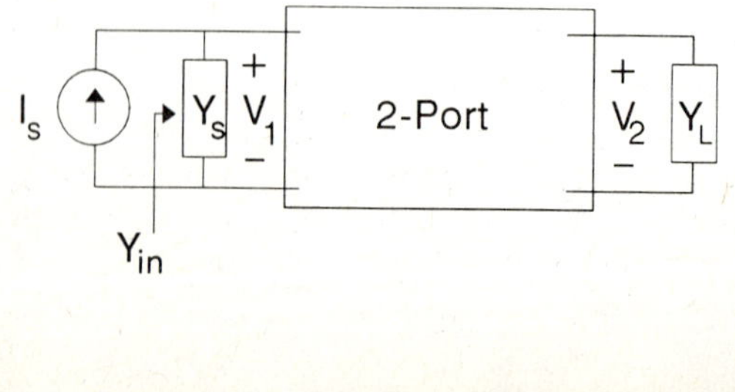

FIGURE 6–25. Circuit for Problem 6–10

TABLE 6–5 Transistor S Parameters.

Freq.	S11		S21		S12		S22	
1.5	0.932	−63.8	5.562	145.1	0.045	60.6	0.493	−28.5
2.0	0.918	−78.9	4.965	133.6	0.055	56.2	0.47	−35.5
2.5	0.875	−87.0	4.451	124.0	0.059	54.0	0.450	−39.0
3.0	0.85	−95.0	3.938	113	0.064	51	0.430	−44.5
3.5	0.8	−101	3.561	105.5	0.066	47	0.410	−48

available gain. Use these values of Γ_L and Γ_S in Eq. (6–11) to verify that your gain is 3 dB below the MSG.

6–17. Design a bandpass amplifier using the transistor whose S parameters are given in Table 6–5. The amplifier is to have a gain of 15 dB from 2 to 3 GHz. The gain is to decrease outside of the band. The design is to be for alumina substrate with $\varepsilon_r = 9.8$ and a thickness of 0.635 mm. The design should be accomplished by first designing for a bilateral match at 3 GHz and then using a CAD optimizer to achieve a flat bandpass.

6–18. Consider three cascaded amplifiers each having the same bandwidth B. The noise power output of the *i*th stage is related to the noise power input of the stage as $P_i = G_i(P_{i-1} + KBT_{ei})$, where G_i and T_{ei} are the power gain and the equivalent noise temperature of the ith stage. By cascading these expressions for noise power outputs and by subsequently substituting for T_{ei} from Eq. (6–32), show that the overall noise factor is given by $F_0 = F_1 + (F_2 - 1)/G_1 + (F_3 - 1)/(G_1G_2)$. The resultant expression for the general case of n cascaded amplifiers is known as Friis' formula.

6–19. Graphically determine the Γ_S required for a NE76184A transistor to have an available gain of 2 dB below the MSG and a noise figure of 1.5 dB at 4 GHz. The transistor S parameters at 4 GHz are $S_{11} = 0.69\angle{-118°}$, $S_{21} = 2.87\angle 78°$, $S_{12} = 0.114\angle 20°$, and $S_{22} = 0.57\angle{-67°}$. The transistor noise parameters at this frequency are $F_{min} = 0.8$ dB, $\Gamma_{opt} = 0.66\angle 82°$, and $r_n = 0.35$. Note that there are two possible solutions for Γ_S. Choose the solution closest to the center of the Smith® chart.

6–20. Determine a Γ_S for the transistor of problem 6–19 that will result in an available gain of 1 dB below the MSG and a noise figure of 1.5 dB. Design the microstrip lines that will produce this gain and noise figure, assuming that the layout will be on a substrate with $\varepsilon_r = 9.8$ and $h = 0.635$ mm. Use a single stub matching, like the topology of Figure 6–4, for both the input and the output circuits.

ENDNOTES

1. J. N. Reddy and M. L. Rasmussen, *Advanced Engineering Mathematics*, p. 32, (New York: John Wiley & Sons, 1982).
2. Donald Woods, "Reappraisal of the Unconditional Stability Criteria for Active 2-Port Networks in Terms of S Parameters," *IEEE Trans. Circuits Systs., CAS-23, 2* (Feb. 1976), 73–81.

3. K. Kurokawa, "Power Waves and the Scattering Matrix," *IEEE Trans. Microwave Theory Tech., MTT-13, 3* (March 1965), 194–202.
4. H. A. Haus and others, "Representation of Noise in Linear Twoports," *Proc. IRE,. 48, 1* (Jan. 1960), 69–73.
5. For a discussion of stationarity and ergodicity, see James L. Melsa and Andrew P. Sage, *An Introduction to Probability and Stochastic Processes,* pp. 229–242, (Englewood Cliffs, NJ, Prentice-Hall, 1973).
6. For a discussion on the use of Thevenin's theorem with noise currents, see C. S. Bull, *Fluctuations of Stationary and Non-Stationary Electron Currents,* Chap. 10, (London: Butterworths, 1966).
7. A. van der Ziel, "Gate Noise in Field Effect Transistors at Moderately High Frequencies", *Proc. IRE, 51, 3* (March 1963), 461–467.
8. W. C. Bruncke and A. van der Ziel, "Thermal Noise in Junction-Gate Field-Effect Transistors," *IEEE Trans. Electron Devices, ED-13, 3* (March 1966), 323–329.
9. Robert A. Pucel, Hermann A. Haus, and Hermann Statz, "Signal and Noise Properties of Gallium Arsenide Microwave Field-Effect Transistors," *Advances in Electronics and Electron Physics,* ed. L. Marton, pp. 229–263, (New York: Academic Press, 1975).
10. Alain Cappy, "Noise Modeling and Measurement Techniques", *IEEE Trans. Microwave Theory Tech., 36, 1* (Jan. 1988), 1–10.
11. Marian W. Pospieszalski, "Modeling of Noise Parameters of MESFET's and MODFET's and Their Frequency and Temperature Dependence," *IEEE Trans. Microwave Theory Tech., 37, 9* (Sept. 1989), 1340–1350.
12. Madhu S. Gupta, Octavius Pitzalis, Jr., Steven E. Rosenbaum and Paul T. Greiling, "Microwave Noise Characterization of GaAs MESFET's: Evaluation by On-Wafer Low-Frequency Output Noise Current Measurement," *IEEE Trans. Microwave Theory Tech., MTT-35, 12* (Dec. 1987), 1208–1217.
13. Hatsuaki Fukui, "Design of Microwave GaAs MESFET's for Broad-Band Low-Noise Amplifiers," *IEEE Trans. Microwave Theory Tech., MTT-27, 7* (July, 1979), 643–650.
14. Octavius Pitzalis, Jr., "Demystify Noise Circuit Modeling and Analysis," *Microwaves & RF, 29, 11* (Nov. 1990), 91–98.

Chapter 7

Wide Band Amplifiers

7.0 INTRODUCTION

Almost with the invention of the triode in the summer of 1906 by Lee DeForest came the need for wide band amplification of electrical signals. In 1939, Wheeler wrote, "For more than twenty years, the design of amplifiers to cover a wide band of frequencies has been a major problem."[1] He determined a formula for the maximum "gain-bandwidth product" of a vacuum tube amplifier using the ratio of the transconductance to stray capacitance very similar to that in use for transistors today. In 1945, Hansen proved that Wheeler's formula gave the theoretical limit on the possible bandwidth for a specified gain of vacuum tube amplifiers.[2] In this article, the term *gain-bandwidth product* was used for the area under that portion of the curve of voltage gain versus frequency where the voltage gain was a constant. Hansen derived this limit in terms of the logarithm of the voltage gain.

At about the same time, and independently of Hansen's work, Bode derived a very general formula that specified that the area given by the integral $\int_0^\infty \ln |1/\rho| \, d\omega$ is limited to the value π/CR, where ρ is the reflection coefficient looking into the parallel combination of the amplifier output circuit shunted by the parasitic capacitance C of the amplifier tube, where R is the Thevenin equivalent resistance of the source driving the output circuit.[3] This was the seminal work leading to the development of further theory concerning the limitations of *equalizer networks.* Equalizer networks are those that match the amplifier to the load (or a subsequent amplifier stage) over the desired frequency range. Bode also discussed the use of feedback to increase the bandwidth of an amplifier stage.

These investigations showed that cascading amplifiers in the usual fashion did not increase the bandwidth, because the overall gain was the product of the gain of the individual stages. However, in 1937 the *distributed amplifier* was patented by William S.

Percival.[4] The theory of this amplifier was first discussed in the general literature in 1948.[5] By adding together the output of several stages as waves traveling along a transmission line (or artificial delay line), this amplifier is capable of overcoming the gain-bandwidth limitations mentioned previously for cascaded amplifiers.

In 1965, Engelbrecht and Kurokawa reported on the design and construction of a microwave wide band balanced amplifier that used a matched pair of transistors.[6] The input to the amplifier was through a 3 dB directional coupler that acts as a power splitter, sending one-half of the power to each transistor. The outputs of the two transistors were recombined through another 3 dB directional coupler. This became the standard design for wide band microwave transistor amplifiers for many years.

These fundamentals are still important today in the design of microwave amplifiers. Wide band amplification requires low intrinsic and parasitic shunt capacitances as expressed in the current gain cut-off frequency f_T which is sometimes referred to as the gain-bandwidth product, and it requires high transconductances and low parasitic inductances. Balanced amplifiers are still widely used for their linearity and power amplification. Distributed amplification is becoming more of a design alternative with tailor made MMIC chips.

The development of epitaxial growth processes of *molecular beam epitaxy (MBE)* and *metal-organic chemical vapor deposition (MOCVD)* in the mid and late 1970s has led to a dramatic increase in the highest frequencies at which transistors can operate. These processes allowed fabrication of transistor *heterojunctions* in which base-emitter, base-collector, gate-channel, and other semiconductor junctions are made with compound layers having different doping levels.[7] This revolution in technology extended frequency ranges above 100 GHz.

7.1 MICROWAVE TRANSISTOR EQUIVALENT CIRCUITS

Although microwave transistors are most accurately described using solid state physics, and although the transistor manufacturers usually specify the S parameters of their products, the design of transistors is often accomplished using lumped constant equivalent circuits. To this purpose, methods for determining lumped constant, equivalent circuits from S parameter measurements have been determined. We present some of the equivalent circuits, but do not discuss their relation to the physical description of conduction through these transistors. The reader wishing to study the physical basis of these equivalent circuits is referred to alternate texts.[8,9]

The Gallium Arsenide metal semiconductor field effect transistor, *GaAs MESFET* was the transistor most frequently used in microwave applications. Its equivalent circuit is shown in Figure 7–1. With the exception of C_{dg}, that portion enclosed in the dotted lines is called the intrinsic portion of the equivalent circuit; it models the conductive and capacitive effects occurring in the channel. The remaining elements are the *extrinsic elements* (sometimes called *parasitic elements*). Frequently, the effect of the capacitance C_{dc} is modeled along with those of C_{dg} and C_{gs}, and the result is shown as two capacitors C_{dg}

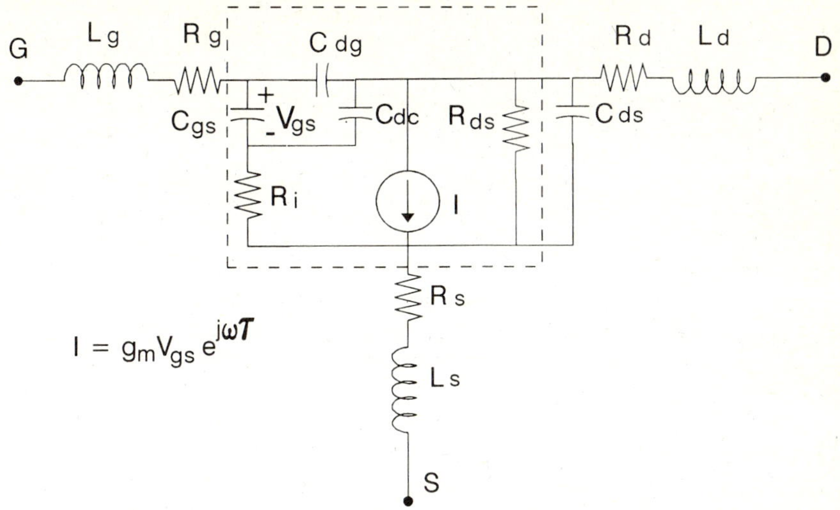

FIGURE 7–1. The Y Parameter Equivalent Circuit for a GaAsFET. The time constant τ models the transit time effects in the channel.

and C_{gs}, while C_{dc} is omitted from the diagram. While C_{gs} is several times larger than C_{ds}, they are typically the same order of magnitude (tenths of a picofarad) and C_{dg} is about an order of magnitude less. The parasitic resistors and R_i are typically of 1 to 10 Ω. The R_{ds} is on the order of hundreds of Ω. The transconductance g_m is the order of 10 to 100 mS.

The development in semiconductor theory and fabrication concerning heterojunctions has resulted in FETs with greatly increased maximum operating frequencies.[10,11] These transistors have compound layers at the channel that greatly increase the mobility and electron density. These are called *heterostructure field effect transistors (HFET)*. This includes the *modulation doped heterojunction field effect transistors (MODFET)* and *high electron mobility transistors (HEMT)*. The equivalent circuit for them is the same as in Figure 7–1. These are capable of very high frequency operation.

These fabrication techniques also have been applied to the fabrication of bipolar junction transistors.[12] These are called *heterojunction bipolar transistors (HBT)*. In this case, the emitter base junction can be made as a heterojunction that can produce a high current gain β or, alternately, a decreased base resistance and junction capacitance. The collector base junction also can be a heterojunction. The layer thicknesses that electrons pass through in the HBT are determined by epitaxial layer growth rather than lithography. With epitaxial growth processes, these layers can be made very thin. This along with graded base regions helps to reduce the transit times for the electrons in the HBT. The approximate equivalent circuit of the HBT is given in Figure 7–2. With the exception of R_{cb}, the resistors in this circuit are typically in the range of 1 to 30 Ω, C_{ce} and C_{ci} are on

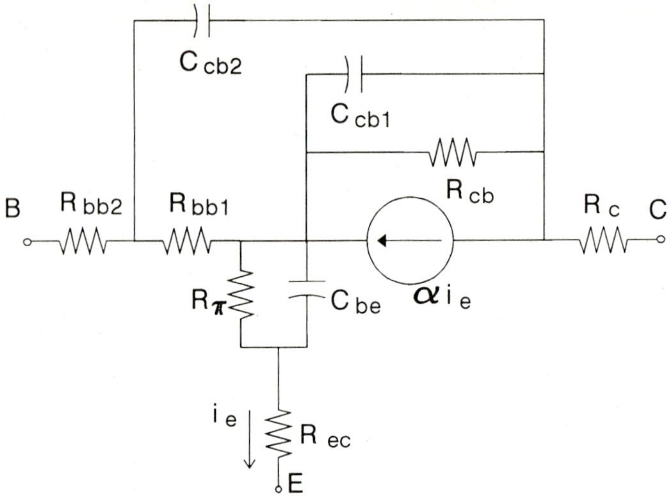

FIGURE 7–2. Heterojunction Bipolar Transistor Equivalent Circuit.

the order of 10 femtofarads (fF), and C_{be} is on the order of several hundred fF. R_{cb} is on the order of tens of KΩ. The α is:

$$\alpha = \frac{\alpha_0 \exp(j\omega\tau_c)}{1 + j\omega\tau_b}\left(\frac{\sin(\omega\tau_c)}{\omega\tau_c}\right),$$

where α_0 is about 0.98, τ_b and τ_c are base and collector transit times that depend upon the device. Of course, all these parameters depend on device materials and dimensions.

7.2 CUT-OFF FREQUENCIES

There are two principal figures of merit for the high frequency operation of transistors. They are the *current gain cut-off frequency* denoted f_T, and the *maximum frequency of oscillation* f_{max}. These can be related to the equivalent circuits given above; however, the relation is approximate because the equivalence between the circuits and the actual transistor operation ceases to be very accurate at the limits of operation of the transistor. In essence, the transistor operates as though the current generator in Figure 7–1 has an inverse frequency dependence at the higher frequency ranges. After simplifying the circuit of Figure 7–1 to that of Figure 7–3, we can solve it for the frequency where the input and output currents are equal in magnitude. For this, we have $|I_{out}/I_{in}| = |g_m - j\omega C_{gd}|/\omega(C_{gs} + C_{dg}) = 1$. Because C_{dg} is small, this becomes Eq. (7–1) as an approximate relation for the current gain cut-off frequency.

$$f_T = g_m/2\pi C_{gs} \tag{7–1}$$

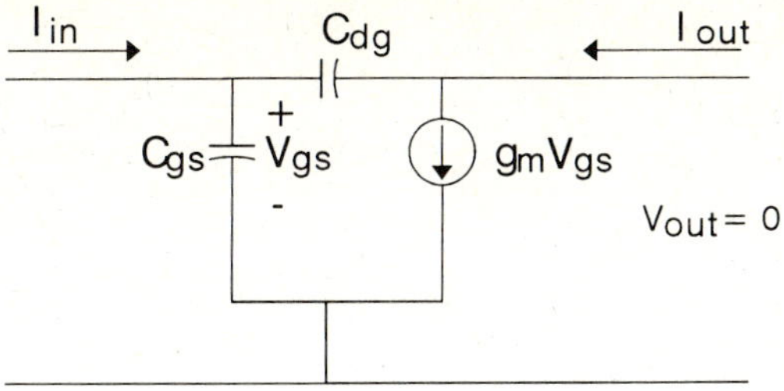

FIGURE 7–3. Approximate Equivalent Circuit for a FET.

The maximum frequency of oscillation f_{max} can be shown to be the frequency at which Mason's unilateral power gain is unity.[13] Usually this parameter is determined experimentally by measurement of the S parameters at frequencies well below f_{max}. Then these parameters are extrapolated to the higher frequency range. Either the MAG or the maximum stable gain is computed from these parameters. The frequency where these are unity is f_{max}. An expression relating f_{max} to the equivalent circuit elements can also be derived from the expression for the MAG in terms of these elements. One version of this parameter is given in Eq. (7–2), where f_T is that of Eq. (7–1) and τ_2 is $2\pi R_g C_{dg}$.[14]

$$f_{max} = f_T/[2\sqrt{G_{ds}(R_g + R_i + R_s)} + f_T \tau_2] \quad (7\text{–}2)$$

Example: NEC GaAs MESFET NE71000. The FET equivalent linear circuit parameters for the NE71000 are given as $R_g = 2.0$, $R_S = 1.5$, $R_i = 1.6$, $C_{gs} = 0.28$, $C_{dg} = 0.033$, $G_{ds} = 1.5$, and $g_m = 48$, where the units are Ohms, picofarads, and millisiemens respectively. From these parameters we have:

$$f_T = 0.048/(2\,\pi\,0.26\;10^{-12}) \cong 30\text{ GHz},$$

$$\tau_2 = 2\,\pi\,2.0 \times 0.033\;10^{-12} = 4.15\;10^{-13}, \qquad \text{and}$$

$$f_{max} = 29.38\;10^9/\left(2\sqrt{.0015(2.0 + 1.6 + 1.5)} + 29.38\;10^9\;4.15\;10^{-13}\right) \cong 100\text{ GHz}$$

The manufacturer lists f_{max} as 90 GHz for this FET.

In the case of the HBT, the current gain cut-off frequency is related to the total emitter to collector transit time τ_{ec} as $f_T = 1/2\pi\tau_{ec}$. Defining the effective base resistance collector capacitance time constant to be τ_{eff}, we have the maximum frequency of oscillation as in Eq. (7–3), where τ_{eff} is sometimes given as $\tau_{eff} = R_B C_{BC}$.

$$f_{max} = 1/[4\pi\sqrt{\tau_{ec}\tau_{eff}}] \quad (7\text{–}3)$$

7.3 CIRCUIT BANDWIDTH LIMITATIONS

The bandwidth limitations of amplifier stages can be expressed in terms of the equivalent circuits of the transistors and their input/output matching circuits. A simplified equivalent circuit for an amplifier stage and its output circuit is shown in Figure 7–4. Bode analyzed this circuit in terms of the reflection coefficient ρ presented to the idealized generator by

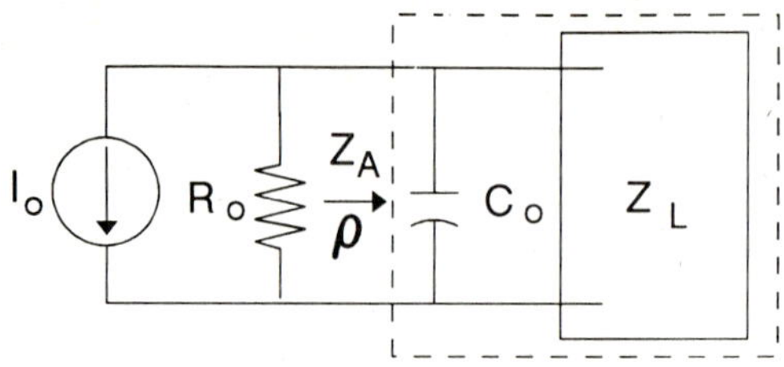

FIGURE 7–4. Idealized Amplifier Circuit. The impedance Z_L is that of the load less any shunt load capacitance. The capacitance C_0 is the output shunt capacitance of the active element combined with parasitic capacitance of the output circuit and the shunt input capacitance of the network matching the load to the amplifier.

C_0 in parallel with network containing the load impedance Z_L, which we assume is a minimum function.[15]

To find the limitation on bandwidth for this circuit, we consider the integral $\oint \ln 1/\rho$ dS over a contour enclosing the right half of the S plane (RHP). The value of this integral will be zero if the integrand has no poles in the right half of the S plane. Examining $1/\rho = (R_0 + Z_A)/(R_0 - Z_A)$, we know that $R_0 + Z_A$ cannot have poles in the RHP because it represents a passive impedance. But $R_0 - Z_A$ can have zeros in the RHP that are poles of $1/\rho$. We label these zeros as S_i. We replace the integral with that shown in Eq. (7–4), where the ratio multiplying $1/\rho$ is a function having zeros at the location of the RHP poles of $1/\rho$.

$$\oint \ln\left((1/\rho)\frac{(S-S_1)(S-S_2)\ldots(S-S_n)}{(S+S_1)(S+S_2)\ldots(S+S_n)}\right) dS \qquad (7\text{–}4)$$

This function is called an *all pass function* whose magnitude is unity everywhere on the $j\omega$ axis. The product of $1/\rho$ and the all pass function does not have poles in the RHP, so the contour integral shown in Eq. (7–4) over a contour enclosing the RHP must be zero. Because of the capacitance C_0, the reflection coefficient magnitude goes to unity as ω goes to infinity. Therefore, as S goes to infinity, the magnitude of the argument of the logarithm in the integrand goes to 1, thus that term in the integrand goes to zero. Therefore, the contribution of the integration on the infinite semicircle in the RHP is simply the phase term. Along the ω axis, the logarithm of the magnitude of the product in this integrand is simply $\ln |1/\rho|$, because the magnitude of the all pass function is unity on the ω axis. The imaginary part of the product of $1/\rho$ times the all pass function is an odd function of ω, so the phase of this product must be an odd function. But, $\ln |1/\rho|$ is an even function of ω on the ω axis. Consequently, Eq. (7–4) can be expressed as in Eq. (7–5), where C_1 is the infinite semicircle in the RHP.

$$2\int_0^{\infty} \ln |1/\rho|\, d\omega + \int_{C_1} \text{phase}\left(1/\rho\frac{(S-S_1)(S-S_2)\ldots(S-S_n)}{(S+S_1)(S+S_2)\ldots(S+S_n)}\right) dS = 0 \qquad (7\text{–}5)$$

To evaluate the integration over C_1 in Eq. (7–5), we note that the integrand consists of the sum of the phases of $1/\rho$, and the all pass function as the radius of the semicircle goes to infinity. Expanding the phase of $1/\rho$ in terms of $1/\omega$ as a Taylor's series about

$1/\omega = 0$, we see that the leading term is $2/(j\omega C_0R_0)$. Similarly, expanding the phase of the pairs of terms $(j\omega - S_i)/(j\omega + S_i)$ in a Taylor's series in $1/\omega$, we find the leading term to be $-2S_i/(j\omega)$ as ω goes to infinity. The phase term then can be expressed as $2\int_{C_1}[1/R_0C_0 - \Sigma\, S_i]\, d\omega / j\omega$. The final result of the integration is Eq. (7–6).

$$\int_0^\infty \ln|1/\rho|\, d\omega = \frac{\pi}{R_0C_0} - \pi\,\Sigma\, S_i \tag{7–6}$$

This result presented by Bode is a restriction on the total area under the curve for $\ln|1/\rho|$ to being equal to or less than π/R_0C_0. Essentially, this is a restriction forced on the gain-bandwidth product of the stage by the parasitic capacitance and generator resistance. If we could somehow match the load to the generator resistance so that $\ln|1/\rho|$ is constant over a frequency band from ω_1 to ω_2 and is zero outside that band, then Eq. (7–6) becomes $(\omega_2 - \omega_1)\ln|1/\rho| \le \pi/(R_0C_0)$. Clearly, this is a restriction on the gain-bandwidth product.

Example: If we take the simple case shown in Figure 7–5, then we can determine explicit expressions for S_i. To do this, we first find Z_A to be:

$$Z_A = \frac{SL + R_L}{S^2LC_{ds} + SC_{ds}R_L + 1},$$

and then from the reflection at the interface $\rho = (Z_A - R_0)/(Z_A + R_0)$, we find $1/\rho$ to be:

$$1/\rho = \frac{S^2LC_{ds} + S(C_{ds}R_L + L/R_{ds}) + 1 + R_L/R_0}{S^2LC_{ds} + S(C_{ds}R_L - L/R_0) + 1 - R_L/R_0}.$$

Now, we find the poles of $1/\rho$ by factoring the denominator. These are:

$$S_i = (1/2)\left[(1/R_0C_{ds} - R_L/L) \pm \sqrt{(1/R_0C_{ds} - R_L/L)^2 - 4(1 - R_L/R_0)/(LC_{ds})}\right].$$

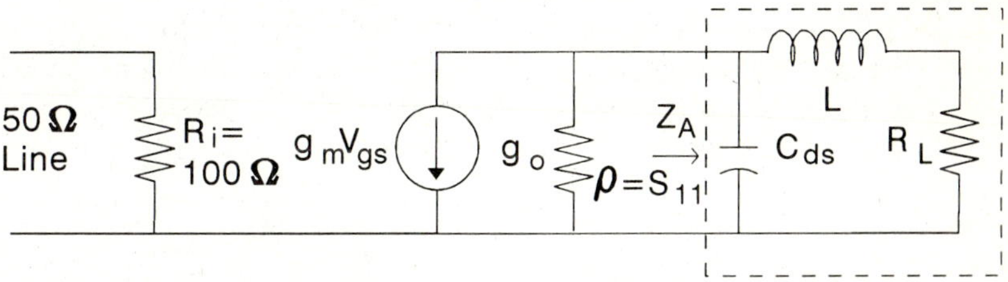

FIGURE 7–5. Simplified Equivalent Circuit of an Amplifier Operating into a Load through an Inductance. The components are: R_i = 100 Ω, g_0 = 5 *mS*, g_m = 50 *mS* and C_{ds} = 0.15 pF. The input is from a 50 Ω line and the load is R_L = 50 Ω.

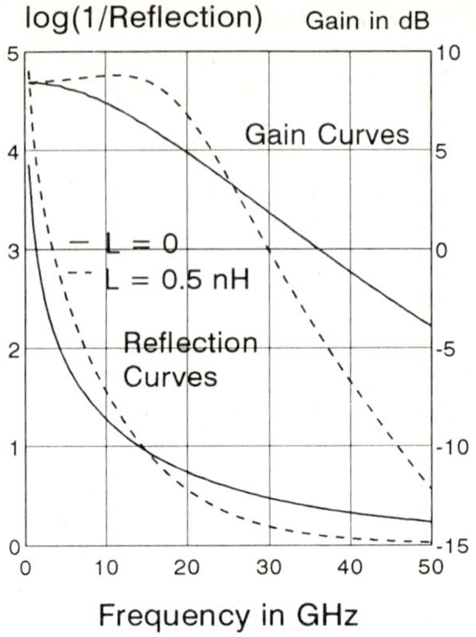

FIGURE 7–6. FET Amplifier Gain Curves and Load Reflection Curves. The dashed lines are for $L = 0.5$ nH and the solid lines are for $L = 0$.

If $R_L/L < 1/R_0C_{ds}$, we note that these poles are in the RHP. Then, $\Sigma\, S_i$ becomes $1/R_0C_{ds} - R_L/L$ and Eq. (7–6) is:

$$\int_0^\infty \ln |1/\rho| = \pi[1/R_0C_{ds} - 1/R_0C_{ds} + R_L/L] = \pi\; R_L/L.$$

When the L/R_L time constant is greater than the R_0C_{ds} time constant, we see that $\pi R_L/L$ is the factor controlling the integral rather than the π/R_0C_{ds} factor. To increase the integral, we must decrease the parasitic inductance. If this inductance is decreased so that $R_L/L > 1/R_0C_{ds}$, then these poles of $1/\rho$ are in the LHP and the factor controlling the integral is π/R_0C_{ds}.

This example has an interesting feature to it. We have calculated the stage gain with $L = 0$ and with $L = 0.5$ nH and ρ for those two inductances. The results are shown in Figure 7–6. The total area under the reflection curve clearly is greater for $L = 0$ than for $L = 0.5$ nH. Similarly, the total area under the gain curve (when plotted on a linear scale) is greater for the $L = 0$ case. However, the "useful area" under the gain curve (i.e., that area in the frequency range where the gain is within 3 dB of the maximum gain) has been increased for the case $L = 0.5$ nH. This is an example of the effect of a *peaking inductance* that produces poles in Z_A at $-R_L/2L \pm(1/2)\sqrt{(R_L/L)^2 - 4/LC_{ds}}$ in the S plane. This causes the magnitude of the load impedance to peak at about 17 GHz, while it causes the overall area under the $\ln |1/\rho|$ curve to decrease.

7.4 FURTHER GAIN BANDWIDTH RESTRICTIONS

Fano did further work on the problem of gain bandwidth restrictions and derived more exact expressions for the equalizer bandwidth limitations that depend on the location and number of the load *transmission zeros.*[16] Transmission zeros are the points in the S plane where the transmission coefficient looking into the load is zero. Following this, Youla derived a theorem relating the coefficients of expansion about the zeros of transmission of the load for various network functions.[17] He showed that, in certain cases, the load could be matched over the entire frequency axis. He further showed a method for realizing the matching networks. These techniques usually require extensive computations and sometimes result in complex networks.

Ku and Petersen have addressed the question of how transistor roll off compensation affects the maximum possible equalizer "gain" for a bandpass amplifier with an upper cut-off frequency ω_2 and a lower cut-off frequency ω_1.[18] Assuming that the roll-off compensation is accomplished in the input matching network with an equalizer gain of $K(\omega/\omega_2)^k$ in the pass band and zero outside of it, and starting with Bode's integral of Eq. (7–6), they derived a relation between ω_2, k, the maximum obtainable K, and the time constant $\tau = R_0C_0$ of Eq. (7–6). For the 6 dB per octave case, where $k = 2$, they arrived at Eq. (7–7), where $u = \omega_1/\omega_2$.

$$\frac{(1-\sqrt{K})^{1-\sqrt{K}} \quad (1+\sqrt{K})^{1+\sqrt{K}}}{(1-u\sqrt{K})^{(1-u\sqrt{K})/u} \quad (1+u\sqrt{K})^{(1+u\sqrt{K})/u}} = \exp(2\pi\tau\omega_2) \tag{7–7}$$

They also showed graphs for the maximum K as a function of the parameters k, τ, ω_1, and ω_2. These expressions only give approximate relations because of the assumptions made in their derivation.

7.5 SINGLE STAGE AMPLIFIER CIRCUITS

Different applications for wide band amplifiers require different types of amplifier circuits. For instance, the front end of a wide band satellite receiver might also require a low noise design. On the other hand, computer pulse circuitry requires a wide band as well as small physical size and relatively high output current capability. Because of different requirements different circuit designs have evolved. Niclas has categorized some of these circuits and compared them.[19] His categories are the reflective match amplifier, the lossy match amplifier, the feedback amplifier, the distributed amplifier, and the active match amplifier. In addition to these, we also have the balanced amplifier. For the first three categories, it is possible to design amplifiers with a single active element, while the last three require at least two active elements. For the present, we discuss those requiring a single active element. Designing these circuits requires input and output matching circuits often referred to as equalizers.

7.6 EQUALIZER CIRCUIT DESIGN

We will design equalizer circuits using lumped element equivalent circuits. These lumped circuit elements can then be simulated with transmission lines where necessary.

$$\begin{pmatrix} V_1 \\ I_1 \end{pmatrix} = \begin{pmatrix} A & B \\ C & D \end{pmatrix} \begin{pmatrix} V_2 \\ -I_2 \end{pmatrix}, \quad \begin{pmatrix} V_1 \\ V_2 \end{pmatrix} = \begin{pmatrix} Z_{11} & Z_{12} \\ Z_{21} & Z_{22} \end{pmatrix} \begin{pmatrix} I_1 \\ I_2 \end{pmatrix}, \quad \begin{pmatrix} I_1 \\ I_2 \end{pmatrix} = \begin{pmatrix} Y_{11} & Y_{12} \\ Y_{21} & Y_{22} \end{pmatrix} \begin{pmatrix} V_1 \\ V_2 \end{pmatrix} \tag{7–8}$$

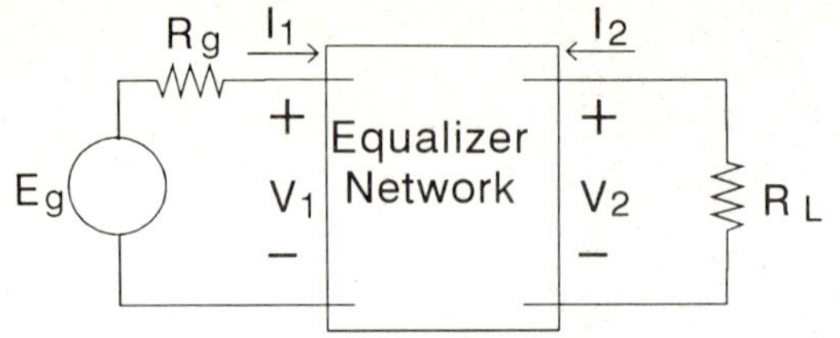

FIGURE 7–7. Linear 2 Port Representing an Equalizer Containing only Reactive Elements.

The circuit shown in Figure 7–7 represented with the relations of Eq. (7–8), where all the circuit parameters are functions of the complex frequency S. The design of an equalizer frequently starts with the specification of the equalizer gain in the form of G_T as a function of S. The object is to determine the open and short circuit impedance functions for the equalizer in terms of the ABCD parameters that depend upon G_T. Then, we use the open and short circuit impedance functions to synthesize the equalizer as a ladder network.

First, we show that A and D are even functions of S and that B and C are odd functions of S. We consider the case when the number 2 port is open circuited so that I_2 is zero, and assume that the equalizer is composed only of reactive circuit elements. From Eq. (7–8), we have the ratio V_2/I_1 equal to both $1/C$ and Z_{21}. Thus, they are equal: $1/C = Z_{21}$. On the $j\omega$ axis, Z_{21} must be a reactance (imaginary function of ω), so we conclude that C is an odd function of S. Similarly, we can short circuit port number 2 and obtain $I_2/V_1 = Y_{21} = -1/B$. Because Y_{21} must be a pure susceptance, we conclude that B must be an odd function of S. For this short circuit case and for $S = j\omega$, the power into port number 1 is $P + jQ = V_1 I_1^*$. Because the circuit has no resistance, P is zero and the product must be imaginary. But $V_1 = -BI_2$ and $I_1 = -DI_2$ for this short circuit case. So, we have $jQ = B\,D^*\,|I_2|^2$. Because B is an odd function of $S = j\omega$, we conclude that D must be an even function of S. We can short circuit port number 1 and determine the input power at port number 2 in the same way as above to lead to the conclusion that A must be an even function of S.

Now, we show that $AD - BC = 1$. This can be shown from the reciprocity theorem or from the conservation of energy. To do this, we choose the case of an arbitrary load and compute the input power for $S = j\omega$. Because the equalizer is purely reactive, we have $P_1 = -P_2$ or $\mathcal{R}e\,(V_1 I_1^*) = -\mathcal{R}e\,(V_2 I_2^*)$. We have $\mathcal{R}e\,(V_1 I_1^*) = \mathcal{R}e\,[(AV_2 - BI_2)(CV_2 - DI_2)^*] = \mathcal{R}e\,(AC^*|V_2|^2 - BD|I_2|^2 - ADV_2 I_2^* - BC^* V_2^* I_2)$. The first two terms in the last expression are purely imaginary and have no real part. In the last term, we can replace BC^* with $-BC$ and note that this product is real. Factoring the real coefficients out of the $\mathcal{R}e$ operator, we have $\mathcal{R}e\,(V_1 I_1^*) = -(AD - BC)\,\mathcal{R}e\,(V_2 I_2^*)$. Because the input power is equal to the output power for a lossless equalizer network, we see that $(AD - BC) = 1$.

Using this relation, we can derive Eq. (7–9).

$$\begin{pmatrix} V_1 \\ V_2 \end{pmatrix} = \frac{1}{C}\begin{pmatrix} A & 1 \\ 1 & D \end{pmatrix}\begin{pmatrix} I_1 \\ I_2 \end{pmatrix}, \qquad \begin{pmatrix} I_1 \\ I_2 \end{pmatrix} = \frac{1}{B}\begin{pmatrix} D & -1 \\ -1 & A \end{pmatrix}\begin{pmatrix} V_1 \\ V_2 \end{pmatrix} \tag{7–9}$$

From Eq. (7–9), we can determine the open and short impedances in terms of the functions A, B, C, and D as shown in Eq. (7–10).

$$Z_{1SC} = B/D,\ Z_{1OC} = A/C,\ Z_{2SC} = B/A,\ Z_{2OC} = D/C \tag{7–10}$$

For Fig. 7–7, we have $G_T = P_L/P_A = (|V_2|^2/R_L)/(E_g^2/4R_g)$. We define the function $H(S)$ to be $H(S) = E_g/(2V_2\sqrt{R_g/R_L})$ so that $H(S)\,H(-S) = 1/G_T$. We can sum voltages around the input loop of Figure 7–7, and using the ABCD parameters, we can express that sum as $E_g = (A + CR_g)V_2 - (B + DR_g)I_2 = (A + CR_g)V_2 + (B + DR_g)V_2/R_L$. Using this in the expression for $H(S)$, we get Eq. (7–11). This expresses the $H(S)$ dependence on the ABCD parameters.

$$H(S) = (1/2)\,(A\sqrt{R_L/R_g} + D\sqrt{R_g/R_L} + B/\sqrt{R_g/R_L} + C\sqrt{R_g/R_L}\,) \tag{7–11}$$

From Eq. (7–11), we get the even and odd parts of H(S) as: $H_e(S) = (1/2)(A\sqrt{R_L/R_g} + D\sqrt{R_g/R_L})$ and $H_o(S) = (1/2)(B/\sqrt{R_gR_L} + C\sqrt{R_gR_L})$. We cannot express any one of the ABCD parameters solely as a function of H_e and H_o. We must define a new function as $K(S) = (1/2)(A\sqrt{R_L/R_g} - D\sqrt{R_g/R_L} + B/\sqrt{R_gR_L} - C\sqrt{R_gR_L})$ that has even and odd parts $K_e(S) = (1/2)(A\sqrt{R_L/R_g} - D\sqrt{R_g/R_L})$ and $K_o(S) = (1/2)(B/\sqrt{R_gR_L} - C\sqrt{R_gR_L})$. Now, we have the relations shown in Eq. (7–12) for the ABCD parameters in terms of these two functions.

$$\begin{aligned} A &= (H_e + K_e)\sqrt{R_g/R_L},\ B = (H_o + K_o)\sqrt{R_g R_L} \\ C &= (H_0 - K_o)/\sqrt{R_g R_L},\ D = (H_e - K_e)\sqrt{R_L/R_g} \end{aligned} \tag{7–12}$$

From Eq. (7–11) for $H(S)$ and its counterpart for $K(S)$, one can easily show that $H(S)\,H(-S) - K(S)\,K(-S) = AD - BC$ which becomes Eq. (7–13).

$$K(S)\,K(-S) = H(S)\,H(-S) - 1 \tag{7–13}$$

$H(S)$ is a rational function, so we define $E(S)$ and $P(S)$ as the numerator and denominator polynomials of $H(S)$: $H(S) = E(S)/P(S)$. From Eq. (7–13), we see that $K(S)$ must have the same polynomial in the denominator as $H(S)$. Letting $F(S)$ be the numerator of $K(S)$ we have $K(S) = F(S)/P(S)$. From Eq. (7–13), we arrive at Eq. (7–14). Given H(S), we know $E(S)$ and $P(S)$. From these and Eq. (7–14), we can find $F(S)$ and then $K(S)$.[20]

$$F(S)\,F(-S) = E(S)\,E(-S) - P(S)\,P(-S) \tag{7–14}$$

We can start with $|S_{21}|^2$ if $\Gamma_S = \Gamma_L = 0$. Note that G_T is $|S_{21}|^2$ (where S_{21} is that of the equalizer) *only* if R_g and R_L are matched to the reference line impedance so that $\Gamma_L = 0 = \Gamma_S$. Otherwise, Eq. (6–10) applies. For the lossless equalizer we use Eq. (3–8): $|S_{21}|^2 = 1 - |S_{11}|^2$. Setting $G_T = |S_{21}|^2$, this becomes $H(S)\,H(-S) - 1 = H(S)\,H(-S)\,|S_{11}|^2$. From Eq. (7–13), we arrive at $K(S)\,K(-S) = H(S)\,H(-S)\,|S_{11}|^2$.

This process for designing the equalizer is:

(1) Specify $G_T = 1/\big(H(S)\,H(-S)\big)$. If $H(S)$ is not a minimum reactance function the step of determining the circuit components from the open and short circuit impedance functions becomes more complex.[21]

(2) From *H(S) H(–S)* determine *H(S)*, and with Eq. (7–14), find *F(S)* then *K(S)*. With high order *H(S)*, this step can be difficult. It generally requires numerical solutions on a computer.

(3) Then use Eq. (7–12) to obtain the ABCD parameters.

(4) Then use Eq. (7–10) to determine the open and short circuit impedances of the equalizer.

(5) From these, the individual circuit components are identified. Because this process relates R_g to R_L, the designer may need to use an impedance transformer to obtain the desired termination resistors.

(6) Finally, the designer usually simulates the circuit on CAD software and resorts to an optimizing routine to reach the final design.

An example of this process is given in section 7–7.

7.7 REFLECTIVE MATCH AMPLIFIER

The design of single transistor wide band reflective match amplifier requires the consideration of three factors. The primary factor is that the gain of the stage will be less than the MAG of the transistor at the upper cut-off frequency ω_2. Usually, it is desirable to make the gain as high as possible at this frequency. The amount by which the stage gain is less than the MAG is related to the bandwidth of the stage, with the gain decreasing as the bandwidth is increased. The second factor is that the roll-off of the transistor $|S_{21}|$ with frequency must be compensated by a corresponding increase in the gain of the input or output matching networks (equalizers) with frequency. A third factor is that if the lower cut-off frequency ω_1 is very low, then direct coupling of the input and output to the source and load is usually needed, otherwise, the coupling capacitors required become prohibitively large.

Mellor and Linvill showed how to design a band pass amplifier stage compensated for transistor roll-off.[22] Their design compensates for transistor roll-off with an equalizer gain that increases with frequency over the band. First, they specify a low pass prototype for the equalizer by selecting a $T(\Omega)$ to use in the $G_T(\Omega) = (|S_{21}(\Omega)|^2$, as shown in section 4–7. Because the equalizer is not terminated in Z_0, its transducer gain is not strictly $|S_{21}(\Omega)|^2$. However, this process produces a design that can be adjusted later with the use of CAD software if necessary. Then, they use a bandpass transformation of the form $\Omega = (1/W)(\omega/\omega_0 - \omega_0/\omega)$ (see section 4–9) in this to find $|S_{21}(\omega)|^2$. From this, they form $H(S)$ $H(-S) = 1/(\omega^{2n}|S_{21}(\omega)|^2)|_{\omega=-js}$ in order to achieve the desired increase of 20 n dB/decade (6 n dB per octave). In this step n must be an integer, because *H(S) H(–S)* must be a rational function. From *H(S) H(–S)*, they determine *K(S) K(–S)* using Eq. (7–14). Then, the network is synthesized from these two functions. One problem with this method is that the designer cannot compensate easily for a roll-off occurring at a rate not corresponding to n being an integer.

Example: The Design of a 3 to 18 GHz Amplifier. For this design, we used the S parameters for the NEC transistor NE71000. Initially, we examined the stability of the transistor by computing input stability circles with the aid of a CAD program. These were computed assuming inductance and resistance in each of the transistor leads as $R_g = 2$, $L_g = 0.05$, $R_s = 1.5$, $L_s = 0.02$, $R_d = 1.5$, and $L_d = 0.05$, Ω and nH respectively. From the stability circles, we determined that loading the gate with a 200 Ω shunt resistor R_i makes the transistor unconditionally stable over the desired frequency range. A plot of the MAG and $|S_{21}|^2$ for this transistor with a 200 Ω shunt input resistor is shown in Fig. 7–8. Because the roll-off is less than 6 dB/octave, our initial equalizer design did not meet the specification. It was adjusted at a later design step to conform to the specification.

We next designed an input equalizer circuit with a 6 dB/octave roll-off compensation. We selected a $T_3(\Omega) = 4\,\Omega^3 - 3\,\Omega$, which is a Chebyshev third order polynomial corresponding to a low pass prototype filter. The equalizer gain is $G_T = \Omega^{2n}/\left(1 + \varepsilon^2 T_3^2(\Omega)\right)$. We used a compensation of $n = 1$ and we selected ε corresponding to 1 dB ripple.[23] For this, ε^2 is 0.2589 and G_T becomes $\Omega^2/[1 + 0.2589(4\Omega^3 - 3\Omega)^2]$. We did not use the band pass transformation, because the low end is limited by any blocking capacitor needed to keep drain bias voltage confined to the transistor. Also, we wanted a low pass filter so that the shunt input capacitance of the transistor could be used as the final capacitance of the equalizer. To this end, we determined the input impedance of the stabilized transistor from its S parameters using the CAD software. This impedance closely resembled a series resistor capacitor combination shunted by the 200 Ω input resistor. The series combination had a resistance of 15 Ω and a capacitor of about 0.5 pF. If we simulate this input impedance with a parallel resistance capacitance combination, the resistance becomes a strong function of frequency varying from 160 Ω at 3 GHz to 36 Ω at 18 GHz. But, the simulated capacitance is relatively constant, varying from 0.58 pF at 3 GHz to 0.3 pF at 18 GHz. So, we used this capacitive effect to simulate the final shunt capacitance of the equalizer.

Substituting $-jS$ for Ω in G_T, we obtained $H(S)\,H(-S) = (4.14\,S^6 + 6.21\,S^4 + 2.33$

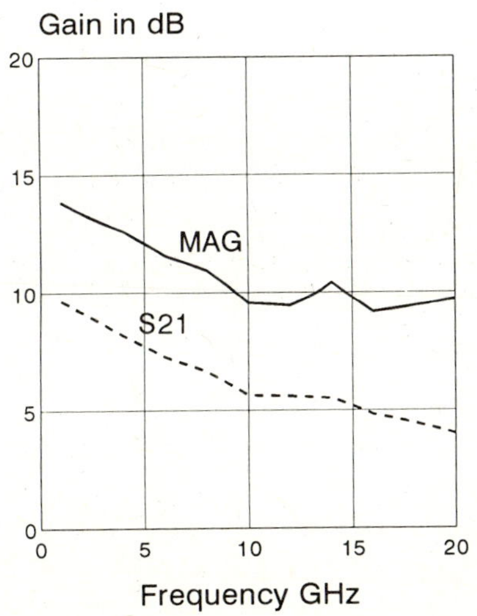

FIGURE 7–8. Gain Functions for the Stabilized NE71000 Transistor.

$S^2 - 1)/S^2$. We let $H(S)$ be $(AS^3 + BS^2 + CS + 1)/S$ to get $H(S)\,H(-S) = [A^2S^6 + (2AC - B^2)S^4 + (C^2 - 2B)S^2 + 1]/S^2$. Then, the coefficients of S^n in $H(S)$ were found by solving $A^2 = 4.14$, $2AC - B^2 = 6.21$, and $C^2 - 2B = 2.33$. From this, we obtained $H(S) = (2.035\,S^3 + 2.01\,S^2 + 2.52\,S + 1)/S$. Using Eq. (7–13), we obtained $K(S)\,K(-S) = (4.14\,S^6 + 6.21\,S^4 + 1.33\,S^2 - 1)/S^2$. From this, we found $K(S) = (2.035\,S^3 + 0.680\,S^2 + 1.64\,S + 1)/S$. Using these in Eq. (7–12), we found:

$$A = \sqrt{\frac{R_g}{R_L}}\left(\frac{4.97\,S^3 + 4.16\,S}{S}\right), \qquad B = \sqrt{R_g R_L}\left(\frac{2.69\,S^2 + 2}{S}\right),$$

$$C = \frac{1}{\sqrt{R_g R_L}}(1.33\,S), \qquad \text{and} \qquad D = \sqrt{\frac{R_L}{R_g}}\,(0.881).$$

Using these in Eq. (7–10), we found:

$$Z_{1oc} = R_g\,(3.06\,S + \frac{1}{0.320\,S}), \qquad Z_{1sc} = R_g\,(3.06\,S + \frac{1}{0.440\,S}),$$

$$Z_{2oc} = R_L\,(\frac{1}{1.51\,S}), \qquad \text{and} \qquad Z_{2sc} = R_L\left(\cfrac{1}{1.51\,S + \cfrac{1}{2.37\,S + \cfrac{1}{0.568\,S}}}\right). \tag{7–15}$$

The prototype circuit of Figure 7–9 was synthesized from these, where the capacitor C_2 and the resistor R_L were simulated by the apparent input admittance of the transistor. The components of Eq. (7–15) were frequency scaled to the cut-off frequency of 18 GHz, and they were impedance scaled to obtain C_2 of 0.4 pF. We chose $C_2 = 0.4$ pF to agree with the transistor apparent input capacitance, which varied from 0.58 to 0.3 pF. From Z_{2oc}, we recognize $C_2 = 1.51/(R_L\omega_2)$. From this, we computed $R_L = 33.4\ \Omega$. This is close to the apparent input shunt resistance at 18 GHz. Using this as R_L in the scaling of the other components, we got $L_1 = 0.7$ nH, $C_1 = 0.15$ pF, and $R_g = 25.9\ \Omega$. We used an external inductance of 0.65 nH. Then, the extrinsic gate lead inductance in series with that added to

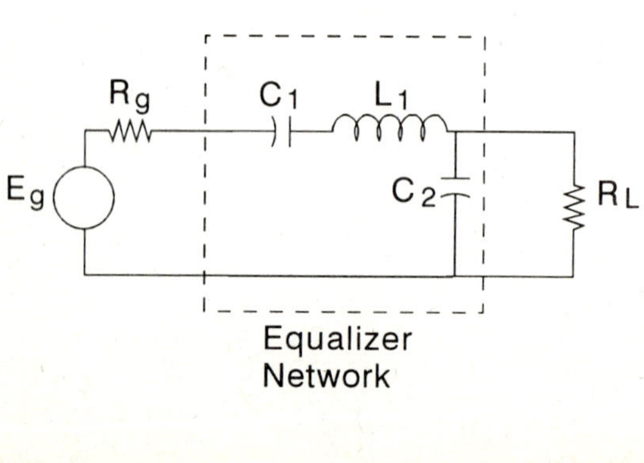

FIGURE 7–9. Equalizer Circuit for a 3 to 18 GHz Amplifier. C_2 is to be the apparent input capacitance of the stabilized transistor.

0.7 nH. The source resistance of 25.9 Ω required impedance transformation from the usual 50 Ω.

To match the output circuit, we calculated the S_{22}, looking back into the drain of the FET with the input equalizer in place. To match that at 18 GHz, we needed about 220 Ω shunted with about a 0.6 nH inductor. At 3 GHz the shunt inductance required was the order of 10 nH. To roughly match the output, we used a shunt load inductance of $L_o = 5$ nH from drain to signal ground, and following that, we inserted a bias blocking capacitance of $C_o = 1$ pF. The circuit is shown in Figure 7–10. The gain of this circuit was determined using the CAD software and is shown in Figure 7–11. It exhibits an increasing gain out to 18 GHz, as was expected, because of the inability to exactly match the transistor roll-off with this method of designing the input equalizer, and because of the approximations used in the design process. However, the positive aspect of this design technique is that we obtained a near match at the high end of the frequency band while determining a topology for the input circuit that allowed the final design to be achieved. Fortunately, this topology also allows for bias insertion through R_i and L_o. We then used the CAD optimizer to adjust C_1, L_1, C_o, and L_o for a flat gain of about 6 dB from 3 GHz to 18 GHz. The resulting circuit elements were: $C_1 = 1.00$ pF, $L_1 = 0.224$ nH, $C_o = 1.47$ pF, and $L_o = 3.7$ nH. The gain curve obtained from simulation on the CAD software with these

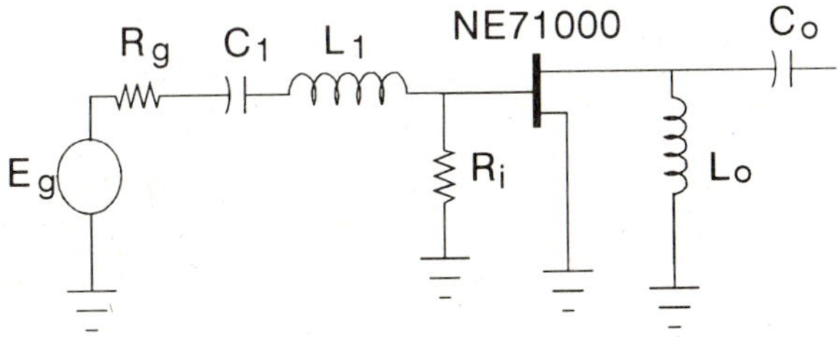

FIGURE 7–10. Reflective Match Amplifier Circuit.

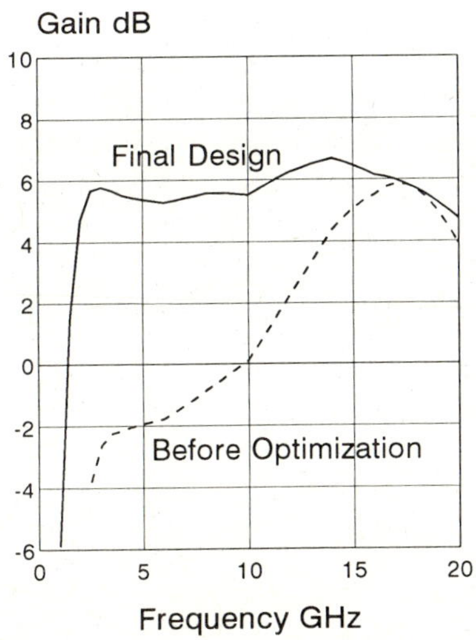

FIGURE 7–11. Amplifier Gain versus Frequency before and after Computer Optimization.

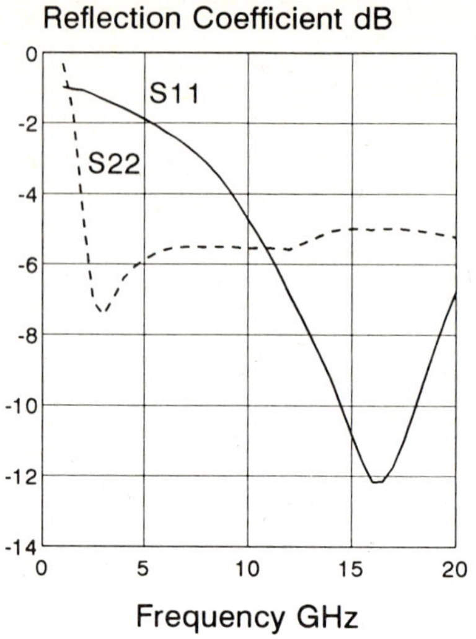

FIGURE 7–12. Reflection Coefficients for a Wide Band Amplifier. The reflection coefficient for the input is S_{11} and S_{22} is that for the output.

values is shown in Figure 7–11. This stays within ± 0.7 dB of 6 dB from 3 GHz to 18.5 GHz. As is typical of reflective match amplifiers, the input reflection coefficient shown for this amplifier in Figure 7–12 is high at the low frequency end and becomes minimum at the high end. The S_{22} for the amplifier is minimum at the low end and stays fairly low over the band.

7.8 BAND PASS REFLECTIVE MATCH WITH RLC LOADS

Series RLC loads and parallel RLC loads can be matched over a band of frequencies using the combination of a Chebyshev prototype low pass filter and a band pass frequency transformation. In principle, the process relies upon the fact that the band pass frequency transformation, $(\omega/\omega_0 - \omega_0/\omega)W$, transforms the series g_n element of the prototype filter in Figure 7–13 into a series LC branch. This branch then can include the series L and C of the load as shown in Figure 7–13. To match a parallel RLC load, the g_n ele-

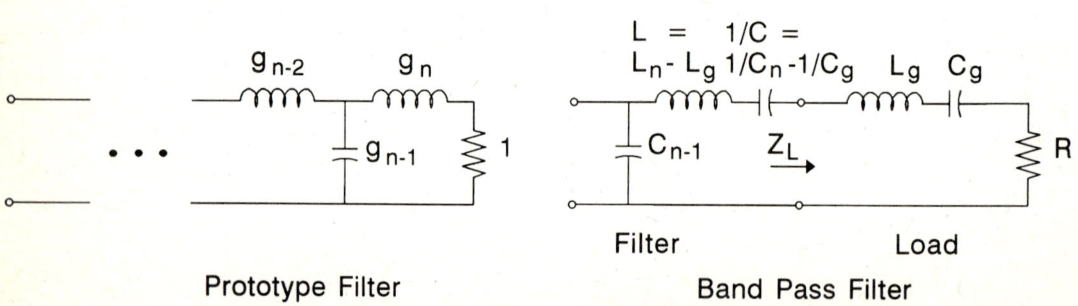

FIGURE 7–13. Band Pass Filter Derived from a Prototype Filter.

ment in the prototype filter must be a shunt element. Then, the band pass frequency transformation changes this element into shunt L and shunt C branches.[24]

7.9 REAL FREQUENCY SYNTHESIS—SINGLE MATCH

The technique we have presented for network synthesis generally requires the specification of a rational function of S representing some network function, such as an input impedance (admittance) of a filter. Usually, this is obtained by assuming a Chebyshev or other type of filter and using the functional representation of that filter. Then, the equalizers or filters are synthesized from these functions. A different method for determining the desired function from which an equalizer can be synthesized was given by Carlin.[25] In this approach, we wish to match a known load Y_L (ω) over a wide frequency band (see Figure 7–14). The Y_L is specified as a graph or set of measurements versus frequency rather than as a rational function of S. To match this over a wide band, we seek a back end equalizer admittance Y_q (ω), which results in the desired equalizer gain G_T (ω). For this "single match" case, the equalizer is assumed to be driven by a source with a purely resistive source impedance that is already matched to the equalizer.

The process of finding Y_q is an iterative one. A first try is made by choosing $Y_{q0}(\omega_i)$ at discrete frequencies ω_i, and the resulting gain is computed at these frequencies. The difference between the desired gain curve and the computed one is examined and the next iteration is made with Y_{q1} (ω_i) selected on the basis of this examination. The Y_{q1} can be chosen to produce a better match to Y_L (ω) in that frequency range where the gain is low by making Y_{q1} more nearly equal to Y_L^* over that range. This usually degrades the match in other parts of the frequency range. This process is continued until a satisfactory $G_T(\omega_i)$ is obtained. The $Y_{qn}(\omega_i)$ corresponding to that must be approximated with a rational function $Y_q(\omega)$. The complexity of the network to be synthesized is dictated by the degree of the $Y_q(\omega)$ chosen to approximate the $Y_{qn}(\omega_i)$. A high degree function will require many reactive elements. One usually uses as low a degree as will meet the design specifications. Finally, the network is synthesized from this $Y_q(\omega)$.

The equations required for this computation are

$$G_T = \frac{4G_g G_L}{|Y_L + Y_q|^2} \quad \text{and} \quad B_q = \frac{1}{\pi}\int_0^{\infty} \frac{dG_q}{d\omega} \ln\left(\frac{|\omega + \omega_o|}{|\omega - \omega_o|}\right) d\omega,$$

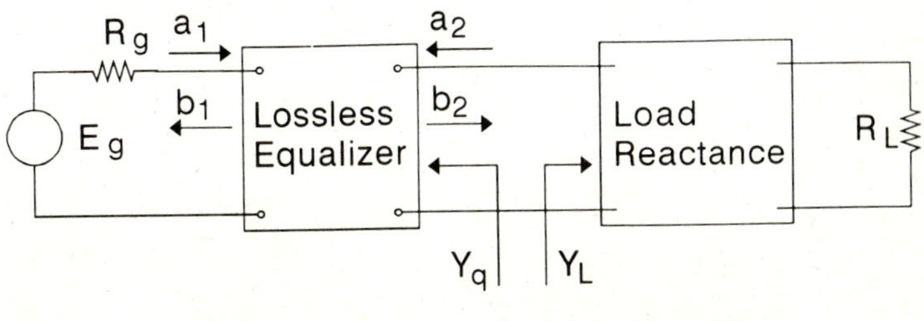

FIGURE 7–14. Equalizer Circuit Configuration.

where $Y_q = G_q + jB_q$. To obtain these, we first examine the expression for G_T given in Eq. (6–10). The equalizer gain is

$$G_T = \frac{1-|\Gamma_S|^2}{|1-\Gamma_S S_{11}|^2}|S_{21}|^2 \frac{1-|\Gamma_L|^2}{|1-\Gamma_L\Gamma_{\text{out}}|^2}.$$

The source driving the equalizer is assumed to be matched so that Γ_S is zero. The output of the equalizer is assumed to be matched to an impedance Y_q^* such that Γ_{out} is zero. If the equalizer is lossless (and includes no internal source) and matched then $P_{\text{in}} = |a_1|^2 = P_{\text{out}} = |b_2|^2$. Determined under these conditions, we have $|S_{21}|^2 = |b_2|^2/|a_1|^2 = 1$. This just shows that the available power out of the lossless equalizer network is the same as the available power into it when the input is matched. Then Eq. (6–10) for G_T becomes $G_T = 1 - |\Gamma_L|^2$, where Γ_L is now the mismatch between Y_L and Y_q, which is given by Eq. (3–18). We have

$$G_T = 1 - \frac{|Y_L^* - Y_q|^2}{|Y_L + Y_q|^2},$$

which reduces to Eq. (7–16).

$$G_T = \frac{4G_q G_L}{|Y_L + Y_q|^2} \tag{7–16}$$

The expression

$$B_q = \frac{1}{\pi}\int_0^\infty \frac{dG_q}{d\omega} \ln\left(\frac{|\omega+\omega_o|}{|\omega-\omega_o|}\right) d\omega$$

relates the imaginary part of the admittance of a minimum reactance function to the real part. This expression was derived by Bode by considering the contour integral $\oint (Y_q - G_q(\infty))\,(1/(S - j\omega_0) - 1/(S + j\omega_0))\,dS$ around the contour enclosing the RHP, but excluding the poles at $\pm j\omega_o$.[26] The integral expression for B_q is used in an approximate way. One assumes that the change in G_q from ω_{i-1} to ω_i is linear, so that $dG_q/d\omega$ can be factored out of the integral. Then, we set $dG_q/d\omega$ to $(G_q(\omega_i) - G_q(\omega_{i-1}))/(\omega_i - \omega_{i-1}) = \Delta G_q(\omega_i)/(\omega_i - \omega_{i-1})$. The remaining integral over that part of the range of integration is given by Eq. (7–17).

$$b_i(\omega) = \int_{\omega_{i-1}}^{\omega_i} \ln\left(\frac{|y+\omega|}{|y-\omega|}\right) dy = (\omega+\omega_i)\ln(\omega+\omega_i) + (\omega-\omega_i)\ln|\omega-\omega_i|$$
$$-(\omega+\omega_{i-1})\ln(\omega+\omega_{i-1}) - (\omega-\omega_{i-1})\ln|\omega-\omega_{i-1}| \tag{7–17}$$

If we approximate the $G_q(\omega)$ as a series of straight lines, the corresponding susceptance function is given by the sum shown in Eq. (7–18).

$$B_q(\omega_k) = \sum_i b_i(\omega_k)\Delta G_q(\omega_i)/[\pi(\omega_i - \omega_{i-1})]. \qquad (7\text{–}18)$$

Carlin used an impedance approach rather than admittance. He used the impedance equivalent to Eqs. (7–16) through (7–18) to approximate Z_q from the given $Z_L(\omega)$. The process was iterative as described above. The initial guess for $R_{q0}(\omega_i)$ was used in Eq. (7–18) (in place of $G_{q0}(\omega_i)$) to find the initial $X_{q0}(\omega_i)$. These were then used in Eq. (7–16) to determine a zero order approximation to $G_T(\omega)$. A comparison of this to the desired $G_T(\omega)$ was made, and the $X_L(\omega)$ was compared to this initial approximation $X_{q0}(\omega_i)$. The $R_{q0}(\omega_i)$ was then refined and the process repeated. Finally, a rational function of ω was fitted to the $Z_{qn}(\omega_i)$ to get $Z_q(\omega)$.

Example: Matching a FET Drain to a Resistive Load. For this example, we chose the NE71000 transistor and assumed that the input circuit was the one from the example in section 7–7. For this example, we reversed the roles of load and source, so that we are using the drain admittance as the "load" admittance discussed previously, and the resistive load to which the transistor is to be matched becomes the source circuit resistance of the previous discussion. The admittance looking back into the drain of the FET is given in Table 7–1 as $Y_d = G_d + jB_d$. We specified that the equalizer gain was to be constant over the frequency band from 3 GHz to 18 GHz and to be as high as possible. To meet this, we initially chose the prototype G_{q0} as shown in Table 7–1. From this, we calculated B_{q0} and G_{T0} using Eq. (7–16) through (7–18). The resulting equalizer gain is high at the low frequency end of the band and low at the high end. Inspecting B_{q0} in this table, we see that it is too small and is of the wrong sign to effect a satisfactory match. To change the sign of B_q (Ω_i), we see from Eq. (7–18) that G_q (Ω_i) must be a decreasing function of frequency Ω. Of course using a decreasing function of Ω for G_q (Ω_i) will

TABLE 7–1 Drain Admittance, Equalizer Admittance and Gain in the Ω Domain

Prototype Radian Frequency	G_d	B_d	G_{q0}	B_{q0}	G_{T0}	G_{qn}	B_{qn}	G_{Tn}	$G_T(\Omega)$
0			0.3	0	1	0.77	0	0.807	0.807
1/6	0.3	0.1	0.3	0.23	0.96	0.75	−0.063	0.815	0.820
2/6	0.3	0.2	0.3	0.048	0.854	0.73	−0.109	0.819	0.855
3/6	0.3	0.3	0.3	0.091	0.702	0.71	−0.148	0.817	0.905
4/6	0.35	0.45	0.35	0.128	0.595	0.71	−0.200	0.838	0.966
5/6	0.4	0.6	0.4	0.133	0.544	0.71	−0.343	0.875	0.914
1	0.45	0.75	0.45	0.107	0.524	0.4	−0.35	0.816	0.760

G_d and B_d are the drain conductance and susceptance normalized to 50 Ω and 18 GHz. G_{qi} and B_{qi} are the normalized equalizer conductance and susceptance for the ith trail. G_{Ti} is the equalizer gain on the ith trail. $G_T(\Omega)$ is the gain of the synthesized equalizer.

cause the match to worsen at either the high or the low frequency end. After adjusting G_q (0) to start at a high value and decrease to a low value, we ran successive iterations of these calculations. After several iterations, we obtained the result listed in Table 7–1 as the nth trial. The equalizer gain variation for this iteration stays within ± 0.2 dB over the band. This was deemed satisfactory so the circuit corresponding to this was synthesized.

We chose to synthesize the equalizer circuit as that shown in Figure 7–15 because the G_{qn} (Ω_i) and the B_{qn} (Ω_i) vary with Ω like the input admittance to a low pass filter. The circuit in Figure 7–15 is a low pass filter and G_{qn} (0) can be readily identified as $1/R$ so that $R = 1.3$. The circuit chosen has $G(\Omega) = R/(R^2 + \Omega^2 L)$. Setting this equal to G_{qn} (Ω) at $\Omega = 1$, we found $L = 1.56$. By requiring $B(1)$ for this circuit to be equal to B_{qn} (1), we got $C = 0.029$. The gain of this equalizer is shown in Figure 7–16, along with that of G_{Tn} (Ω_i). Clearly a more complex equalizer could be designed to give a more constant gain function, but the trade between circuit complexity and a more constant gain did not seem reasonable. The equalizer circuit elements were scaled in frequency and impedance to get $R = 65\ \Omega$, $L = 0.69$ nH and $C = 5.1$ fF.

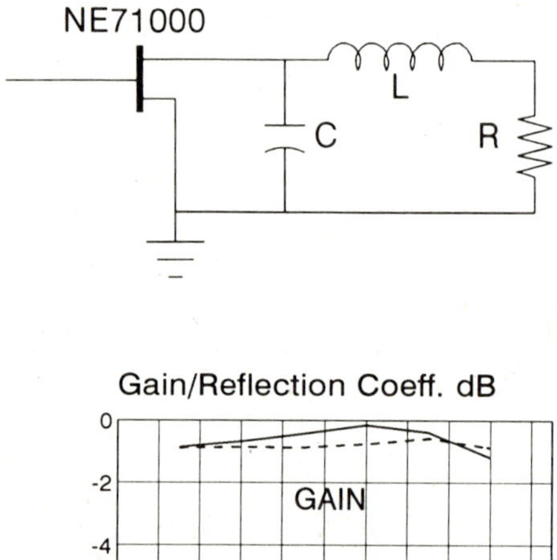

FIGURE 7–15. Circuit for Matching FET Drain.

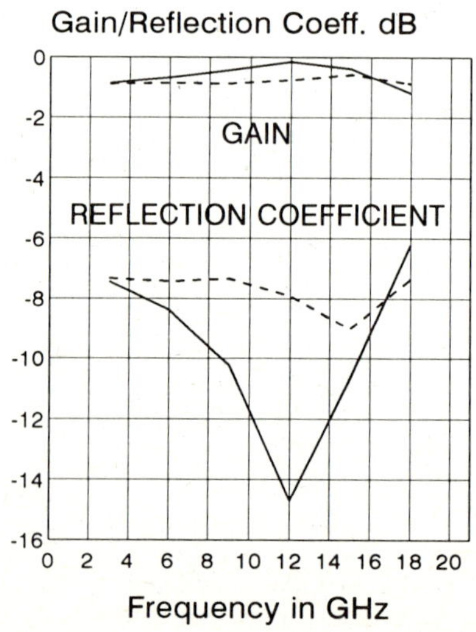

FIGURE 7–16. Equalizer Gain and Reflection. The solid line is that of the two-element circuit. The dotted line is that determined numerically.

7.10 INPUT AND OUTPUT EQUALIZATION WITH REAL FREQUENCY SYNTHESIS

The equalization calculation of the previous section cannot very well be applied sequentially to the input and then the output of an amplifier when the transistor is not unilateral, because of the interaction of these circuits. For that case, Carlin and Komiak refined the real frequency technique to apply it to the simultaneous equalization of both the input and output of an amplifier.[27] Their basic approach was the same as in section 7–9, except that they used a gain equation meant for the entire circuit. They first equalized the input to obtain a flat gain and then adjusted the design for the desired overall gain. This technique was also used by Jung and Wu who added consideration of the stability of the stage to the process.[28] We base this discussion on the Jung and Wu approach. The variables to be adjusted for a constant gain are the Y_S and Y_L of Figure 7–17. These are adjusted after comparing them to Y_{in} and Y_{out}. To improve the gain at any frequency, it is necessary to improve the match between Y_s and Y_{in} or between Y_L and Y_{out}. The equalizer admittances are assumed to be minimum susceptance, which can be realized as low pass ladder networks. The inductors L_G and L_D are not required for this process if the design is for a low pass amplifier. These inductors are incorporated to allow a band pass design while still keeping Y_S and Y_L as minimum susceptance functions. The gain equation is Eq. (7–19), where y_{ij} are the normalized Y parameters of the circuit containing the transistor and the shunt inductors L_G and L_D.

$$G_T = \frac{4|y_{21}|^2 g_s\, g_L}{|(y_{11}+y_s)(y_{22}+y_L)-y_{21}y_{12}|^2} \tag{7–19}$$

In this equation, the y_S, y_L, g_S, and g_L are those of the equalizers normalized to 50 Ω. This equation can be derived as shown in problem 7–6. The stability of the stage is computed using Eq. (7–20), where θ is the angle of the product $y_{12}y_{21}$. Equation (7–20) is the Llewellyn stability criterion.[29]

$$\frac{2(g_s+g_{11})(g_L+g_{22})}{|y_{12}\ y_{21}|(1+\cos\theta)} = L_2 > 1 \tag{7–20}$$

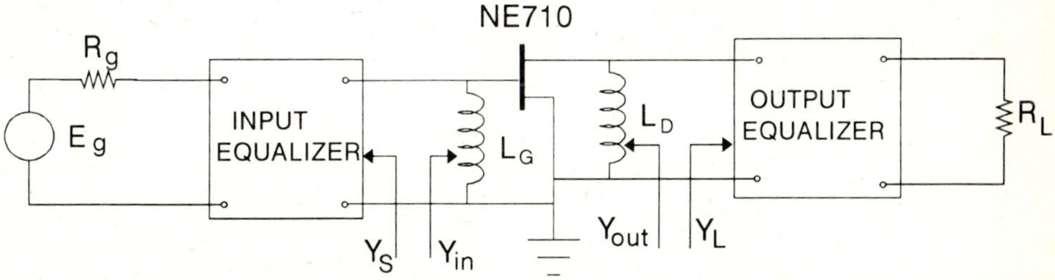

FIGURE 7–17. Input and Output Equalization of an Amplifier Circuit.

This is derived from the criterion that $\mathcal{R}e(y_{in}) > 0$ for stability. These equations along with Eq. (7–18) and a computer optimization program were used by Jung and Wu to determine the variation of $Y_s(\omega) = G_s(\omega) + jB_s(\omega)$ and $Y_L(\omega) = G_L(\omega) + jB_L(\omega)$ needed to produce the desired band pass characteristic. Because of the nature of Eq. (7–18), the procedure requires that the frequency range for these calculations must start at $\omega = 0$. Because of the stability requirement, this range must go at least to the frequency where the transistor becomes unconditionally stable.

Example: Design of a 4 to 8 GHz Band Pass Amplifier. The FET used in this example is the NE710 biased at $V_{DS} = 3$ volts and $I_{DS} = 30$ mA. The S parameters given by the manufacturer were converted to y parameters over the frequency range from 2 to 26 GHz using the computer program in the chapter 3 appendix. Inductors of 0.5 nH were assumed to shunt the gate and drain of the FET, and a new set of y parameters incorporating these inductors was calculated. Stability calculations showed that the FET was potentially unstable up to 23 GHz. Gain calculations showed that the maximum stable gain at 8 GHz (given by $|S_{21}|/|S_{12}|$) was 13.8 dB. From these, we concluded that the frequency range for our calculations had to be from 0 to at least 23 GHz, and the maximum gain we could expect had to be less than 13.8 dB. We wrote a computer program requiring an input of $g_s(\omega_i)$ and $g_L(\omega_i)$ with which we calculated $G_T(\omega_i)$ using Eq. (7–19), the stability factor using Eq. (7–20), $b_s(\omega_i)$ and $b_L(\omega_i)$ using Eq. (7–18), the $y_{in} = y_{11} - y_{12}\,y_{21}/(y_{22} + y_L)$, and the $y_{out} = y_{22} - y_{12}\,y_{21}/(y_{11} + y_s)$. The lower case letters such as g_s indicate values normalized to 50 Ω. We did not include an optimization routine. Instead, we relied upon visual inspection of the input and output match for the equalizers by comparing $y_s(\omega_i)$ to $y_{in}(\omega_i)$ and $y_L(\omega_i)$ to $y_{out}(\omega_i)$. For the first series of computations, we tried to achieve a fairly high constant gain by varying only the input match. We assumed that $g_L(\omega_i)$ was unity for the entire frequency range. We chose values for $g_s(\omega_i)$ and calculated $b_s(\omega_i)$, $g_{in}(\omega_i)$, $b_{in}(\omega_i)$, $G_T(\omega_i)$, and the stability factor. After comparing the calculated y_{in} to the resulting y_s, we adjusted $g_s(\omega_i)$ to improve the input match. After several trials, we were able to achieve 11 dB gain across the band without attempting to match $g_L + jb_L$ to $g_{out} + jb_{out}$. Then, we adjusted both $g_s(\omega_i)$ and $g_L(\omega_i)$ simultaneously to increase the gain across the band while maintaining stable operation. The resulting admittances, stability factor L_2, and gain are shown in Table 7–2. Note that g_s and g_L are not asymptotic to zero. Shunt resistances were required to maintain stability at frequencies above the band. The asymptotic values of g_s and g_L at high frequencies are equal to the inverse of these shunt resistances. From Table 7–2, we see that the gain varies from 11.6 dB to 12.9 dB across the band. The extra one dB of gain achieved by matching the output came at a cost of a reduced stability margin.

After obtaining the data for $y_L(\omega)$ and $y_s(\omega)$ that correspond to the desired gain, we synthesized the input and output equalizers. The equalizer admittances approximated the $y_L(\omega)$ and $y_s(\omega)$ data, as shown in Figure 7–18. For synthesis, we assumed that each equalizer would be a low pass T section with series inductors and a shunt capacitor. These sections have an input admittance with a pair of complex conjugate zeros and three poles. We adjusted the location of the complex poles and zeros to give break fre-

TABLE 7–2 Optimized Equalizer Admittances and Amplifier Gain.

Freq. GHz	g_s	b_s	g_L	b_L	Stability	Gain dB
0	2	0	1.3	0	—	—
2	1.8	–0.51	1.3	–0.231	65.6	9.94
3	1.7	–0.747	1.3	–0.401	42.6	10.31
4	1.4	–0.876	1.15	–0.543	23.2	11.6
5	1.3	–0.907	1.05	–0.611	16.8	12.18
6	1.2	–0.999	0.95	–0.661	10.7	12.62
7	1.1	–1.96	0.9	–0.728	8.5	12.96
8	0.6	–1.21	0.8	–0.864	4.3	12.34
9	0.45	–0.995	0.45	–0.835	1.9	8.64
11	0.4	–0.785	0.4	–0.628	1.5	4.56
13	0.35	–0.680	0.35	–0.553	1.1	1.30
15	0.35	–0.614	0.35	–0.521	1.3	–0.68
17	0.35	–0.605	0.3	–0.503	2.2	–2.74
19	0.30	–0.598	0.3	–0.485	3.2	–4.44
21	0.30	–0.596	0.3	–0.506	1.4	–5.19

The admittances are normalized to 50 Ω.

quencies on a Bode plot of $g(\omega)$ which correspond to the break frequencies of the data. Then, we corrected the component values by trial and error to obtain the desired admittances. The resulting admittance functions and circuits are shown in Figures 7–18 and 7–19.

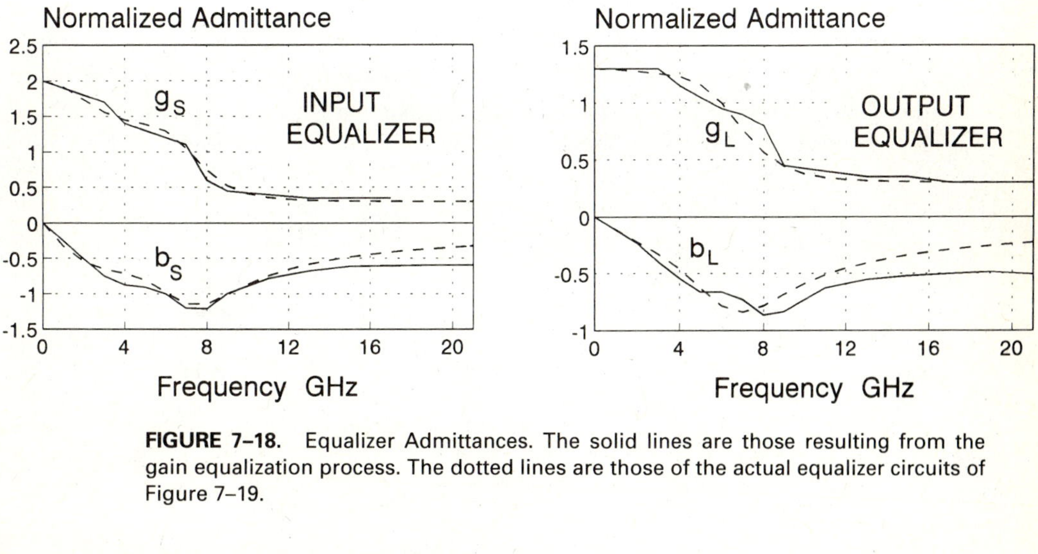

FIGURE 7–18. Equalizer Admittances. The solid lines are those resulting from the gain equalization process. The dotted lines are those of the actual equalizer circuits of Figure 7–19.

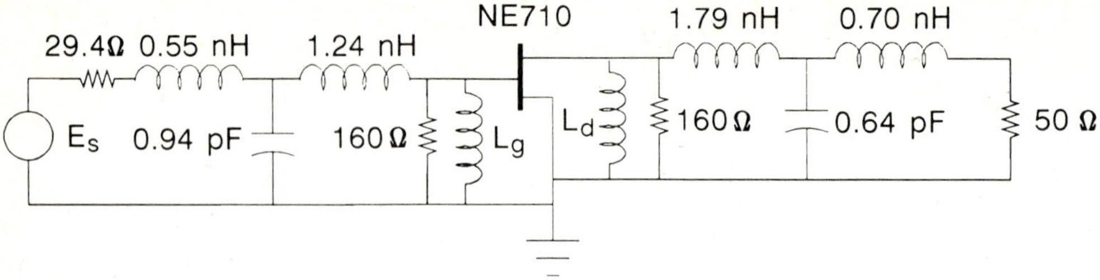

FIGURE 7–19. 4 to 8 GHz Bandpass Amplifier.

7.11 LOSSY MATCH AMPLIFIERS

Ultra wide band amplifiers with frequency ranges greater than 3 octaves can be made by inserting resistors in series between the shunt inductors (or shunt transmission lines) and signal ground in a reflective match amplifier. This amplifier then becomes a resistance capacitance coupled amplifier at the very low frequency range. A diagram of the signal equivalent circuit used for this type of amplifier is shown in Figure 7–20. The change in design procedure is to include the shunt branch admittances y_g and y_d (normalized to Y_0) as part of the transistor S parameters. These admittances must be asymptotic to shunt conductances g_g and g_d at the low frequency end of the band f_1. The equalizer design is done as before, with the exception that low pass filters must be used that reduce to short circuits, effectively connecting the coupling capacitors C_1 and C_2 directly to the shunt conductances at f_1. The low frequency equivalent circuit of the amplifier is also shown in Figure 7–20.

If the transistor is unconditionally stable, the design algorithm is to

(1) Compute the MAG at the high frequency end of the band f_2. This will put a limit on the gain obtainable for the design.

(2) Determine g_g and g_d required to obtain the gain at f_1. Then, these fix the C_1 and C_2 needed by the requirements that $X_{C1} \ll R_g$ and $X_{C2} \ll R_d$ at f_1.

(3) Design the equalizer networks to obtain maximum gain at f_2, commensurate with the requirements on the input and output reflection coefficients Γ_{in} and Γ_{out}, and with the requirements on the bandwidth.

(4) Optimize the resulting design with CAD software.

Niclas discussed the design technique required at the low frequency end of the band.[30] He showed that matching the input and the output at this frequency does not lead to maximum transducer gain. To derive his result, we use the low frequency equivalent circuit in Figure 7–20 and assume that the reactances of the capacitors are negligible. The transducer gain is $G_T = P_L/P_A$.

$$P_L = |I_L|^2 R_L = \left(\frac{g_m V_{gs} G_L}{G_L + G_d}\right)^2 R_L = \frac{g_m^2 R_L}{(1+g_d)^2}\left(\frac{|I_s| R_s}{1+g_g}\right)^2 \tag{7–21}$$

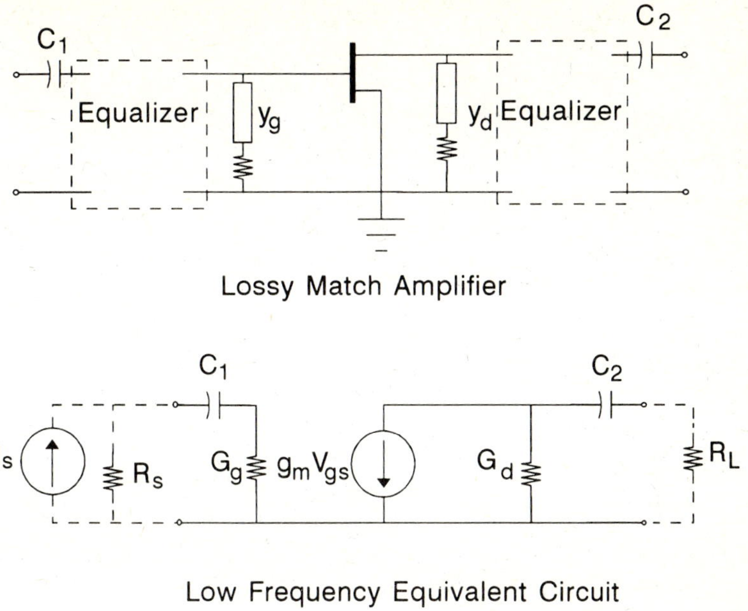

FIGURE 7–20. Signal Equivalent Circuit for a Lossy Match Amplifier. The grounds shown are signal grounds. The low frequency equivalent circuit is shown below.

The load power is given in Eq. (7–21), where $g_g = G_g/G_S$ and $g_d = G_d/G_L$. We also have $P_A = |I_S/2|^2 R_S$. The resulting transducer gain is given in Eq. (7–22). We see that neither of the matching conditions $g_g = 1$ nor $g_d = 1$ maximizes G_T.

$$G_T = \frac{4\, g_m^2\, R_L\, R_S}{(1+g_d)^2(1+g_g)^2} \tag{7–22}$$

From Eq. (7–22), one can derive Eq. (7–23), which expresses this gain in terms of the S parameters (see problem 7–8). Again, we observe that G_T is not maximized when either S_{11} or S_{22} is minimized.

$$G_T = (1 + S_{11})^2(1 + S_{22})2\, g_m^2\, R_L\, R_S/4 \tag{7–23}$$

Example: Wide Band Amplifier. Niclas designed such an amplifier to cover a frequency range from 0.1 to 8 GHz—a frequency range of more than 6 octaves with a 2 stage gain of about 12 dB. One stage of this amplifier circuit is shown in Figure 7–21. One feature of this circuit that should stand out immediately is that the bypass and coupling capacitors are large, on the order of tens of pF. They must have a low impedance at all signal frequencies. Another feature is that the reactive elements are transmission lines. Those which are in series with the signal path become electrically short at the low frequency end of the band. The open circuited stubs that act as shunt capacitors become

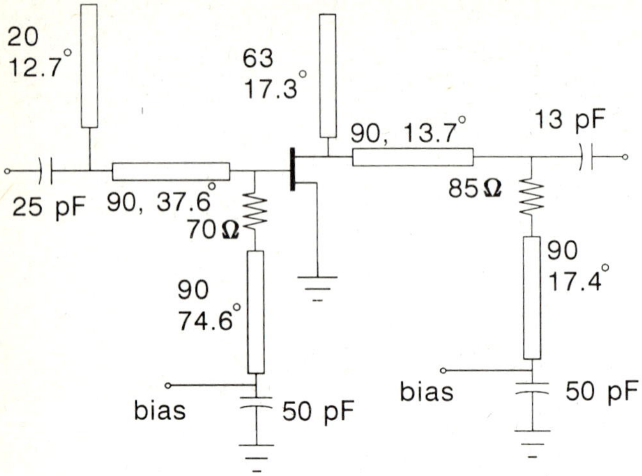

FIGURE 7–21. Lossy Match Amplifier Circuit. Capacitors are in pF, resistors in Ω, transmission line impedances in Ω and lengths are in electrical degrees at 8 GHz[30]. © 1982 IEEE.

open circuits at the low frequency end. The shunt lines terminated in a resistor become purely resistive at the low frequency end of the band. So, the equivalent circuit at low frequencies is simply an R-C coupled amplifier.

7.12 BALANCED AMPLIFIERS

As can be seen from Figures 7–12 and 7–16, the input and output match for single stage wide band amplifier circuits tends to be poor. This occurs because the primary design goal is a constant gain over a wide band. One remedy for this is to use a balanced amplifier topology. This topology is shown in Figure 7–22. One Lange coupler is used at

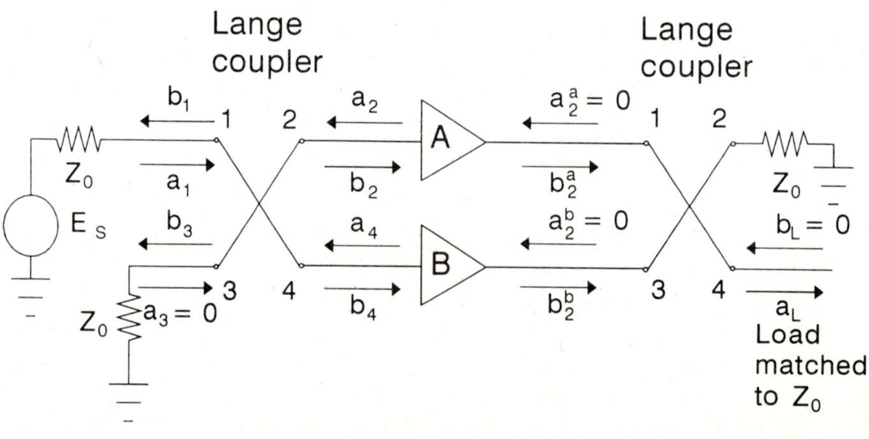

FIGURE 7–22. Balanced Amplifier Schematic.

the input, and another is used at the output. The input coupler splits the power into two outputs at ports 2 and 4 for input to the amplifiers shown as A and B. The output coupler combines the amplified outputs of A and B and supplies it to the load. Because of the properties of the coupler, reflections at the inputs to amplifiers A and B are combined and appear at the ports 1 and 3 of the input Lange coupler. If the amplifiers have equal S_{11} parameters at the center frequency of a 3 dB Lange coupler, then zero reflection occurs at port 1 and all the reflected power is dissipated in the resistor attached to port 3. In this way, the input reflection coefficient of the Lange coupler is reduced, regardless of the reflection coefficient of the individual amplifiers A and B. Another advantage is that the power output of the balanced amplifiers is greater than that of either amplifier alone.

To analyze the circuit of Figure 7–22, we use the Lange coupler S parameters from Eq. (5–35) and shown as Eq. (7–24) here, where $\theta = f\pi/2f_0$ with f_0 being the center frequency of the Lange couplers.

$$\begin{aligned}
S'_{21} = S'_{12} = S'_{43} = S'_{34} &= j\frac{C\sin\theta}{\sqrt{1-C^2}\cos\theta + j\sin\theta} \\
S'_{31} = S'_{13} = S'_{24} = S'_{42} &= 0 \\
S'_{41} = S'_{14} = S'_{23} = S'_{32} &= \frac{\sqrt{1-C^2}}{\sqrt{1-C^2}\cos\theta + j\sin\theta}
\end{aligned} \tag{7–24}$$

The transducer gain for the matched balanced amplifier is given by $G_T = |S_{21}|^2 = |a_L/a_1|^2$. We use the superscripts a and b for the S parameters of the amplifiers A and B. To simplify the results, we assume that the ports 1 and 3 of the input Lange coupler are matched as are ports 2 and 4 of the output Lange coupler. A more complete analysis not using this assumption was given by Engelbrecht and Kurokawa.[31] At the output, we have $a_L = S'_{14} b_2^a + S'_{34} b_2^b$, where $b_2^a = S_{21}^a S'_{21} a_1$ and $b_2^b = S_{21}^b S'_{41} a_1$. From this, we get $G_T = |S'_{41}|^2 |S'_{21}|^2 |S_{21}^a + S_{21}^b|^2$ for the transducer gain. Substituting into this from Eq. (7–24) and assuming $S_{21}^a = S_{21}^b$, we get the transducer gain of Eq. (7–25).

$$G_T = \frac{4C^2(1-C^2)\sin^2\theta\,|S_{21}^a|^2}{[(1-C^2)\cos^2\theta + \sin^2\theta]^2} \tag{7–25}$$

Similarly, we can determine the input reflection coefficient to be that of Eq. (7–26).

$$S_{11} = \frac{-C^2 S_{11}^a \sin^2\theta + (1-C^2) S_{11}^b}{(1-C^2)\cos^2\theta + \sin^2\theta} \tag{7–26}$$

Usually, balanced amplifiers are designed to use 3 dB Lange couplers however, Walker has shown that a wider bandwidth can be obtained using a tighter coupling.[32] A plot of Eq. (7–25) is shown in Figure 7–23 for the case where amplifiers A and B are identical and have $S_{21}^a = 1$ across the frequency band. This graph shows that the 3 dB

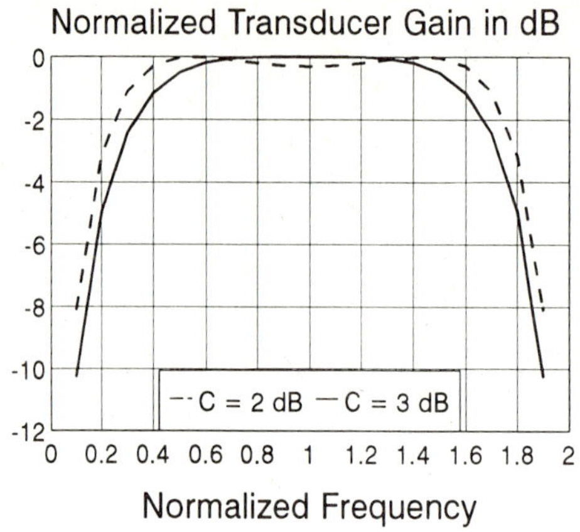

FIGURE 7–23. Gain Curves for Balanced Amplifiers. These curves were calculated assuming that the amplifiers had unity gain across the band. The resulting loss in gain is that due to the Lange couplers.

Lange couplers restrict the 1 dB bandwidth of a balanced amplifier to 2 octaves, regardless of how wide band amplifiers A and B are. Similarly, the 2 dB Lange couplers restrict the 1 dB bandwidth of the balanced amplifier to about 2.5 octaves. A closer coupling than 2 dB is difficult to achieve with conventional Lange couplers. A coupling closer than 2 dB can degrade the gain in the center of the band to an undesirable amount.

Figure 7–24 was computed using Eq. (7–26) and assuming amplifiers A and B are identical and have total reflection at the input: $|S^b_{11}| = |S^a_{11}| = 1$. The actual input match will be better than that shown in Figure 7–24 by the dB amount of the input reflection coefficients of the amplifiers. From Figure 7–24, we see that the input match becomes worse in the center of the frequency band as the coupling is increased above 3 dB; however, the match is improved at the edges of the band by this increase.

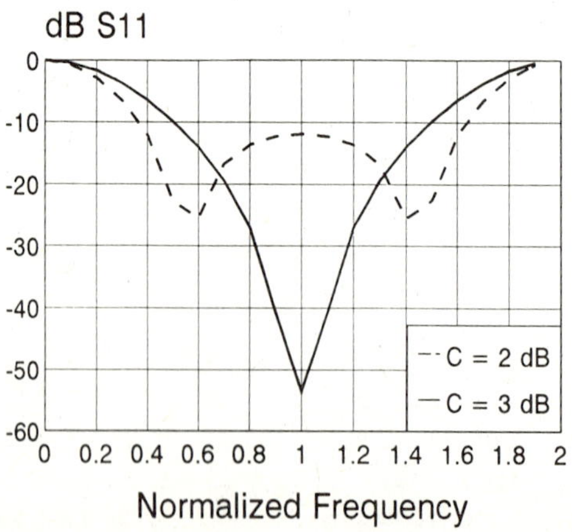

FIGURE 7–24. Balanced Amplifier Input Reflection Coefficient.

7.13 MILLER EFFECT

When designing amplifiers, one often uses an approximation to account for capacitive feedback internal to the transistor. This capacitive feedback results in an increased apparent capacitance at the input of the transistor. This effect is referred to as the *Miller capacitance.* We refer to Figure 7–25 as an approximate equivalent circuit of a bipolar transistor. The circuit equations for this circuit are shown in Eq. (7–27).

$$\begin{pmatrix} I_1 \\ 0 \end{pmatrix} = \begin{pmatrix} (1-\alpha)G_e + SC & -SC \\ \alpha G_e - SC & SC + G_L \end{pmatrix} \begin{pmatrix} V_1 \\ V_2 \end{pmatrix} \tag{7–27}$$

If we solve this for the voltages V_2 and V_1 and take their ratio, we have the voltage gain $A = V_2/V_1$. This is given in Eq. (7–28). We can solve Eq. (7–27) for I_1 as a function of V_1 and V_2.

$$A = \frac{-\alpha G_e + SC}{G_L + SC} \tag{7–28}$$

Substituting for V_2 from Eq. (7–28), we obtain the input admittance in the form of $Y_{in} = I_1/V_1$ and rearrange that solution as $Y_{in} = G_e(1 - \alpha) + SC - SCA$. Rewriting that, we have Eq. (7–29) where we have identified the input conductance $G_{in} = (1 - \alpha)G_e$ and capacitance $C_{in} = C(1 - A)$.

$$Y_{in} = G_{in} + SC_{in} = (1 - \alpha)G_e + SC(1 - A) \tag{7–29}$$

Because A is negative (at least in the low frequency range) and usually has a magnitude much greater than unity, we see that the apparent input capacitance is much greater than C. This effect is used at low frequencies to obtain very long time constant circuits by adding capacitors in parallel with C. At microwave frequencies, the result of this effect will be a decrease in the upper frequency limit of amplifiers due to the shunt Miller input capacitance.

One way to eliminate the Miller effect is to use an *emitter follower.* Emitter followers are amplifiers where the collector is at signal ground and a resistor is between the emitter and the signal ground. The output is taken from the emitter and is a voltage nearly

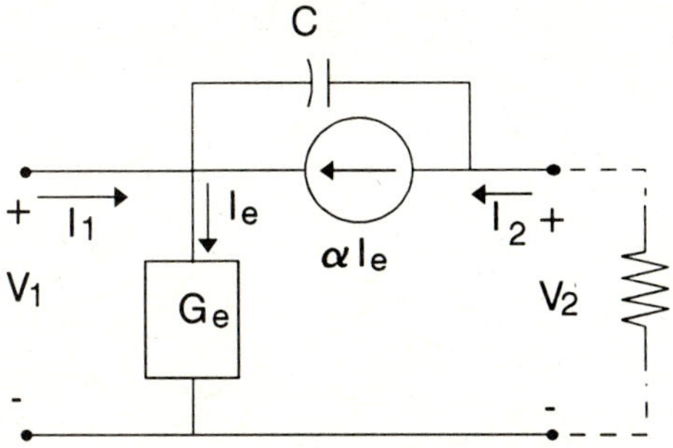

FIGURE 7–25. Transistor Equivalent Circuit for Determining the Miller Effect. The capacitance C models intrinsic collector to base capacitance. The current gain is assumed to be constant for this circuit.

equal to the base input voltage. In that case, the voltage gain from the base to the collector is zero, and the resulting effective input capacitance is simply the collector to base capacitance in parallel with any parasitic capacitance from the base to signal ground. Although no voltage gain is realized, there is a current gain that occurs because the emitter circuit acts as a low impedance source for succeeding stages. If the transit time of the transistor is not the factor limiting the upper frequency response, then a considerable increase in the upper frequency response is obtained because the effective input capacitance has been greatly reduced.

7.14 THE DARLINGTON PAIR

A method for improving the bandwidth obtainable with a given transistor type is to couple a pair of that type of transistors as in Figure 7–26. The combination is called a *Darlington pair* and it can be modeled as a single transistor with an increased β and an increased f_T. The Darlington pair has an equivalent β greater than the β of the individual transistor, and it has an equivalent current gain cut-off frequency about twice that of the individual transistor. To derive an expression for the equivalent β, we start with $I_{c1} = \beta_1 I_b$ for the currents of the first transistor in Figure 7–26 and obtain $I_b = (1 + \beta_1)\, I_{e1} = (1 + \beta_1)\, I_{b2}$. We substitute $I_{b2} = (1 + \beta_2)\, I_e$ to obtain $I_b = (1 + \beta_1)(1 + \beta_2)\, I_e$. This becomes Eq. (7–30). From this, we see that the equivalent β is considerably increased.

$$I_b = (1 + \beta')\, I_e \qquad \text{where} \qquad \beta' = \beta_1 + \beta_2 + \beta_1\, \beta_2 \tag{7–30}$$

To find a relation for the current gain cut-off frequency of the Darlington pair circuit, we start with the relation for a single transistor $\alpha = \alpha_o/(1 + jf/f_\alpha)$ and substitute it into $\beta = \alpha/(1 - \alpha)$ to obtain Eq. (7–31), where $\beta_o = \alpha_o/(1 - \alpha_o)$ and $f_\beta = f_\alpha(1 - \alpha_o)$.

$$\beta = \beta_o/(1 + jf/f_\beta) \tag{7–31}$$

The current gain cut-off frequency for the single transistor occurs at $|I_c/I_b| = 1$, which is $|\beta| = 1$. From this and Eq. (7–31), we get a cut-off frequency for the single transistor of $f_T = f_\beta \sqrt{\beta_o^2 - 1} \cong \beta_o f_\beta$.

Now, we consider the current gain cut-off frequency for the Darlington pair by finding the frequency f_T where $|\beta'| = 1$ for the β' of Eq. (7–30). At $f = f_T$ with $\beta_1 = \beta_2 = \beta$, we have $|\beta'| = |\beta\,(2 + \beta)| = 1$, where β is given by Eq. (7–31).

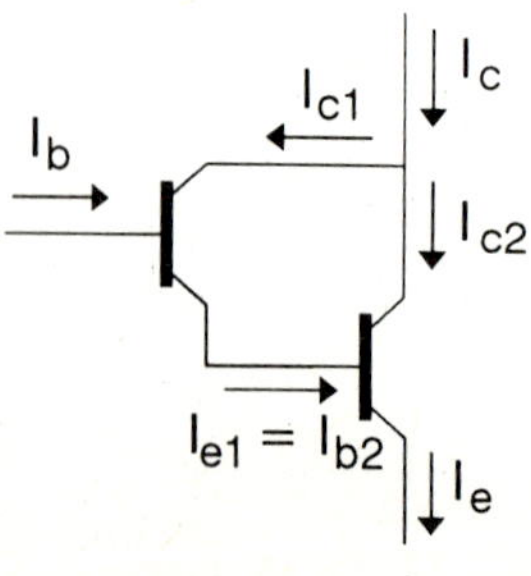

FIGURE 7–26. Darlington Pair Connection.

Assuming β_o is real, this can be written as

$$\frac{\beta_o}{\left|(1+jf_T/f_\beta)^2\right|}\left|2+\beta_o+j2f_T/f_\beta\right|=1 \qquad \text{or}$$

$$\left(\frac{\beta_o}{\left|1+(f_T/f_\beta)^2\right|}\right)^2\left((2+\beta_o)^2+4(f_T/f_\beta)^2\right)=1.$$

Letting $x = (f_T/f_\beta)^2$, this becomes $x^2 + x(2 - 4\beta_o^2) + 1 - \beta_o^2(2 + \beta_o)^2 = 0$. If $\beta_o \gg 1$, then we can approximate this equation as $x^2 - x\,4\beta_o^2 - \beta_o^4 = 0$ and solve for Eq. (7–32).

$$f_T \cong \beta_o f_\beta \sqrt{2+\sqrt{5}} \cong 2\,\beta_o f_\beta \qquad (7\text{–}32)$$

In this relation we see that the Darlington pair has a current gain cut-off frequency approximately twice that of the individual transistors in the pair.

7.15 FET FEEDBACK AMPLIFIER

Feedback in amplifier circuits is normally used in a parallel mode or in a series mode. The parallel mode in a FET amplifier with a grounded source is usually a circuit branch connecting the drain to the gate. The series mode feedback in such an amplifier is usually accomplished with a voltage drop across an impedance inserted between the source terminal and the signal ground. The parallel mode feedback is best described using the *augmented Y matrix* for the circuit. This matrix is the sum of the Y parameter matrix describing the FET and the Y parameter matrix for the feedback circuit. In the series mode, the Z parameter matrices are used. The data describing the FET are usually supplied in the form of S parameter variation with frequency, so the analysis of these feedback circuits requires conversion between S and Y matrices. Because of the complexity of the circuits, these analyses are usually done with CAD software. Pavio discussed several practical aspects of feedback amplifier design.[33]

At very low frequencies, the transit time and capacitive effects internal to the FET are negligible, and the feedback effects can be determine using a purely resistive circuit as shown in Figure 7–27. The low frequency augmented Y matrix for Figure 7–27 is given in Eq. (7–33).

$$(Y)=\begin{pmatrix} g_{fb}, & -g_{fb} \\ 50\,g_m - g_{fb}, & g_{ds}+g_{fb}\end{pmatrix}, \qquad g_{fb}=50\,G_{fb}, \qquad g_{ds}=50\,G_{ds} \qquad (7\text{–}33)$$

The S parameters can be determined from this using Eq. (3–13); they are given in Eq. (7–34).

$$\begin{aligned} S_{11} &= [1 + g_{ds} - g_{fb}\,(g_{ds} + 50\,g_m)]/\Delta \\ S_{22} &= [1 - g_{ds} - g_{fb}\,(g_{ds} + 50\,g_m)]/\Delta \\ S_{21} &= 2(g_{fb} - 50\,g_m)/\Delta,\ S_{12} = 2\,g_{fb}/\Delta \\ \Delta &= 1 + g_{ds} + g_{fb}\,(2 + g_{ds} + 50\,g_m) \end{aligned} \qquad (7\text{–}34)$$

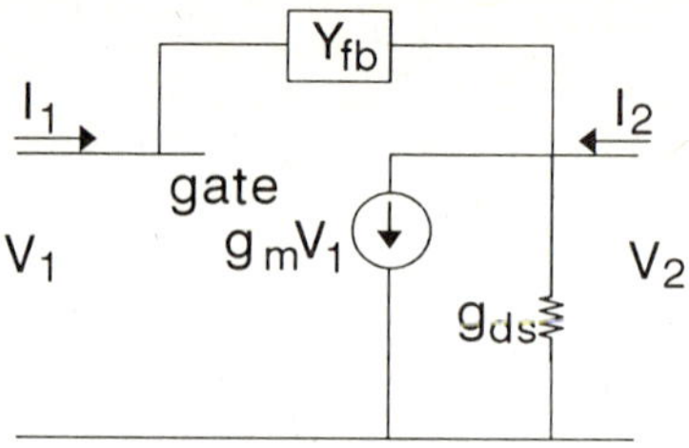

FIGURE 7–27. Low Frequency Equivalent Amplifier Circuit with Feedback.

These can be compared with the case for no feedback by letting g_{fb} be zero in Eq. (7–34). Both the input and output reflection coefficients, as given by S_{11} and S_{22}, are decreased by the feedback.[34] This equates to an improved match for the case with feedback. However, the transducer gain, which is $|S_{21}|^2$, is also decreased by the feedback. At the higher frequency range, the circuit equations become more complex, but the net result is that the gain versus frequency curve for this parallel resistive feedback is relatively flat, out to about where it would intersect the nonfeedback gain curve. At this frequency, it becomes asymptotic to that curve. For an example see Figure 7–29. For this design procedure, one must select a FET that has a MAG better than the desired gain at the high frequency end of the band (keeping in mind the stability criteria), and then use Eq. (7–34) to determine the feedback resistance that produces the desired gain at the low end of the band. After the required feedback is determined, the input and output circuits can then be designed to improve the match at the high frequency end of the band.

Example: Wide Band Feedback Amplifier. Niclas reported on a wide band amplifier with parallel feedback.[35] The signal equivalent circuit of this amplifier before adding the input and output match is shown in Figure 7–28. This circuit differs somewhat from that discussed above. This circuit uses a feedback branch with a series resistance and inductance. Also, a peaking inductance (see section 7–3) has been added between the drain of the FET and the feedback branch. The components for this circuit are shown in Table 7–3. The feedback resistor was 160 Ω and the feedback inductor was 0.45 nH. The peaking inductance L_D was 0.6 nH. The CAD simulated performance of this circuit is shown in Figures 7–29 through 7–31. In Figure 7–29, we see that the FET with only resistive feedback (no inductance in the feedback branch and no peaking inductance) has a transducer gain that is about 4 dB at the low frequency end. This agrees with the 4.65 dB computed using values from Table 7–3 in Eq. (7–34). This gain curve is also asymptotic to the gain curve of the FET by itself (no feedback or peaking inductance). Additionally, the gain of the FET without feedback and the effect of the peaking inductance on this gain are shown. The peaking inductance causes an increased gain (over that of the FET without the L_D) in the frequency band from about 8 GHz to 17 GHz. The gain curve for the composite circuit exhibits the effect of the combination of the feedback and the peaking inductance. It is flat over the lower frequency range and peaks well above the curve for a purely resistive feedback over the upper frequency range. The various components combine to yield a nearly constant gain out to 14 GHz.

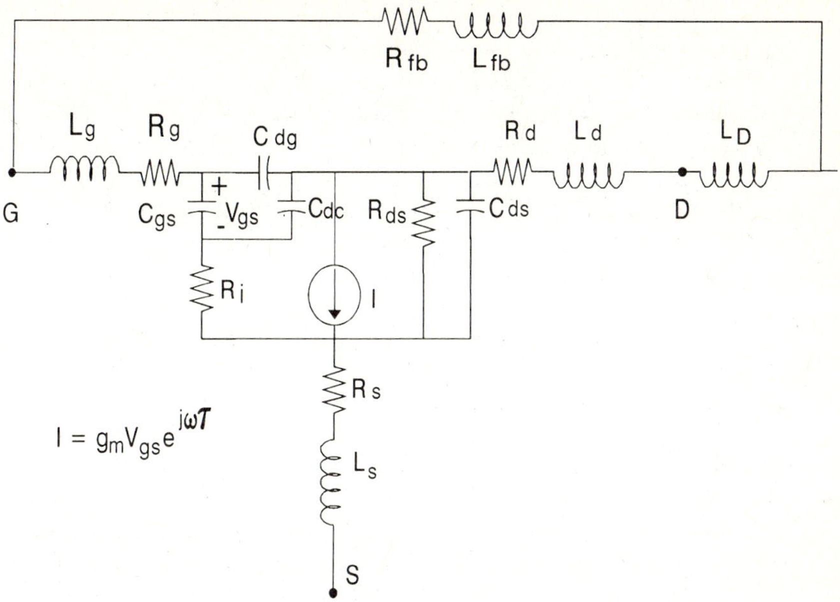

FIGURE 7–28. FET with a Parallel Feedback Branch of a Series Resistance and Inductance. A peaking inductance L_D has been added. The component values are shown in Table 7–1.[34] ©1980 IEEE.

Figures 7–30 and 7–31 show the input and output match of the circuit in Figure 7–28. These figures show that the feedback has improved the match at the input and output over most of the range. Using Eq. (7–34), we computed S_{11} = −18.6 dB and S_{22} = −24.2 dB for zero frequency, which agree with those shown in Figures 7–30 and 7–31. So, the feedback circuit has improved both the input and the output match considerably at the low frequency end. At the high frequency end, the peaking inductance improves the output match. However, this inductance combined with the feedback branch degrades the input match at the high end. This occurs because the feedback becomes positive rather than negative at the high frequencies. This results from the added phase shift due to L_D

TABLE 7–3 Component Values for the Circuit in Figure 7–28.

Intrinsic Elements		Extrinsic Elements	
g_m = 54 mS.	C_{dc} = 32 fF	L_s = 0.151 nH	R_g = 1.5 Ω
τ = 3.5 psec.	R_i = 4.4 Ω	C_{ds} = 82 fF	L_g = 0.19 nH
C_{gs} = 0.67 pF	G_{ds} = 5 mS	R_d = 2 Ω	R_s = 0.9 Ω
C_{gd} = 17 fF		L_d = 0.143 nH	

and L_{fb}. Niclas and his coauthors show phasor diagrams that document this change to positive feedback at the upper frequencies.

A 50 pF capacitor was put into the feedback branch to block the dc bias voltages. This capacitor caused the gain to decrease for frequencies below 1 GHz. The input match was improved with a simple matching circuit on the input. Then, a series transmission line was required to maintain a good output match. Niclas and his coworkers built two models of this amplifier. The widest bandwidth achieved was about 5 octaves.

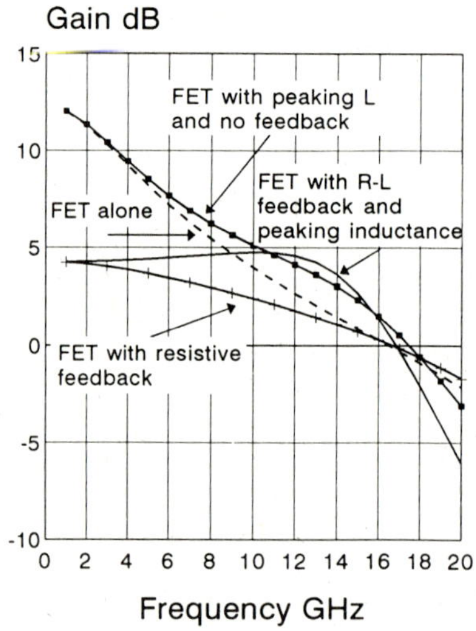

FIGURE 7–29. FET Gain Curves with and without Feedback Circuits.

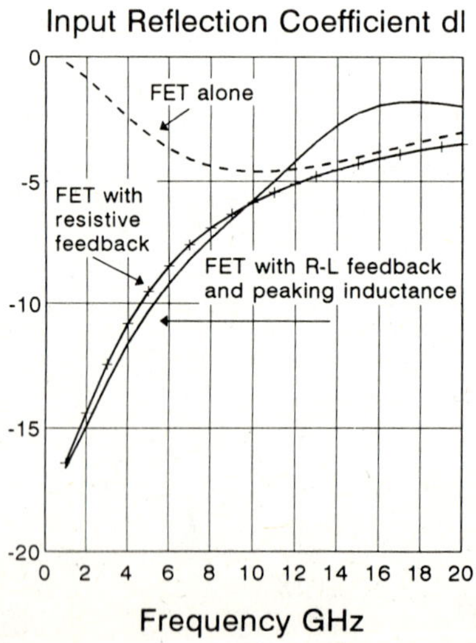

FIGURE 7–30. S_{11} for the Amplifier of Fig. 7–28 Showing the Effects of Feedback and Peaking Inductance.

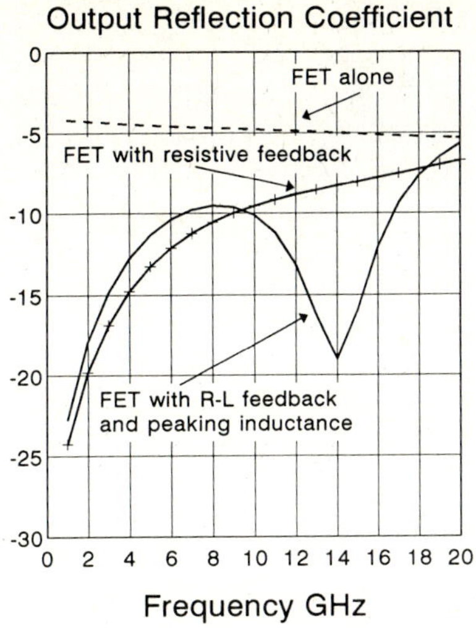

FIGURE 7–31. S_{22} for the Amplifier of Fig. 7–28 Showing the Effects of Feedback and Peaking Inductance.

7.16 FEEDBACK AMPLIFIER DESIGN BY GRAPHICAL MEANS

An approximate method for designing amplifiers having parallel feedback was presented by Perez and Ortega.[36] Their approach is to determine the feedback branch components best able to result in the desired gain curve over the pass band or best able to result in the desired match at the input and output. To do this, they show that, to the first order, the overall S_{21} and the overall S_{ii} of the feedback amplifier can be determined graphically from the S_{21} and S_{ii} of the amplifier without the feedback branch.

To show this relation, we start with Eq. (3–13) and determine the S parameters from the (normalized) y parameters for an arbitrary network. We obtain Eq. (7–35).

$$\begin{aligned} S_{11} &= (1 - y_{11} + y_{22} - y_{11}y_{22} + y_{12}y_{21})/\Delta \\ S_{22} &= (1 - y_{22} + y_{11} - y_{11}y_{22} + y_{12}y_{21})/\Delta \\ S_{21} &= -2y_{21}/\Delta,\ \Delta = (1 + y_{11})\,(1 + y_{22}) - y_{12}y_{21} \end{aligned} \tag{7–35}$$

From this, we get the relations shown as Eq. (7–36).

$$\begin{aligned} 2/S_{21} &= y_{12} - (1 + y_{11} + y_{22} + y_{11}y_{22})/y_{21} \\ 2S_{11}/S_{21} &= -y_{12} - (1 - y_{11} + y_{22} - y_{11}y_{22})/y_{21} \\ 2S_{22}/S_{21} &= -y_{12} - (1 + y_{11} - y_{22} - y_{11}y_{22})/y_{21} \end{aligned} \tag{7–36}$$

The y parameters of the feedback branch of Figure 7–27 are given in Eq. (7–37).

$$(y_f) = \begin{pmatrix} y_{fb} & -y_{fb} \\ -y_{fb} & y_{fb} \end{pmatrix} \tag{7–37}$$

We let the transistor have an admittance matrix given by (y) and S parameters denoted by S_{ij}. The transistor with the feedback has an augmented y matrix (y^T) and the S parameters S^T_{ij}. Then, we form the difference between the relations of Eq. (7–36) for the transistor with feedback and the relations of the Eq. (7–36) for the transistor without feedback. When we use the approximation $y^T_{21} = y_{21} - y_{fb} \cong y_{21}$, these differences become Eq. (7–38).

$$\begin{aligned}
2S^T_{11}/S^T_{21} - 2S_{11}/S_{21} &= y_{fb}\left(1 + \frac{y_{11} + y_{22} + y_{fb}}{y_{21}}\right) \\
2S^T_{22}/S^T_{21} - 2S_{22}/S_{21} &= y_{fb}\left(1 + \frac{y_{11} + y_{22} + y_{fb}}{y_{21}}\right) \qquad (7\text{–}38) \\
2/S^T_{21} - 2/S_{21} &= -y_{fb}\left(1 + \frac{2 + y_{11} + y_{22} + y_{fb}}{y_{21}}\right)
\end{aligned}$$

If the fractions inside the brackets of Eq. (7–38) are negligible with respect to unity, then these equations become Eqs. (7–39), (7–40), and (7–41).

$$2/S^T_{21} - 2/S_{21} \cong -y_{fb} \qquad (7\text{–}39)$$

$$2S^T_{11}/S^T_{21} - 2S_{11}/S_{21} \cong y_{fb} \qquad (7\text{–}40)$$

$$2S^T_{22}/S^T_{21} - 2S_{22}/S_{21} \cong y_{fb} \qquad (7\text{–}41)$$

We see that, to the first order, the difference between these parameters with and without feedback is simply the admittance of the feedback branch. The graphical method is to calculate and plot values for these parameters without feedback, then add or subtract the y_{fb} necessary to obtain the desired value for the parameters.

Example: Gain Adjustment with Parallel Feedback. We take the amplifier of the previous example to show how the feedback branch tends to flatten the gain curve of the stage over the desired pass band. The graph used is shown in Figure 7–32. For this graph, we calculated the locus of values of $2/S_{21}$ for $\text{Log}(|S_{21}|^2) = 0.4, 0.6, 0.8$ and 1. These correspond to transducer gains of 4, 6, 8, and 10 dB. These values result in the circular plots on Figure 7–32. Then, we computed the S parameters for the transistor and peaking inductance in Figure 7–28, but not including the feedback branch. A plot of the resulting $2/S_{21}$ for the frequency range from 1 to 15 GHz is the one shown as the solid line in Figure 7–32. We see that this plot corresponds to a change of gain from perhaps 12 dB at 1 GHz to about 3 dB at 15 GHz. According to Eq. (7–39), we then subtracted from that curve the value of y_{fb} for a feedback branch of 160 Ω in series with 0.45 nH. The dotted line is the result. We see that the gain variation is now from about 8 dB to 4 dB. So, the feedback branch has reduced the gain variation by a considerable amount. In practice, one would compute the dB gain circles and the $2/S_{21}$ curve, then determine what y_{fb} would be necessary to obtain the desired gain variation. Comparing the gain curve of Figure 7–32 with that of Figure 7–29, we see that the approximations we have made to obtain Eq. (7–39) result in a substantial error. However, these approximations would be more valid for a bipolar transistor than for this FET.

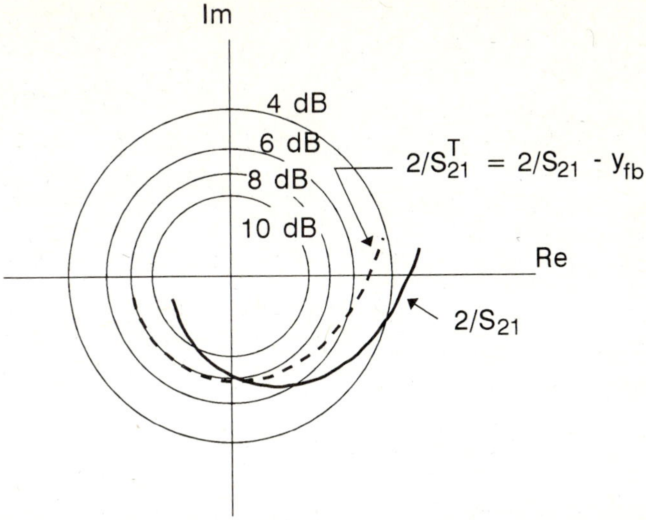

FIGURE 7–32. Graphical Design of Feedback Amplifiers. The circles are constant gain circles and the solid line is for $2/S_{21}$ from 1 to 15 GHz. The dotted line is $2/S_{21} - y_{fb}$ ranging from 1 to 15 GHz. At any frequency the difference between these two curves is the feedback admittance.

7.17 HBT FEEDBACK AMPLIFIER

The effect of combined series and parallel feedback in a wide band HBT amplifier stage can be described in a way similar to that in section 7–15 by treating the effects at a very low frequency where the capacitive and inductive components can be ignored. The gain is set at the low frequency end of the band by choosing the feedback resistors.

For this discussion, we use the approximate equivalent circuit shown in Figure 7–33. The feedback occurs through R_f (parallel mode) and R_E (series mode). The circuit equations in matrix form are given in Eq. (7–42). Equation (7–42) can be solved for the voltage gain $A = V_2/V_1$ given in Eq. (7–43).

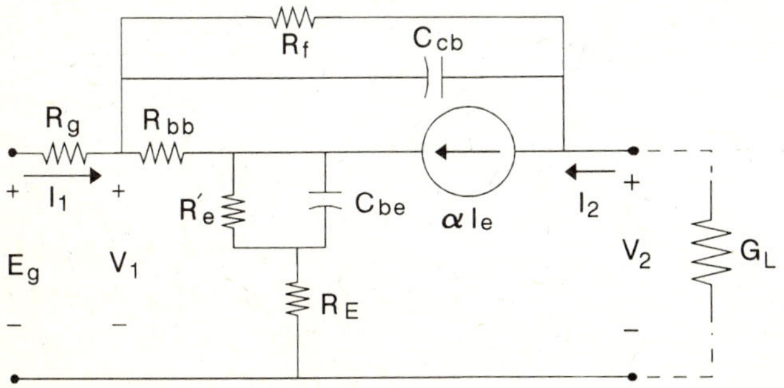

FIGURE 7–33. Approximate Equivalent Circuit of an HBT with Feedback.

$$\begin{pmatrix} I_1 \\ 0 \\ 0 \end{pmatrix} = \begin{pmatrix} Y_f + G_{bb}, & -G_{bb}, & -Y_f \\ -G_{bb}, & G_{bb} + Y_e(1-\alpha), & 0 \\ -Y_f, & \alpha\, Y_e, & Y_f + G_L \end{pmatrix} \begin{pmatrix} V_1 \\ V_{be} \\ V_2 \end{pmatrix} \quad \text{or} \quad (I) = (Y_1)(V) \tag{7–42}$$

$$A = V_2/V_1 = \frac{-\alpha\, G_{bb} Y_e + Y_f (G_{bb} + Y_e(1-\alpha))}{[G_{bb} + Y_e(1-\alpha)][Y_f + G_L]} \tag{7–43}$$

For the low frequency case (where $Y_f = G_f$, $Y_e = G_e = 1/(R_E + R'_e)$ and $\alpha = \alpha_0$), this can be rewritten as in Eq. (7–44), where $R_e = R_E + R'_e$.

$$A = \frac{1 - \dfrac{\alpha_0\, R_f/R_{bb}}{1 - \alpha_0 + R_e/R_{bb}}}{1 + R_f/R_L} \tag{7–44}$$

The input impedance is $Z_{\text{in}} = V_1/I_1 = [-G_{bb}\, \alpha\, Y_e + Y_f\, (G_{bb} + Y_e\, (1 - \alpha))]/\Delta_y$, where Δ_y is the determinant of the Y matrix of Eq. (7–42). This can be rewritten for the low frequency case as shown in Eq. (7–45).

$$Z_{in} = (R_L + R_f) \frac{R_e + R_{bb}(1-\alpha_0)}{R_e + R_L + (R_f + R_{bb})(1-\alpha_0)} \tag{7–45}$$

Similarly, the formula for output impedance for low frequency case can be derived for this circuit starting with $Z_{\text{out}} = V_2/I_2$ and $E_g = 0$. It is shown as Eq. (7–46).

$$Z_{\text{out}} = (R_g + R_f) \frac{R_e + (R_{bb} + R_x)(1-\alpha_0)}{R_e + R_g + R_{bb}(1-\alpha_0)} \quad \text{for} \quad R_x = 1/(1/R_f + 1/R_g) \tag{7–46}$$

The transducer gain is $G_T = P_L\,/P_A$, where $P_L = |V_2|^2/R_L = |A\,V_1|^2/R_L$ and $P_A = |E_g|^2/4R_g$. Substituting $V_1 = E_g\; Z_{\text{in}}/(R_g + Z_{\text{in}})$ into this, we can get the formula for the transducer gain as shown in Eq. (7–47).

$$G_T = \frac{|A|^2\, 4\, G_L\, G_g}{|Y_{in} + G_g|^2} \tag{7–47}$$

One interesting possibility is that the feedback resistors can be used to match the input and output impedances of the circuit.[37] To investigate this, we take the low frequency case and assume $R_g = Z_0 = R_L$. Then, we require $Z_{\text{in}} = Z_0 = Z_{\text{out}}$ in Eqs. (7–45) and (7–46). We then solve Eq. (7–45) for $R_e R_f = Z_0^2 - (1 - \alpha_0)(Z_0 - R_{bb})R_f$ or $R_e R_f \cong Z_0^2$. Similarly, we solve Eq. (7–46) for $R_e R_f = Z_0^2 + (1 - \alpha_0)(R_{bb} R_f + Z_0 R_x + R_f R_x)$ or $R_e R_f \cong Z_0^2$. To obtain both an input and an output match, we must choose R_e and R_f so their product is Z_0^2. Increasing R_e or decreasing R_f to match the input and output will cause a loss of gain as can be seen from an inspection of Eq. (7–44).

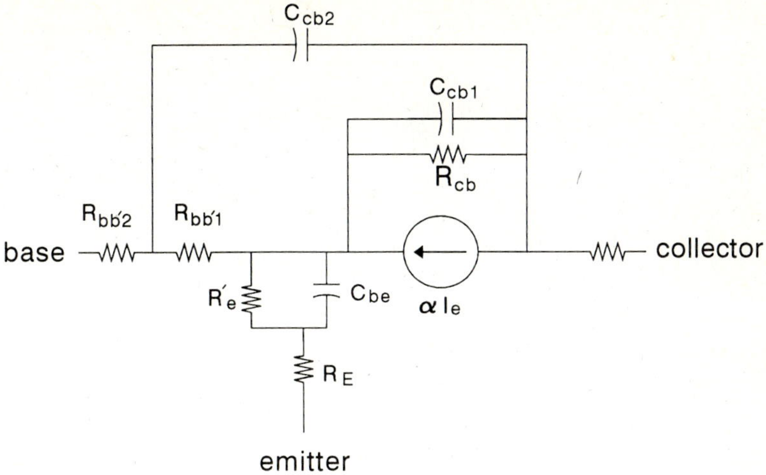

FIGURE 7–34. HBT Equivalent Circuit.[38] © 1991 IEEE.

Example: An HBT used for feedback amplifiers was reported by Rodwell and others.[38] Figure 7–34 is the equivalent circuit of their transistor. Values for the constants of Figure 7–34 are given in Table 7–4. This circuit is approximated by that of Figure 7–33 with $R_{bb} = 50\ \Omega$, $R_E = 9.5\ \Omega$, $R'_e = NKT/qI_e = 2.2$ (assuming $I_e = 15$ mA), $C_{be} = 124\ fF$ and $C_{cb} = 30\ fF$. At dc, the capacitors are open circuits and we apply the relation $R_e R_f = Z_0^2$, where $R_e = R_E + R'_e$ to match the input and the output. Using $R_e = 11.7\ \Omega$, a Z_0 of $50\ \Omega$ results in R_f of $214\ \Omega$. Substituting these into Eqs. (7–45), (7–44), and (7–47) gives

$$Z_{\text{in}} = (50 + 214)(11.7 + 50(0.02))/(11.7 + 50 + (214 + 50)(0.02)) = 50\ \Omega$$

$$A = [1 - 0.98(214/50)/(0.02 + (11.7/50))]/(1 + (214/50)) = -2.94$$

$$G_T = |-2.94|^2 4\ (0.02)^2/|0.02 + 0.02|^2 = 8.63, \text{ which is } 9.36 \text{ dB.}$$

If R_e is increased to $15\ \Omega$, R_f must be decreased to $167\ \Omega$ if we wish to maintain the matched input and output. Then, the G_T drops to 6.54 dB. If we do not require matched input and output and use $R_f = 480\ \Omega$ along with $R_e = 11.7\ \Omega$, the transducer gain increases to 12.9 dB. So, the transducer gain increases with less feedback despite the input

TABLE 7–4 Circuit Elements for the HBT in Figure 7–34.

$R_{bb1} = 35\ \Omega$	$C_{be} = 124$ fF	
$R_{bb2} = 15\ \Omega$	$C_{cb1} = 7$ fF	
$R_{ex} = 9.5\ \Omega$	$C_{cb2} = 23$ fF	$\alpha = \dfrac{\alpha_0\, e^{-j\omega\tau_c}}{1 + j\omega\tau_b}\left(\dfrac{\sin(\omega\tau_c)}{\omega\tau_c}\right)$
$R_c = 6.6\ \Omega$	$\alpha_0 = 0.98$	
$R_{cb} = 25\ \Omega$	$\tau_b = 0.96$ ps	
$R_e = NkT/qI_E$; $N = 1.35$	$\tau_c = 0.36$ ps	

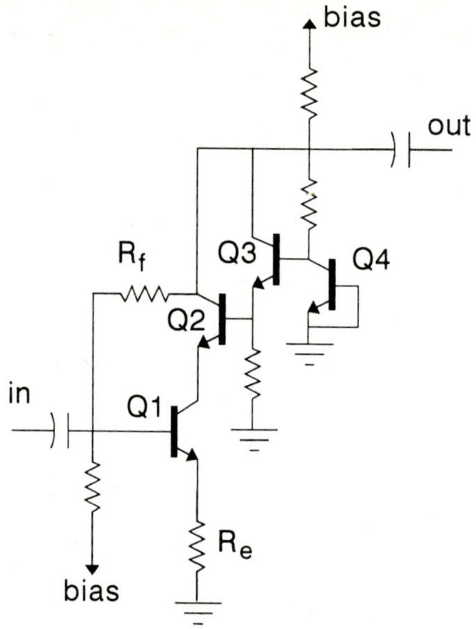

FIGURE 7–35. Cascode Amplifier for Minimizing Miller Capacitance. (IEEE 1991)

and output mismatches. Rodwell and his coworkers simulated their circuit on CAD software with an R_f of 208 Ω and predicted a 14.4 GHz upper cut-off frequency and a transducer gain of 9.3 dB. They constructed this amplifier and measured an upper 3 dB cut-off frequency of 14.1 GHz.

To increase the bandwidth of this HBT amplifier stage, Rodwell and his coworkers connected the HBT in a *cascode* arrangement shown in Figure 7–35. For this circuit, the transistors Q_3 and Q_4 are an active bias network for the transistor Q_2, which keep the base of Q_2 at signal ground without requiring large bypass capacitors. With this arrangement, the Miller effect is no longer limiting the bandwidth because the input stage with Q_1 approximates an emitter follower. They realized an upper cut-off frequency with this network of 33 GHz using the same HBT that was used in the common emitter feedback circuit having a 14.1 GHz upper cut-off frequency.

7.18 DISTRIBUTED AMPLIFICATION

The bandwidth limitation encountered when cascading amplifiers is overcome when amplifier signals are added together through the connection of the amplifier outputs to a transmission line as shown in Figure 7–36. This connection is called a *distributed amplifier.* In this amplifier, the amplified signals from several stages are added without also adding the output capacitances of the stages. In this way, we increase the overall gain without the bandwidth limitation usually caused by a concurrent increase in the capacitance. For this to work, the signals from each stage arriving at the load must have the

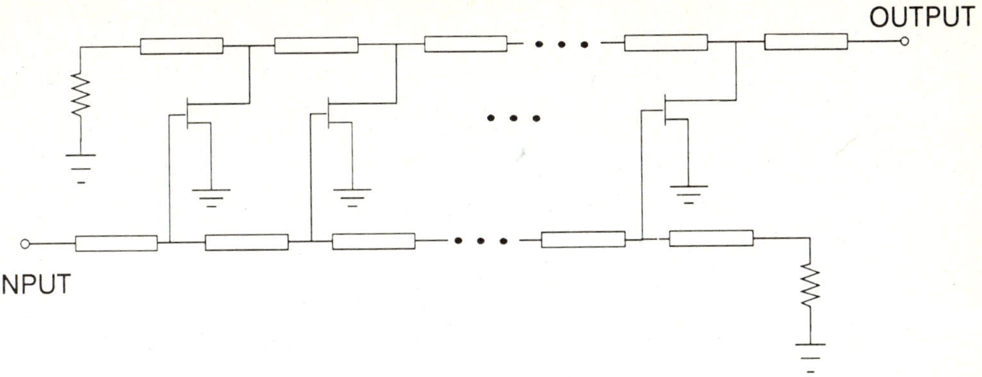

FIGURE 7–36. Distributed Amplifier Topology.

same phase. Because the outputs of the individual stages are separated from the load by different lengths of transmission line, this requires that the input signals to the stages have the correct phase to match these different delays. To achieve this phase relation, the signal inputs for the stages are taken from a signal traveling along another transmission line. The proper phase relation is obtained when the phase delay on the input line between the stages is equal to the phase delay on the output line between stages.

This simple explanation must be considerably enhanced when analyzing the distributed amplifier circuit. The presence of the input, output, and mutual impedances of the active elements alters the circuit operation so that the input and output circuits cannot be accurately modeled as simple transmission lines. The most accurate analysis yet reported is that done by computer modeling. The shortcoming of this type of analysis is that it does not yield simple design formulas for determining an initial design. Several approximate analyses applicable to the distributed amplifier have been reported. Each of these is capable of describing some aspect of the distributed amplifier operation.

One such analysis is to include the input and output admittances of the active elements as though they were part of the transmission line admittances and to assume the reverse mutual admittances y_{12} of the active elements are negligible. For this analysis, the active element admittances are divided by the length of line between the active elements to yield an admittance per unit length and the result is added to the line admittance to find an effective line admittance per unit length Y_e. Combining the transmission line equations for this composite line $dV/dz = -Z_e I$ and $dI/dz = -Y_e V$, we have $d^2V/dz^2 = Y_e Z_e V$. From this, we see that the propagation constant of this composite line is given by $\gamma^2 = Y_e Z_e$. Ayasli and his coworkers have used this approach to derive a gain equation for the distributed amplifier.[39] Here, we follow their approach. The kth active element of the distributed amplifier shown in Figure 7–37 contributes a current $g_m V_{gk}$ to the output line. This current divides into two parts that go up and down the line. The part toward the load propagates to the load and appears at the load with a factor of $\exp(-\gamma_d \ell_d (n-k+1))$, where $\gamma_d = \alpha_d + j\beta_d$ is the output line propagation constant and ℓ_d is the length of output

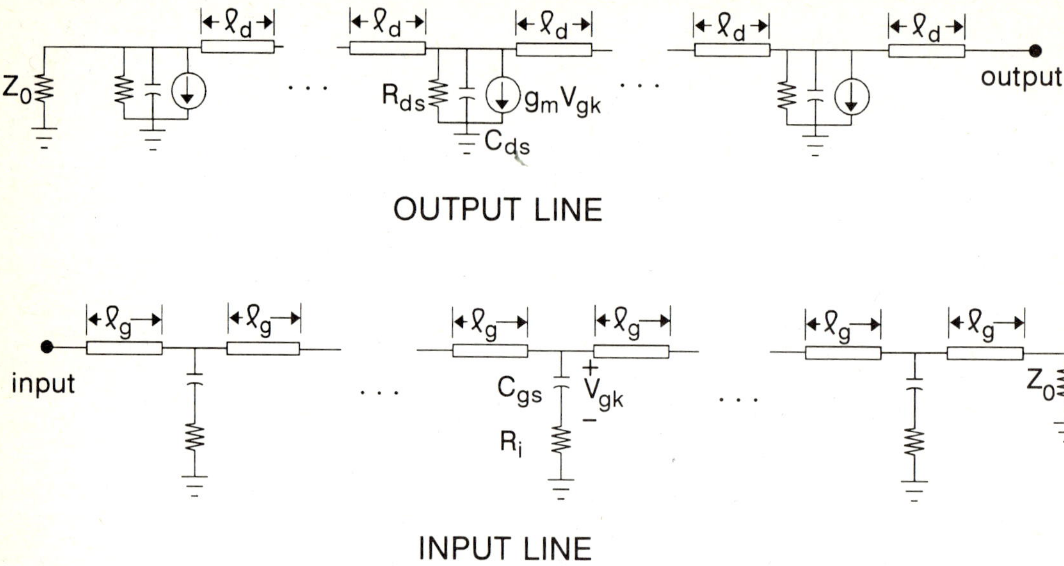

FIGURE 7–37. Input and Output Lines for a FET Distributed Amplifier.

line between any two active elements. Summing the contribution of each active element, we obtain an output load current of

$$I_o = (g_m/2)\sum_{k=1}^{n} V_{kg} \exp\left(-(n-k+1)(\alpha_d \ell_d + j\beta_d \ell_d)\right).$$

We let $a_d = \alpha_d \ell_d$, $a_g = \alpha_g \ell_g$ and $\beta_d \ell_d = \beta_g \ell_g = \theta$, where the subscript d is for the output line and the subscript g is for the input line. Because β_d is usually not equal to β_g, the output line is normally a meander line to get $\ell_d > \ell_g$. Here, we have assumed the input and output phase delays per section are equal. The quantities a_d and a_g are the attenuation per section of the output and input lines. Then, the load current becomes

$$I_o = (g_m/2)\exp\left((-n-1)(a_d + j\theta)\right)\sum_{k=1}^{n} V_{kg} \exp\left(k(a_d + j\theta)\right).$$

The input line voltage at the kth stage is $V_{kg} = V_{\text{in}} \exp(-k(a_g + j\theta))$. Substituting this into the load current expression, we have

$$I_o = (V_{in} g_m/2)\exp\left((-n-1)(a_d + j\theta)\right)\sum_{k=1}^{n} \exp(k(a_d - a_g)).$$

The sum in this expression is a geometric progression that can be expressed in closed form as $(e^A - e^{nA})/(1 - e^A)$, where $A = a_d - a_g$. Replacing the sum with this expression, we can rewrite the load current as

$$|I_o| = (g_m|V_{in}|/2)\left|\frac{\exp(-na_d - a_g) - \exp(-na_g - a_d)}{1 - \exp(a_d - a_g)}\right|.$$

If $a_d \cong a_g \ll 1$ and if $N = n + 1$, then this expression becomes

$$|I_o| \cong (g_m|V_{in}|/2)\left|\frac{\exp(-Na_d) - \exp(-Na_g)}{a_g - a_d}\right|.$$

To find a gain expression, we write the load power as $P_o = |I_o|^2 Z_0$ and the input power as $P_{in} = |V_{in}|^2/Z_0$. The gain is $G = P_o/P_{in}$ and can be written as Eq. (7–48) in terms of the attenuation per section of the output and input circuits for the stages.

$$G \cong (g_m Z_0/2)^2\left(\frac{\exp(-Na_d) - \exp(-Na_g)}{a_g - a_d}\right)^2 \quad (7\text{–}48)$$

Because the exponents are not equal in Eq. (7–48), the gain will not be monotonically increasing with N (or n). There is an optimum number of stages that will provide the maximum gain obtainable for given a_g and a_d (see problem 7–16). Attempting to increase the gain above this by increasing the number of stages will cause the overall gain to decrease. The physical explanation for this is that the signal traveling down the input line attenuates to the point where the stages receiving this signal produce an insignificant output. Then, these stages attenuate the signal on the output line more than they contribute to the output signal. Because of the frequency dependence of α_g, this occurs more noticeably at the high frequencies. This causes an increased loss of gain at the high frequency end of the band. For frequencies approaching dc, the amplifier reduces to a parallel configuration where the composite g_m is Ng_m and the gain equation becomes the same as for the lossy match amplifier given in Eq. (7–22) with g_m replaced by Ng_m. For $R_L = R_S = Z_0$ and $g_d = g_g = 1$ in that equation, we obtain the gain formula of Eq. (7–49).

$$G_T = (N g_m Z_0/2)^2 \quad (7\text{–}49)$$

An analysis treating the distributed amplifier with the filter theory discussed in section 4–1 was presented by Beyer and others.[40] Using the formulas for a sequence of T sections cascaded in a ladder network, they derived formulas for the amplifier gain and gain bandwidth product. In this analysis, the input and output ladder networks ("lines") have a cut-off frequency given by the same formula as that for a T section in section 4–1 and given here as Eq. (7–50), where L is the series inductance of the lines connecting the individual stages in the amplifier and C is the effective shunt capacitance at the active elements of the amplifier.

$$1 - \omega_c^2 LC/4 = 0 \quad (7\text{–}50)$$

The capacitance must include the effect of the active element capacitance plus the inherent line capacitance. We see that the bandwidth of the distributed amplifier can be limited either by the active element cut-off frequencies or by the input or output "line" (ladder network) cut-off frequencies.

Another approach is to use the results of an analysis of a true traveling wave amplifier. Jutzi and Kohn performed the analysis for a traveling wave amplifier with continuously coupled transmission lines.[41,42] Unlike most other investigations, this analysis included the effects of capacitive feedback from the output line to the input line as well as the forward transconductance. Their propagation constant has the same form as Eq. (5–12); they determined that there would be both fast and slow waves (even and odd modes). They also determined that, with sufficient capacitive coupling between the lines, an exponential amplification occurs causing a peak in the gain versus frequency curve at high frequencies. This peak can help to extend the high frequency end of the pass band.

Payne published an early article on distributed amplification where he considered the amplifier as a ladder network of lumped circuits, much the same as the filter networks of section 4–1.[43] However, he analyzed the network by multiplying the ABCD matrices for the individual stages. Along with other results, he showed that the gain will fluctuate nearly periodically as a function of frequency about some average value. He also showed that these fluctuations are smaller when the loading of the output line by the active elements is increased.

Example: Four Section FET Distributed Amplifier. For this example, we used the approximate methods of analysis to determine the gain and cut-off frequencies of a distributed amplifier with 4 FETs as shown in Figure 7–38. Then, we modeled it on CAD software to obtain plots of gain and input reflection coefficient versus frequency. The amplifier has a GaAs substrate with ε_r = 12.8 and a thickness of 0.1 mm and is to be matched to 50 Ω. The FET equivalent circuit parameters are given in Table 7–5. These were carefully chosen to exhibit some of the results discussed above. Using Eq. (7–49), we determined that the gain at dc would be 9 (9.54 dB). Using Eqs. (2–15) and (2–17)

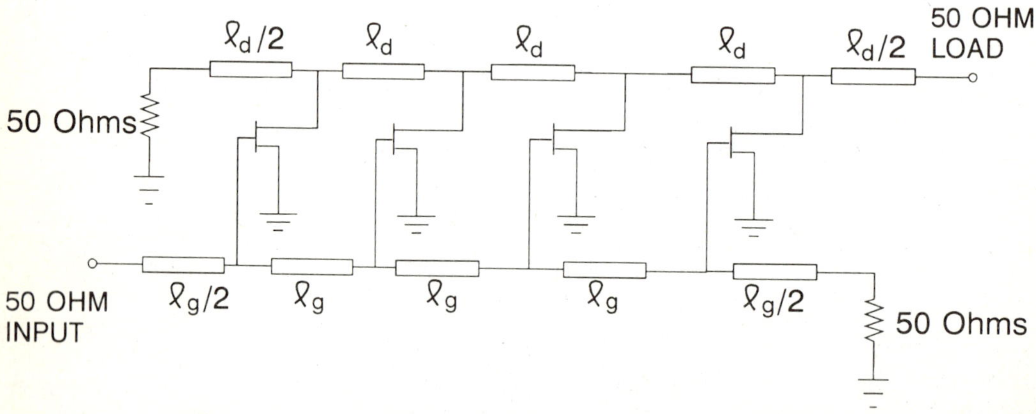

FIGURE 7–38. A 4-Section FET Distributed Amplifier. The FET equivalent circuit parameters are given in Table 7–5.

TABLE 7–5 Equivalent Circuit Elements for Distributed Amplifier FET.

$C_{gs} = 0.2$ pF	$C_{ds} = 0.07$ pF
$R_i = 7\ \Omega$	$R_{ds} = 300\ \Omega$
$C_{dg} = 0.02$ pF	$g_m = 0.03$ S
$f_T = 50$ GHz	$C_{dc} = 0$
$T = 1$ psec.	

repetitively, we determined that microstrip lines with a width of 0.075 mm would have an impedance of 49.7 Ω and an effective dielectric constant of 8.305.

To find the cut-off frequency for the output line, we need the inductance per section and capacitance per section of this line. The inductance per unit length of the drain line is given by $L_d = Z_0/V_p = (Z_0/C)\sqrt{\varepsilon_{eff}} = (49.7/3)10^{-8}\sqrt{8.305}$ or $L_d = 477$ nH/m. Similarly, from $C_d = 1/(Z_0V_p) = \sqrt{\varepsilon_{eff}}/(Z_0C) = \sqrt{8.305}/(49.7 \times 3 \times 10^8)$, we get the capacitance per unit length of the drain line C_d equal to 193.4 pF/m. This capacitance is in parallel with the shunt elements of the FET drain, as shown in Figure 7–37. We assumed that the output line sections would be $\ell_d = 1$ mm long and computed the effective shunt capacitance per unit length as $C_{ed} = 193.4 + C_{ds}/\ell_d = 193.4 + .07/.001 = 263.4$ pF/m. The cut-off frequency of this line from Eq. (7–50) is $f_c = 1/(\pi\sqrt{L_eC_e})$, where L_e and C_e must be the lumped element values for a section. We have $f_c = 1/(\pi\sqrt{477 \times 263.4 \times 10^{-21} \times \ell_d^2} = 28.4$ GHz. To find $\gamma_d = Z_eY_e$, we must have Z_e and Y_e. The effective conductance per unit length is $G = 1/(R_{ds}\ell_d) = 10/3$. Then, $Y_e = G + j\omega C_{ed}$. We have $\gamma_d^2 = j\omega L_d(G + j\omega C_{ed})$.

To find the cut-off frequency and γ_g for the input line, we follow a similar set of calculations. The difference here is that the gate equivalent circuit elements are in series with each other, and this series branch is in parallel with the shunt capacitance of the input line as is shown in Figure 7–37. The admittance of this branch is $Y = j\omega C_{gs}/(1 + j\omega C_{gs}R_i) \cong \omega^2C_{gs}^2R_i + j\omega C_{gs}$. This approximation is valid to within about 5% at 20 GHz. From a calculation of the phase constants of the input and output lines, we determined that the input line section between active elements should be about $\ell_g = 0.8$ mm (note that the calculation depends on the outcome of the calculation for the input line shunt capacitance and series inductance, so several iterations were used). The input section shunt capacitance is equal to $193.4 \times 0.0008 + 0.2$ pF = 0.3547 pF. The series inductance per section is 477×0.0008 nH = 0.3816 nH. The input line cut-off frequency is $f_c = 1/(\pi\sqrt{3.816 \times 0.3547 \times 10^{-11}}) = 27.4$ GHz. The propagation constant for the input line is given by $\gamma_g^2 = j\omega L_d(\omega^2C_{gs}^2R_i + j\omega(C_{gs} + C_g\ell_g))/\ell_g$, where C_g is the gate line shunt capacitance per unit length. Using these expressions for the propagation constants, we determined $\alpha_d + j\beta_d = 64.7 + j154.4$ and $\alpha_g + j\beta_g = 0.909 + j182.2$ at 2 GHz. At 20 GHz, we found $\alpha_d + j\beta_d = 71.2 + j1403$ and $\alpha_g + j\beta_g = 90.79 + j1824$. When the line lengths of 1 and 0.8 mm are used, the phase delay per section on the output line is nearly equal to the phase delay per section on the input line at 20 GHz.

The circuit was simulated on CAD software for several different sets of FET equivalent circuit parameters. Some of the results are shown in Figures 7–39 and 7–40. Examining these figures, we see that the gain exhibits the periodic variations versus fre-

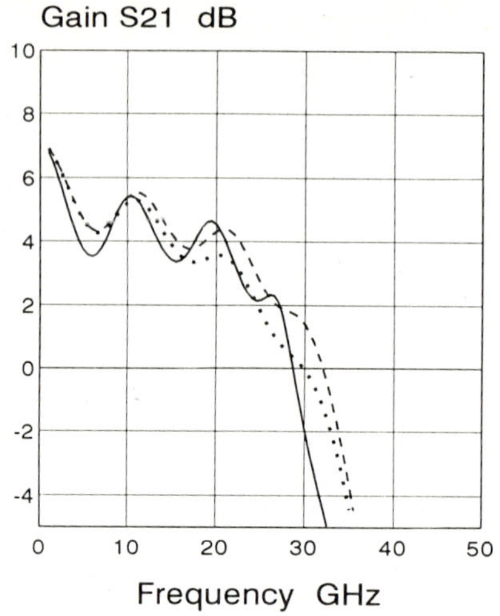

FIGURE 7–39. Gain versus Frequency for a 4-Section FET Distributed Amplifier. The solid line curve is that for the amplifier with the parameters of Table 7–5. The dotted line curve is that when R_i is increased to 10 Ω and the C_{dg} is set to zero. The dashed line is that for $R_i = 7$ Ω and C_{dg} is zero.

quency described by Payne. We also see that increasing the gate circuit loading by increasing R_i to 10 Ω causes this variation to be less pronounced and the gain drops off at a lesser frequency. We also see that the presence of an 0.02 pF drain to gate capacitance decreases the cut-off frequency of the lines, because it presents an increased shunt capacitance to both lines and causes a slight increase in gain at the frequencies just below cut-off. In the extreme case of capacitive feedback and very low attenuation, this effect is pronounced and corresponds to the exponential increase described by Jutzi and Kohn. The input reflection coefficient versus frequency is shown in Figure 7–40.

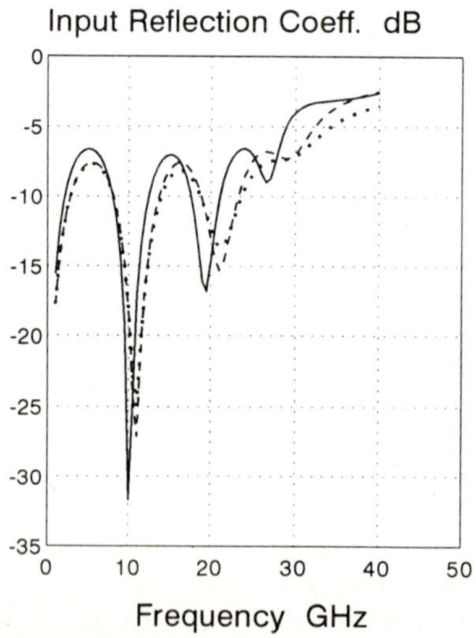

FIGURE 7–40. 4-Section Distributed Amplifier $|S_{11}|$ on a Log Scale. The solid line curve is that for the amplifier with the parameters of Table 7–5. The dotted line curve is that when R_i is increased to 10 Ω and the C_{dg} is set to zero. The dashed line is that for $R_i = 7$ Ω and C_{dg} is zero.

7.19 DISTRIBUTED AMPLIFIERS

Because of the combined effects of the variation of both the input line and output line attenuation with frequency and with the output line signal voltage variation from section to section, the signal power contributed by each active element in a distributed amplifier varies considerably with frequency and with location in the amplifier. Walker has shown in the simplified lossless line case that the active element in the output line next to the load contributes nearly one-half of the total power over most of the frequency range.[44] Because the power distribution is not uniform in the conventional distributed amplifier, a modification of the basic distributed amplifier design has been proposed. To maximize the power output of a distributed amplifier, Ayasli and his coworkers proposed capacitively coupling the input line to the active elements in such a way as "tailor the excitation" to each active element.[45] Besides increasing the power output capability in the case where FETs are the active elements, this process has the added benefit that one might more nearly equalize the shunt capacitance presented by the active elements to the input line and so allow the input and output lines to have equal line lengths between the sections. Prasad and his coworkers have published design procedures for this capacitively coupled distributed amplifier.[46]

The gain limitation caused by the limitation on the number of active elements required to produce maximum gain in a distributed amplifier has led to an amplifier circuit called a *matrix amplifier.*[47] The configuration is shown in Figure 7–41. In this amplifier,

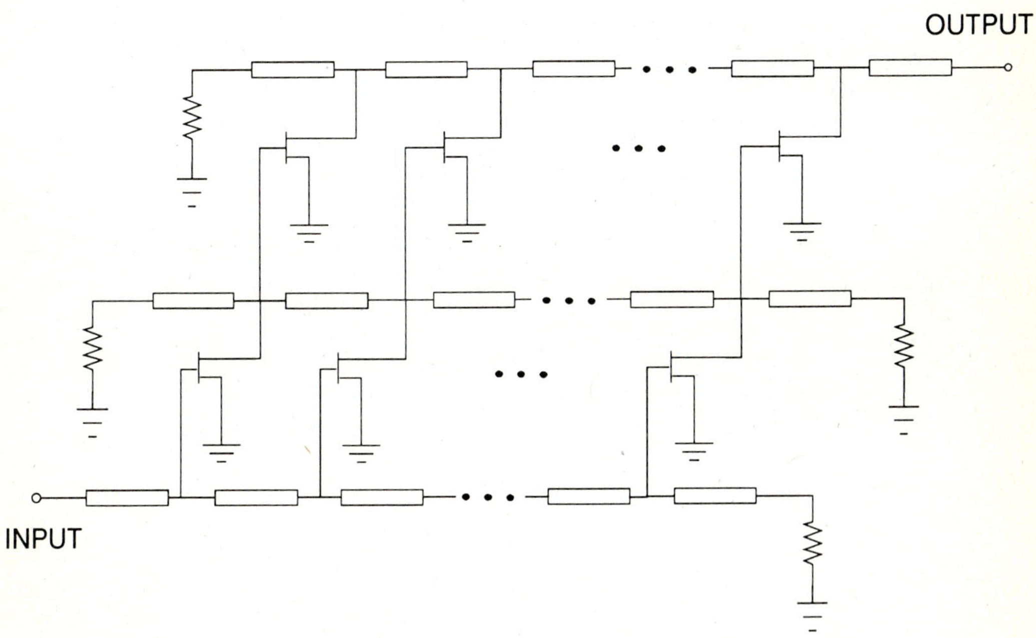

FIGURE 7–41. Two Tier Matrix Amplifier.

the input line is the same as with a distributed amplifier, but there are two (or more) tiers of active elements with the "output line" of one tier being the input line to the next tier. Then, the output line of the top tier feeds the output of the amplifier. Simply stated, the device combines the additive property of distributed amplifiers along with the multiplicative property of cascaded amplifiers. In this way, one can maintain the large bandwidth property of the distributed amplifier and still obtain higher gain by adding more active elements in the second tier of the matrix amplifier. As Niclas and Pereira state, the analysis of this device is a formidable task best undertaken by computer simulation. The initial design still must be accomplished using formulas for the design of a distributed amplifier. Then, the final design can be reached by computer optimization of individual components in the amplifier. Niclas and Pereira reported on the design and performance of a 3 tier matrix amplifier.[48] In this report they give some design formulas.

The distributed amplifier has an inherent bidirectionality to it. The idle terminations of the input and output lines can be converted into an alternate input and output. The amplifier can then be used as an amplifying directional coupler with rather poor rejection in the unwanted direction. Chu and his coworkers have found that a matrix amplifier with the idle line terminated at RF produces a satisfactory bidirectional amplifier.[49]

Reports have been published of distributed amplifiers using HFETs operating up to as high as 100 GHz. One report was of a seven-section 5 to 100 GHz distributed amplifier using coplanar waveguide as the transmission medium.[50]

The active stages in the distributed or matrix amplifier need not be a single active element. One matrix distributed amplifier was built with a Darlington pair of HBT transistors as the active stage.[51] Another 5 to 60 GHz distributed amplifier was built using a cascode connection of a pair of HFETs.[52]

Clearly, distributed amplifiers and their variations is an active field with many avenues to be pursued.

PROBLEMS

7–1. Find f_T and f_{max} for the FET with the parameters listed in Table 7–3.

7–2. Assume $C_{dg} = C_{dc} = 0$ in the FET of Figure 7–1, and assume this FET has $G_{ds} = 5$ mS, $C_{ds} = 0.082$ pF, and $R_L = 50\ \Omega$. With these assumptions, find an approximate value of a peaking coil which will give a maximum in Z_A at 10 GHz when used in the circuit of Figure 7–5.

7–3. Using the reciprocity theorem, $I_1/V_2|_{V_1=0} = I_2/V_1|_{V_1=0}$, with the circuit of Figure 7–7, prove that $AD - BC = 1$.

7–4. A lossless equalizer network has $H(S) = (4 + 5S + 2S^2 + S^3)/4$ and $K(S) = -(3 + S^3)/4$, where all the elements have been normalized to $R_g = R_L = 1$ and $\Omega = 1$. Synthesize the network using Eqs. (7–10) and (7–12) to find the normalized element values. Scale these element values to $R_g = R_L = 50\ \Omega$ and $\Omega = 10^{10}$.

7–5. Use ε^2 corresponding to $\frac{1}{2}$ dB ripple in the example of section 7–7 and determine $H(S)$ and $K(S)$.

7–6. Start with a current source I_S and source admittance y_S feeding a 2 port that is loaded with an admittance y_L and derive the equation

$$\begin{pmatrix} I_S \\ 0 \end{pmatrix} = \begin{pmatrix} y_{11} + y_S & y_{12} \\ y_{21} & y_{22} + y_L \end{pmatrix} \begin{pmatrix} V_1 \\ V_2 \end{pmatrix},$$

where V_1 and V_2 are the input and output voltages of a 2 port and y_{ij} are its admittance parameters. Derive Eq. (7–19) by starting with the load power as $|V_2|^2 g_L$, where V_2 is found from the solution of the matrix equation above.

7–7. Show that the input admittance to a 2 port with a load of y_L is given by the relation shown in the example of section 7–10.

7–8. Show that Eq. (7–23) follows from Eq. (7–22) by using $S_{11} = (1 - g_g)/(1 + g_g)$ and $S_{22} = (1 - g_d)/(1 + g_d)$, where $g_g = R_S/R_g$ and $g_d = R_L/R_d$.

7–9. At frequencies below 1 GHz, the stabilized FET of the example in section 7–7 has an input impedance approximately like a 0.28 pF capacitor shunted by the 200 Ω stabilizing resistor and an output impedance approximately like a C_{ds} = 0.03 pF shunted by G_d = 1.5 mS. Assume that the dc transconductance is 70 mS and that y_{12} is negligible. Given the MAG curve of Figure 7–8 and the Eq. (7–22), find values for the G_g and G_d of Figure 7–20 that can be used as the starting point for a lossy match amplifier design FET that will have an upper cut-off frequency of 10 GHz.

7–10. Graphically determine the curve for the transducer gain versus frequency and input reflection coefficient $|S_{11}|$ versus frequency if a balanced amplifier is made with 2 dB Lange couplers whose center frequency is 10 GHz and with matched amplifiers whose gain and $|S_{11}|$ curves are those of Figures 7–11 and 7–12.

7–11. Derive Eq. (7–26) assuming that the $S_{12}^a = S_{12}^b = 0$.

7–12. Kobayashi and his coworkers used a modified Darlington pair as shown in Figure 7–42, where R_{bias} was adjusted such that $I_x = I_{e2}$. Assuming that the two transistors have equal βs, show that at signal frequencies the overall β of the combination is approximately twice the β of the individual transistors.

7–13. The NE68100 bipolar transistor has the S_{21} parameter values given in the Table 7–6. Using the graphical method of Fig. 7–32 determine the approximate gain $|S_{21}|^2$ this transistor would have over a frequency range from 2 to 6 GHz with a parallel feedback path of 100 Ω in series with an 0.8 nH inductor.

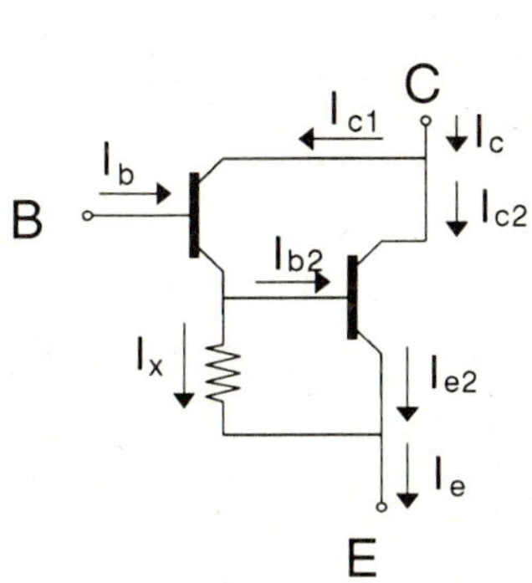

FIGURE 7–42. Circuit for Problem 7–12

TABLE 7–6 S Parameters for NE68100 Bipolar Transistor.

Freq. GHz	S11		S21		S12		S22	
2	0.68	180	4.78	80	0.04	55	0.4	–27
3	0.68	173	3.53	72	0.05	61	0.4	–29
4	0.68	167	2.8	65	0.06	65	0.39	–33
5	0.68	162	2.34	58	0.07	67	0.39	–36
6	0.69	157	2.02	52	0.08	68	0.39	–41
7	0.69	153	1.76	46	0.09	67	0.39	–46
8	0.7	149	1.57	40	0.1	67	0.39	–52
9	0.7	144	1.42	35	0.11	68	0.39	–57

The parameters are given as magnitude and angle with the angle in degrees.

7–14. Determine the output impedance for the example in section 7–17 using Eq. (7–46). What will be the reflection coefficient caused by this impedance feeding 50 Ω?

7–15. Using the conversion program in the chapter 3 appendix, convert the S parameters of Table 7–6 to y parameters over the frequency range of 2 to 6 GHz. Determine the augmented y matrix when the NE 68100 has a parallel feedback of $Y_{fb} = 1/(R_{fb} + j\omega L_{fb})$ with $R_{fb} = 100\ \Omega$ and $L_{fb} = 0.8$ nH. For the frequencies of 2, 4, and 6 GHz, calculate the stage gain with feedback using the components of this augmented matrix in Eq. (7–19). Compare your results with the solution of problem 7–13.

7–16. Starting with the gain equation of Eq. 7–48, find a formula for the number of stages that will give maximum gain for a distributed amplifier with specified per section losses a_d and a_g.

7–17. Use the expression found in the solution of problem 7–16 to estimate the number of active elements that will yield the maximum gain with $a_d = 0.07$ and $a_g = 0.3$.

ENDNOTES

1. Harold A. Wheeler, "Wide-Band Amplifiers for Television," *Proc. IRE, 27, 7* (July 1939), 329–338.
2. W. W. Hansen, "On Maximum Gain-Band Width Product in Amplifiers," *J. Applied Physics, 16,* (Sept. 1945), 528–534.
3. Hendrik W. Bode, Ph.D., *Network Analysis and Feedback Amplifier Design,* 360–366, (New York: D. Van Nostrand Company, 1945).
4. William S. Percival, British Patent 460, 562.
5. Edward L. Ginzton, William R. Hewlett, John H. Jasberg, and Jerre D. Noe, "Distributed Amplification," *Proc. IRE, 36, 8* (Aug. 1948), 956–969.
6. R. S. Engelbrecht and K. Kurokawa, "A Wide-Band Low Noise L-Band Balanced Transistor Amplifier," *Proc. IEEE, 53, 3* (March, 1965), 237–247.
7. An excellent summary and compilation of papers is given in *Compound Semiconductor Transistors: Physics and Technology,* ed. Sandip Tiwari, (New York: The Institute of Electrical and Electronics Engineers, 1993).

8. S. M. Sze, *Semiconductor Devices Physics and Technology,* (New York: John Wiley & Sons, 1985).
9. Peter H. Ladbrooke, *MMIC Design GaAs FETs and HEMTs,* (Norwood, MA: Artech House, 1989).
10. Timothy J. Drummond, William Ted Masselink, and Hadis Morkoç, "Modulation-Doped GaAs/(Al,Ga)As Heterojunction Field-Effect Transistors: MODFETs", *Proc. IEEE, 74, 6* (June 1986), 773–822.
11. Hadis Morkoç, Boris Sverdlov, and Guang-Bo Gao, "Strained Layer Heterostructures, and their Applications to MODFET's, HBT's, and Lasers," *Proc. IEEE, 81, 4* (April 1993), 493–556.
12. Herbert Kroemer, "Heterostructure Bipolar Transistors and Integrated Circuits," *Proc. IEEE, 70, 1* (January 1982), 13–25. Also see Morkoç, and others "Strained Layer . . .", 522–530.
13. This parameter is not the unilateral gain used in chapter 6. For a discussion of this parameter, see Madhu S. Gupta, "Power Gain in Feedback Amplifiers, A Classic Revisited," *IEEE Trans. Microwave Theory Tech., 40, 5* (May 1992), 875.
14. Charles A. Liechti, "Microwave Field-Effect Transistors," *IEEE Trans. Microwave Theory Tech., MTT-24, 6* (June 1976), 289.
15. Bode, Network Analysis . . . , 363–367. The development here follows Bode's presentation.
16. R. M. Fano, "Theoretical Limitations on the Broadband Matching of Arbitrary Impedances," *Jour. Franklin Inst., 249,* (January 1950), 57–83; (February 1950), 139–54.
17. D. C. Youla, "A New Theory of Broad-band Matching," *IEEE Trans. Circuit Theory, 11, 3* (March 1964), 30–50.
18. Walter H. Ku and Wendell C. Peterson, "Optimum Gain-Bandwidth Limitations of Transistor Amplifiers as Reactively Constrained Active Two-Port Networks," *IEEE Trans. Circuits Systems, CAS-22, 6* (June 1975), 523–533.
19. Karl B. Niclas, "Multi-Octave Performance of Single-Ended Microwave Solid-State Amplifiers," *IEEE Trans. Microwave Theory Tech., MTT-32, 8* (August 1984), 896–908.
20. see Gabor C. Temes and Sanjit K. Mitra, *Modern Filter Theory and Design,* 79–80, (New York: John Wiley & Sons, 1973).
21. see R. Saal and E. Ulbrich, "On the Design of Filters by Synthesis", *IRE Trans. Circuit Theory, CT-5, 4* (December 1958), 284–327.
22. Douglas J. Mellor and John G. Linvill, "Synthesis of Interstage Networks of Prescribed Gain Versus Frequency Slopes," *IEEE Trans. Microwave Theory Tech., MTT-23,* no. 12 (December 1975), 1013–1020.
23. The selection of the ripple factor, $\varepsilon,^2$ need not be arbitrary. It represents a degree of freedom that can be used to advantage. See R. Levy, "Explicit Formulas for Chebyshev Impedance-Matching Networks, Filters and Interstages," *Proc. IEE, Electronics, 111, 6* (June 1964), 1099–1106.
24. See the foregoing reference: Levy, "Formulas for Impedance-Matching . . .". *Proc. IEE, 1964.*
25. Herbert J. Carlin, "A New Approach to Gain-Bandwidth Problems," *IEEE Trans. Systems, CAS-24, 4* (April 1977), 170–175.
26. Bode, *Network Analysis* . . . , 305–315. This integral also is known as the Hilbert transform.
27. Herbert J. Carlin and James J. Komiak, "A New Method of Broad-Band Equalization Applied to Microwave Amplifiers," *IEEE Trans. Microwave Theory Tech., MTT-27, 2* (February 1979), 93–99.
28. Wen-Lin Jung and Jingshown Wu, "Stable Broad-Band Microwave Amplifier Design," *IEEE Trans. Microwave Theory Tech., 38, 8* (August 1990), 1079–1085.
29. Mohammed Shauib Ghausi, *Principles and Design of Linear Active Circuits,* (New York: McGraw-Hill Book Company, 1965), 68.

30. Karl B. Niclas, "On Design and Performance of Lossy Match GaAs MESFET Amplifiers," *IEEE Trans. Microwave Theory Tech., 30, 11* (November 1982), 1900–1907.
31. R. S. Engelbrecht and K. Kurokawa, "A Wide-Band Low Noise L-Band Balanced Transistor Amplifier," *Proc. IEEE, 53, 3* (March 1965), 237–247.
32. John L. B. Walker, "Improving Operation of Classic Broadband, Balanced Amplifiers," *Microwaves & RF, 26, 9* (September 1987), 175–180.
33. Anthony M. Pavio, Jr., "A Network Modeling and Design Method for a 2–18-GHz Feedback Amplifier", *IEEE Trans. Microwave Theory Tech., MTT-30, 12* (Dec. 1982), 2212–2216.
34. This analysis is derived from that described by Karl B. Niclas, Walter T. Wilser, Richard B. Gold and William R. Hitchens in "The Matched Feedback Amplifier: Ultrawide-Band Microwave Amplification with GaAs MESFET's," *IEEE Trans. Microwave Theory Tech., MTT-28, 4* (April 1980), 285–294.
35. Niclas and others, "The Matched Feedback Amplifier: . . . ."
36. Felix Perez and Vicente Ortega, "A Graphical Method for the Design of Feedback Networks for Microwave Transistor Amplifiers: Theory and Applications," *IEEE Trans. Microwave Theory Tech., MTT-29, 10* (Oct. 1981), 1019–1027.
37. Eric Ulrich, "Use Negative Feedback to Slash Wideband VSWR," *Microwaves, 17* (October 1978), 66–70.
38. Mark Rodwell, J. F. Jensen, W. E. Stanchina, R. A. Metzger, D. B. Rensch, M. W. Pierce, T. V. Kargodorian, and Y. K. Allen and others in "33-GHz Monolithic Cascode AlInAs/GaInAs Heterojunction Bipolar Transistor Feedback Amplifier," *IEEE J. Solid-State Circuits, 26, 10* (Oct. 1991), 1378–1382.
39. Yalçin Ayasli, Robert L. Mozzi, James L. Vorhaus, Leonard D. Reynolds, and Robert A. Pucel, "A Monolithic GaAs 1–13-GHz Traveling-Wave Amplifier," *IEEE Trans. Microwave Theory Tech., MTT-30, 7* (July 1982), 976–981.
40. James B. Beyer, S. N. Prasad, Robert C. Becker, James E. Nordman and Gert K. Hohenwarter, "MESFET Distributed Amplifier Guidelines," *IEEE Trans. Microwave Theory Tech., MTT-32, 3* (March 1984), 268–275.
41. W. Jutzi, "Uniform Distributed Amplifier Analysis with Fast and Slow Waves," *Proc. IEEE, 56, 1* (January 1968), 66–67.
42. Gerhard Kohn, "Distributed Field-Effect Amplifiers," *Proc. IEEE, 56. 6* (June 1968), 1136–1137.
43. Delmar V. Payne, "Distributed Amplifier Theory," *Proc. IRE, 41, 6* (June 1953), 759–762.
44. J. L. B. Walker, "Some Observations on the Design and Performance of Distributed Amplifiers," *IEEE Trans. Microwave Theory Tech., 40, 1* (January 1992), 164–168.
45. Yalçin Ayasli, Steven W. Miller, Robert Mozzi, and Larry K. Hanes, "Capacitively Coupled Traveling-Wave Power Amplifier," *IEEE Trans. Microwave Theory Tech., MTT-32, 12* (December 1984), 1704–1709.
46. S. N. Prasad, James B. Beyer, and Ik-Soo Chang, "Power-Bandwidth Considerations in the Design of MESFET Distributed Amplifiers," *IEEE Trans. Microwave Theory Tech., 36, 7* (July 1988), 1117–1123.
47. Karl B. Niclas and Ramon R. Pereira, "The Matrix Amplifier: A High-Gain Module for Multioctave Frequency Bands," *IEEE Trans. Microwave Theory Tech., MTT-35, 3* (March 1987), 296–306.
48. Karl B. Niclas and Ramon R. Pereira, "On the Design and Performance of a 6–18 GHz Three-Tier Matrix Amplifier", *IEEE Trans. Microwave Theory Tech., 37, 7* (July 1989), 1069–1077.
49. S. L. G. Chu, M. J. Schindler, A. M. Bertrand, and T. Tsukii, "A Novel Broadband

Bidirectional Matrix Amplifier," *GaAs IC Symp. Tech. Dig.* 1991, 315–318, (New York: The Institute of Electrical and Electronic Engineers, 1991).

50. R. Majidi-Ahy, M. Riaziat, C. Nishimoto, M. Glenn, S. Silverman, S. Weng, Y. C. Pao, G. Zdasiuk, S. Bandy, and Z. Tan, "5–100 GHz InP CPW MMIC 7-Section Distributed Amplifier," *IEEE 1990 Microwave Millimeter-Wave Monolithic Circuits Symp. Dig.*, ed. Ho-Chung Huang, 31–34, (Piscataway, NJ: The Institute of Electrical and Electronics Engineers, 1990).
51. Kevin W. Kobayashi, Reza Esfandiari, Majig E. Hafizi, Dwight C. Streit, Aaron K. Oki, Liem T. Tran, Donald K. Umemoto, and Mike E. Kim, "GaAs HBT Wideband Matrix Distributed and Darlington Feedback Amplifiers to 24 GHz," *IEEE Trans. Microwave Theory Tech., 39, 12* (December 1991), 2001–2008.
52. C. Yuen, Y. C. Pao, and N. G. Bechtel, "5–60 GHz High-Gain Distributed Amplifier Utilizing InP Cascode HEMTs," *GaAs IC Symp. Tech. Dig.* 1991, 319–322. (Piscataway, NJ: The Institute of Electrical and Electronics Engineers, 1991).

Index